# Clean Energy

The text discusses various technologies and methods that help in generating clean and renewable forms of energy. It also covers the use of fuels in energy conversion media such as internal combustion engines, and compares these with existing fossil fuels. This book also focuses on green hydrogen revolution and the urban heat island effect.

This book

- covers topics such as sustainable biodiesel production, photovoltaic (PV) device optimization and performance enhancement, maximizing energy production, and improving the overall efficiency, reliability, and performance of solar energy systems.
- explains energy-efficient green buildings, blue–green infrastructure, high building energy demand, solar chimney, urban green spaces in reducing the urban heat island effect, urban air quality, building resilience to airborne pathogens, and challenges related to urban green infrastructure.
- discusses the carbon sequestration potential of urban parks, the evolution of carbon capture, utilization, and storage (CCUS) technology, hydrogen production technologies, production and utilization of hydrogen as a future fuel for electric vehicles, and the evaluation of carbon emissions from various vehicles.
- presents solar-based thermal energy systems, optimizing solar air heater performance, and solar-integrated hydrogen production and utilization in India.
- illustrates agro-residue-derived biochar for environmental management and competing alternative fuel technologies for commercial vehicles in India.

The book is primarily written for senior undergraduates, graduate students, and academic researchers in the fields of mechanical engineering, environmental engineering, energy engineering, electrical engineering, electronics and communications engineering, automotive engineering, and renewable energy.

**Advances in Manufacturing, Design and Computational Intelligence Techniques**

*Series Editor:*

Ashwani Kumar

*Senior Lecturer, Mechanical Engineering, at Technical Education Department, Uttar Pradesh, Kanpur, India*

The book series editor is inviting edited, reference and text book proposal submission in the book series. The main objective of this book series is to provide researchers a platform to present state-of-the-art innovations, research related to advanced materials applications, cutting-edge manufacturing techniques, and innovative design and computational intelligence methods used for solving nonlinear problems of engineering. The series includes a comprehensive range of topics and their applications in engineering areas such as additive manufacturing, nanomanufacturing, micromachining, biodegradable composites, material synthesis and processing, energy materials, polymers and soft matter, nonlinear dynamics, dynamics of complex systems, MEMS, green and sustainable technologies, vibration control, AI in power station, analog-digital hybrid modulation, advancement in inverter technology, adaptive piezoelectric energy harvesting circuit, contactless energy transfer system, energy-efficient motors, bioinformatics, computer-aided inspection planning, hybrid electrical vehicle, autonomous vehicle, object identification, machine intelligence, deep learning, control-robotics-automation, knowledge-based simulation, biomedical imaging, image processing and visualization. This book series compiled all aspects of manufacturing, design and computational intelligence techniques from fundamental principles to current advanced concepts.

Hybrid Metal Additive Manufacturing: Technology and Applications
*Edited by Parnika Shrivastava, Anil Dhanola, and Kishor Kumar Gajrani*

Thermal Energy Systems: Design, Computational Techniques, and Applications
*Edited by Ashwani Kumar, Varun Pratap Singh, Chandan Swaroop Meena, and Nitesh Dutt*

Clean Energy: Technology, Advances, and Applications
*Edited by Tikendra Nath Verma, Thokchom Subhaschandra Singh, Upendra Rajak, Prerana Nashine, Gaurav Dwivedi, and Ashwani Kumar*

https://www.routledge.com/Advances-in-Manufacturing-Design-and-Computational-Intelligence-Techniques/book-series/CRCAIMDCIT?publishedFilter=alltitles&pd=published,forthcoming&pg=1&pp=12&so=pub&view=list?publishedFilter=alltitles&pd=published,forthcoming&pg=1&pp=12&so=pub&view=list

# Clean Energy
## Technology, Advances, and Applications

Edited by
Tikendra Nath Verma,
Thokchom Subhaschandra Singh,
Upendra Rajak, Prerana Nashine,
Gaurav Dwivedi, and Ashwani Kumar

CRC Press
Taylor & Francis Group
Boca Raton London New York

CRC Press is an imprint of the
Taylor & Francis Group, an **informa** business

Designed cover image: shutterstock

First edition published 2025
by CRC Press
2385 NW Executive Center Drive, Suite 320, Boca Raton FL 33431

and by CRC Press
4 Park Square, Milton Park, Abingdon, Oxon, OX14 4RN

*CRC Press is an imprint of Taylor & Francis Group, LLC*



ISBN: 9781032861043 (hbk)
ISBN: 9781032861036 (pbk)
ISBN: 9781003521341 (ebk)

DOI: 10.1201/9781003521341

Typeset in Sabon
by codeMantra

# Contents

# Aim and Scope

*Clean Energy: Technology, Advances, and Applications* is designed to be a comprehensive resource for both novice and experienced learners in the ever-expanding field of clean energy. This book serves as a springboard for young researchers in renewable energy, alternative fuels, and clean energy technologies. It provides a foundational understanding of current methodologies employed in clean energy generation. This book functions as a valuable reference point. Researchers can utilize the book to benchmark their ongoing research endeavors against the findings presented here. It also delves into uncertainty analysis, a crucial aspect of scientific research, providing readers with the tools to implement similar practices within their own studies.

*Clean Energy: Technology, Advances, and Applications* takes readers on a comprehensive journey through the world of clean energy, equipping them with the knowledge and tools necessary to address global sustainability challenges. The foundational chapter establishes the critical role clean energy plays in achieving a sustainable future. It explores the challenges posed by traditional energy sources and highlights the benefits of transitioning to clean alternatives. It dives deep into various renewable energy sources. This book examines the mechanisms and technologies behind solar, wind, geothermal, and hydrogen energy production, empowering readers to understand how these resources can be harnessed for clean energy generation. The book chapter presents innovative approaches for creating energy-efficient buildings. It outlines methods to optimize energy consumption and reduce environmental impact from construction to demolition. This book highlights the concept of blue–green infrastructure, which integrates natural elements like water features and vegetation into urban design. The book also explores the benefits of blue–green infrastructure in promoting sustainability, mitigating environmental challenges, and enhancing urban living.

This book unpacks the phenomenon of urban heat islands, where cities experience higher temperatures than surrounding rural areas. It explores the causes and potential mitigation strategies for this growing urban challenge.

In addition, it delves into the world of CCUS technologies and examines methods for capturing carbon emissions from various sources, exploring their potential for reuse or safe storage to reduce their environmental impact. It also explores emerging fuel technologies for electric vehicles and examines advancements in carbon footprint mapping, hydrogen fuel cells, and other innovations, paving the way for a cleaner transportation future. It establishes the critical link between vehicle emissions and urban air quality, highlighting the detrimental impact of transportation emissions on human health and the environment and emphasizing the urgency for clean energy solutions. Additionally, it provides valuable insights into the production and utilization of biodiesel, a clean-burning fuel derived from renewable biological sources and explores the potential of biodiesel as a sustainable alternative to traditional fossil fuels.

By weaving together these diverse facets of clean energy, *Clean Energy: Technology, Advances, and Applications* provides a holistic understanding of the field and its potential to revolutionize the way we power our world.

**Editors**

**Tikendra Nath Verma, Thokchom Subhaschandra Singh, Upendra Rajak, Prerana Nashine, Gaurav Dwivedi, and Ashwani Kumar**

# Preface

Clean energy and its ever-evolving technologies offer a vital lifeline for our planet's future. By harnessing renewable sources like solar, wind, geothermal, and hydrogen energy, we can curb our reliance on fossil fuels and the associated greenhouse gas emissions that fuel climate change. Clean energy technologies not only generate power but empower sustainable practices. Imagine buildings that minimize energy use, cities that integrate nature to combat urban heat islands, and electric vehicles powered by clean fuels – these advancements are within reach. Embracing clean energy technologies is not just about generating power; but it is about redefining our relationship with the environment and building a sustainable future for generations to come. *Clean Energy: Technology, Advances, and Applications* delves into the cutting-edge world of clean energy technologies, providing readers with a comprehensive understanding of current research and applications.

**Chapter 1** serves as a roadmap, introducing recent trends in clean energy fuel technologies and outlining ongoing research endeavors across various renewable energy domains. It delves into the various techniques for producing biodiesel, exploring advancements that enhance its sustainability. **Chapters 2–4** shift the focus to solar energy, a critical renewable source. **Chapters 2–4** explores deep into advancements designed to enhance PV system performance. **Chapter 2** explores the potential of artificial neural networks for optimizing PV systems, while **Chapter 3** explores the promise of tin-based perovskites for creating efficient and stable PV devices. **Chapter 4** builds upon this foundation by investigating the application of super-hydrophobic surfaces for solar cells, a technology with the potential to further improve efficiency.

**Chapters 5–8** explore the application of clean energy principles in the building sector. **Chapter 5** provides a comprehensive review of energy-efficient green buildings, considering environmental, economic, and technical factors. **Chapter 6** investigates the use of solar chimneys for ventilation in green buildings, a strategy for promoting energy efficiency. Recognizing the growing challenge of urban heat islands, **Chapter 7** explores the development of blue–green infrastructure as a potential solution to mitigate high building energy demands. Finally, **Chapter 8** examines the role of urban green spaces in reducing the urban heat island effect, highlighting the multifaceted benefits of integrating nature into our built environment. **Chapter 9** studies load uncertainty analysis.

**Chapters 10 and 11** delve deeper into the realm of carbon management, exploring cutting-edge technologies for mitigating climate change. **Chapter 10** highlights the exciting potential of urban green spaces for carbon capture, presenting a specific case study on how green parks contribute to carbon sequestration within urban areas. The findings offer valuable insights into the role of urban planning and green infrastructure in mitigating climate change. **Chapter 11** takes a broader perspective on carbon management technologies, tracing the evolution of CCUS technologies. It examines the current state of the art in CCUS, exploring various methods for capturing carbon emissions from industrial sources and power plants. The chapter then takes a forward-looking approach, exploring promising advancements and potential future developments in CCUS, offering a glimpse into a cleaner and more sustainable future.

The latter portion of this book highlights solar energy and alternative fuels, specifically biodiesel, as promising clean energy sources. **Chapter 12** explores various techniques for producing biodiesel, exploring advancements that enhance its sustainability. **Chapter 13** broadens the conversation by examining alternative fuel options for commercial vehicles within the Indian context. This comparative analysis sheds light on the potential of biodiesel relative to other emerging technologies, providing valuable insights for stakeholders in the Indian transportation sector.

Dedicated to the "green hydrogen revolution," **Chapters 14–16** explore the immense potential of hydrogen as a clean and sustainable fuel source. **Chapter 14** serves as a launching pad, exploring the current landscape of hydrogen production technologies. It sheds light on established methods while envisioning advancements on the horizon. **Chapter 15**, with a specific focus on India, examines the exciting possibilities of integrating solar energy with hydrogen production, exploring the potential of this approach in furthering India's clean energy goals. **Chapter 16** bridges the gap between hydrogen production and its application in electric vehicles, reviewing methods for utilizing hydrogen as a clean fuel source for EVs, paving the way for a cleaner transportation future. By dissecting these critical aspects of green hydrogen, these chapters equip readers with a comprehensive understanding of this transformative technology and its potential to reshape the energy landscape.

**Chapters 17–20** take a practical approach, transitioning from foundational concepts to real-world applications. These chapters delve into critical areas where clean energy solutions are making a tangible impact. **Chapter 17** tackles the pressing issue of transportation emissions, employing the analytical hierarchy process – a robust decision-making tool – to comparatively evaluate carbon emissions from different vehicle types. This analysis empowers readers to understand the environmental impact of various transportation options and make informed choices. In **Chapter 18**, the focus shifts to biochar, a charcoal-like substance derived from agricultural waste. This chapter explores conversion techniques for producing biochar and its diverse applications in environmental management. Biochar's potential for soil enhancement, carbon sequestration, and pollutant remediation makes it a valuable tool for promoting sustainability. **Chapter 19** addresses the growing concern of indoor air quality, exploring the development of a model for assessing indoor air quality and its resilience to airborne pathogens. Understanding these dynamics is crucial for designing and maintaining healthy and sustainable buildings. The concluding **Chapter 20** brings the conversation full circle by focusing on urban air quality. It presents a comparative study of pollution levels in different urban environments. By analyzing these variations, readers gain insights into the factors impacting urban air quality and the importance of clean energy solutions in mitigating air pollution.

These application-based chapters showcase the transformative power of clean energy technologies in addressing real-world challenges, paving the way for a cleaner and healthier future.

Editor

**Tikendra Nath Verma, Thokchom Subhaschandra Singh, Upendra Rajak, Prerana Nashine, Gaurav Dwivedi, and Ashwani Kumar**

# Acknowledgment

The journey that led to the fruition of this book, *Clean Energy: Technology, Advances, and Applications*, has been paved with the invaluable contributions of numerous individuals and entities. We are immensely grateful to CRC Press for providing us with a platform to share our knowledge on clean energy. Their unwavering support and insightful guidance throughout the entire publishing process were instrumental in shaping this book into a valuable resource. From initial concept to final publication, their commitment to excellence significantly enhanced the overall quality of the work. Our sincere appreciation goes to the numerous contributors and reviewers who generously shared their expertise. Their diverse perspectives enriched every chapter, contributing to the depth and breadth of the content. Their illuminating feedback ensured the accuracy and comprehensiveness of this book, making it a vital reference for the scholarly community.

We dedicate this book to the passionate and dedicated individuals whose tireless efforts are driving advancements in clean energy. Their relentless pursuit of innovation inspires us all. We hope that this work serves as a source of inspiration and a valuable reference for the scholarly community, furthering the collective effort toward a sustainable future. Finally, we honor the collaborative spirit that unites us all in the pursuit of excellence and progress within the dynamic intersection of engineering and technology. This book stands as a testament to the power of collaboration, and we are grateful to have been a part of this collective endeavor.

# About the Editors

**Tikendra Nath Verma** earned a Bachelor of Engineering (BE) degree in mechanical engineering with first division from Pt. Ravi Shankar Shukla University, Raipur, Chhattisgarh, in 2006. He completed his master's degree in thermal engineering in 2009 at the Department of Mechanical Engineering, Maulana Azad National Institute of Technology, Bhopal, Madhya Pradesh. He earned a PhD from the National Institute of Technology, Raipur, Chhattisgarh. He joined the Institute of Technology, Korba, Chhattisgarh, as an assistant professor in July 2013, serving there until September 2015. He then joined the National Institute of Technology Manipur as an assistant professor in September 2015, where he worked until May 2020. In June 2020, he joined the Department of Mechanical Engineering at the Maulana Azad National Institute of Technology (MANIT), Bhopal, as an assistant professor. He is currently an associate professor at MANIT, Bhopal. He has three ongoing projects supported by various funding agencies, including the Department of Science and Technology-Core Research Grant (DST-CRG), Technical Education Quality Improvement Programme (TEQIP)-III, Seed Grant, and DST Manipur. He has published more than 213 research articles (11 in Science Citation Index (SCI)/Science Citation Index Expanded (SCIE) and 35 in Scopus) in various national and international journals, book chapters, and conferences (51) with high impact factors. His work has been cited 5194 times, with an h-index of 44 and an i10 index of 97. He has successfully guided six PhD and 22 MTech students. He has organized several programs as a convener and organizing secretary for different Faculty Development Programmes (FDPs), workshops, and seminars. He holds six granted patents and has filed six patents in India and Australia. He serves as a guest editor for the *Environmental Science & Pollution Research, Materials Today: Proceedings, and Lecture Notes in Mechanical Engineering. He is also the editor of Vegetable Oil Volume 1: Biofuel Technology and Volume 2: Future of Different Seeds Oil.* Recently, he was featured in the "Top 2% scientist of the world from India in all fields" by a survey conducted by Stanford University. He is an active member of the Institution of Engineers (India) (IEI), Fluid Mechanics and Fluid Power, and the Indian Society of Heat and Mass Transfer.

He also organized four KARYASHALA (High-End Workshops) under the DST SERB Scheme (each 5 lakhs).

**Thokchom Subhaschandra Singh** earned a Bachelor of Engineering (BE) in mechanical engineering with Distinction and a Master of Engineering (ME) in thermal engineering with Distinction from Anna University, Chennai. He earned a Doctor of Philosophy (PhD) in mechanical engineering from the National Institute of Technology Manipur. He is currently an assistant professor of mechanical engineering at the National Institute of Technology Manipur and has over 8 years of teaching experience. He has completed projects funded by DST Manipur and TEQIP-III, NIT Manipur. He has published numerous research articles in reputed international and national journals and presented his work at international and national conferences. He has supervised more than 15 MTech students in thermal and fluids engineering at NIT Manipur. He is also associated with the IEI Manipur State Centre as the Honorary Secretary (2023 onward) and has previously served as the Joint Honorary Secretary of the State Centre of the IEI (2018–2021 and 2021–2023). He has published book chapters with reputed publishers and has reviewed many articles in reputed journals. He has been granted three Australian patents and one Indian design patent. He successfully established "Fuel Production and Testing Laboratory" in the Department of Mechanical Engineering at NIT Manipur in 2020. He holds professional memberships in reputed organizations such as the American Society of Mechanical Engineers (ASME), Institute of Engineers (IE) (India), International Association of Engineers (IAENG), Society of Automotive Engineers (SAE), and Indian Society of Mechanical Engineers (ISME).

**Upendra Rajak** earned a Bachelor of Engineering (BE) degree in mechanical engineering with first division from Rajiv Gandhi Proudyogiki Vishwavidyalaya (RGPV), Bhopal, Madhya Pradesh, in 2010. He completed a master's degree in hydropower engineering in 2013 from the Department of Civil Engineering, Maulana Azad National Institute of Technology, Bhopal, Madhya Pradesh. He served as an assistant professor at the Vidhyapeeth Institute of Science & Technology, Bhopal, from July 2013 to July 2016. He earned a PhD from the National Institute of Technology Imphal, Manipur. He joined the Rajeev Gandhi Memorial College of Engineering and Technology, Nandyal, as an assistant professor in February 2020 and is currently an assistant professor (Grade II) in the Department of Mechanical Engineering. His research interests include computational fluid dynamics, heat and mass transfer, and renewable energy and alternative fuels in internal combustion engines. He has published more than 100 papers in SCI- and Scopus-indexed journals, authored two books, and contributed to seven book chapters in reputed publishers. He also serves as a potential reviewer for reputed journals such as *Energy, Fuel, SN Applied Sciences, International Journal of Thermal Engineering, and Energy Sources, Part A: Recovery, Utilization, and Environmental Effects.*

**Prerana Nashine** earned a Bachelor of Technology (BTech) degree in mechanical engineering from the National Institute of Technology, Raipur, Chhattisgarh. She completed a master's degree in energy systems and pollution from the Department of Mechanical Engineering, National Institute of Technology, Raipur, Chhattisgarh. She earned a PhD from the National Institute of Technology, Rourkela. She has 8 years of teaching experience and 1.5 years of industrial experience. She worked as a mechanical engineer at DSC Limited. She worked as an assistant professor in the Department of Mechanical Engineering at Manipur Technical University, Imphal. From November 2022, she has been an assistant professor in the PSS Central Institute of Vocational Education (a constituent unit of the National Council of Educational Research and Training (NCERT) under the Ministry of Education, Government of India). She has published 37 research articles (25 in SCI, 8 in Scopus, and 4 in University Grant Commission (UGC)-approved/non-SCI-approved journals) in various national and international journals with high impact factors. She has also organized programs as a convener and organizing secretary for different FDPs, workshops, and seminars. Additionally, she has served as a potential reviewer for reputed journals.

**Gaurav Dwivedi** completed a BTech in mechanical engineering from I.K. Gujrat Punjab Technical University (IKGPTU), Jalandhar; an MTech in energy systems from IIT Roorkee; and a PhD in renewable energy (biofuel) from IIT Roorkee, India. He has 12 years of teaching experience in mechanical and energy engineering, during which he taught various subjects such as IC engine, thermodynamics, fluid machinery, and bioenergy at various institutes including VIT University, Vellore, and Amity University, Noida. Currently, he is an assistant professor in the Energy Centre at the Maulana Azad National Institute of Technology, Bhopal, India. His research interests include biodiesel production methodology, enhancement of fuel properties, and the application of biofuel and hydrogen in engine operations. He has published more than 150 papers in various international journals and presented his work at various international conferences, with more than 80 papers indexed in SCI and Scopus. Additionally, he has authored four books with international publishers and contributed to 12 book chapters. He received the Young Scientist Award at NIT Trichy for his contributions to the field of renewable energy. He is an active member of the IEI, Solar Energy Society of India, and IEEE. He also served as a guest editor for international journals including *Sustainable Energy Technology and Assessment, Materials Today: Proceedings*, and the *Journal of Traffic and Transportation*. He is currently managing three ongoing projects with a combined funding of more than 45 lakhs, sponsored by DST and MPCST.

He has also been featured in the "Top 2% scientist of the world from India (all fields)" for the years 2020, 2021, 2022, and 2023, according to a survey conducted by Stanford University.

**Ashwani Kumar** holds a PhD in mechanical engineering, specializing in mechanical vibration and design. Currently, he serves as a senior lecturer in mechanical engineering at the Technical Education Department of Uttar Pradesh, India, under the Government of Uttar Pradesh, a position he has held since December 2013. Prior to this, he worked as an assistant professor in the Department of Mechanical Engineering at Graphic Era University, Dehradun, India from July 2010 to November 2013. With over 14 years of combined experience in research, academia, and administration, he has held various roles, including coordinator for All India Council for Technical Education (AICTE)-Extension of Approval, nodal officer for the Pradhan Mantri Kaushal Vikas Yojana for Technical Institutes (PMKVY-TI) Scheme (Government of India), internal coordinator for the CDTP Scheme (Government of Uttar Pradesh), industry-academia relation officer, assistant center superintendent in the Institute Examination Cell, zonal officer for conducting the Joint Entrance Examination (Diploma), and Sector Magistrate for State Assembly Elections held in Uttar Pradesh. As an academician and researcher, he serves as the series editor for five book series: *Advances in Manufacturing, Design and Computational Intelligence Techniques, Renewable and Sustainable Energy Developments, Smart Innovations and Technological Advancements in Mechanical and Materials Engineering, Energy Systems, and Computational Intelligence and Biomedical Engineering*. Furthermore, he is the guest editor for a special issue titled "*Sustainable Buildings, Resilient Cities, and Infrastructure Systems*" in Buildings (ISSN: 2075-5309, I.F. 3.8). He serves as editor-in-chief of the *International Journal of Materials, Manufacturing and Sustainable Technologies* (ISSN: 2583-6625) and as editor of the *International Journal of Energy Resources Applications* (ISSN: 2583-6617). Additionally, he acts as a guest editor and editorial board member for eight international journals and a review board member for 20 prestigious international journals indexed in SCI/SCIE/Scopus, including *Applied Acoustics, Measurement, Journal of Engineering Science and Technology (JESTEC)*, and *Arabian Journal for Science and Engineering (AJSE)*. He has an extensive publication record, with more than 100 articles in journals, book chapters, and conferences. He has authored, co-authored, and edited more than 40 books in mechanical, materials, and renewable energy engineering. He holds three patents and has been recognized for his contributions to academia and research, earning the title of Best Teacher for excellence in both fields. Actively engaged in academic pursuits, he has successfully supervised 15 BTech, MTech, and PhD theses and serves as an external doctoral committee member of S.R.M. University, New Delhi. He is associated with various international conferences as an invited speaker, session chair, advisory board member, review board member, and program committee member. He has delivered numerous invited talks in webinars, FDPs, and workshops. His current research interests include AI and ML in mechanical engineering, smart materials and manufacturing techniques, thermal energy storage, building efficiency, renewable energy harvesting, and sustainable mobility. Dr. Kumar's ORCID ID is 0000-0003-4099-935X, and his Google Scholar web link is https://scholar.google.com/citations?hl=en&user=KOILpEkAAAAJ.

# Contributors List

**Meena Agrawal**
Energy Centre
Maulana Azad National Institute of Technology Bhopal
Bhopal, India

**Mohd Anas**
Department of Electrical Engineering
JSS Academy of Technical Education
Noida, India

**Farzan Ashraf**
Department of Mechanical Engineering
Jamia Millia Islamia
New Delhi, India

**Sonal Santosh Bagade**
Department of Physics
Maulana Azad National Institute of Technology
Bhopal, India

**Sachin Kumar Banait**
Energy Centre
Maulana Azad National Institute of Technology
Bhopal, India

**Prashant Baredar**
Energy Centre
Maulana Azad National Institute of Technology
Bhopal, India

**Sudhansu B. Barik**
Department of Physics
Maulana Azad National Institute of Technology
Bhopal, India

**Vibhor Bhardwaj**
Department of Architecture and Planning
Maulana Azad National Institute of Technology
Bhopal, India

**Shubhangi Choudhary**
Department of Architecture and Planning
Maulana Azad National Institute of Technology
Bhopal, India

**Deepika**
Department of Environmental Engineering
Delhi Technological University
New Delhi, India

**Ashish Dewangan**
Department of Mechanical Engineering
Galgotias College of Engineering & Technology
Greater Noida, India

**Aishwarya Dwivedi**
Department of Architecture and Planning
Maulana Azad National Institute of Technology
Bhopal, India

**Azhar Equbal**
Department of Mechanical Engineering
Jamia Millia Islamia
New Delhi, India

**Arun Kumar Singh Gangwar**
Department of Mechanical Engineering
Integral University
Lucknow, India

**Mariette Rose George**
Department of Architecture and Planning
Maulana Azad National Institute of Technology
Bhopal, India

**Bharat Girdhani**
Energy Centre
Maulana Azad National Institute of Technology Bhopal
Bhopal, India

**Anil Kumar Haritash**
Department of Environmental Engineering
Delhi Technological University
New Delhi, India

**Sourabh Kaithwas**
Energy Centre
Maulana Azad National Institute of Technology Bhopal
Bhopal, India

**Sakhi Katre**
Energy Centre
Maulana Azad National Institute of Technology
Bhopal, India

**Sujeet Kesharvani**
Energy Centre
Maulana Azad National Institute of Technology
Bhopal, India

**M. Emran Khan**
Department of Mechanical Engineering
Jamia Millia Islamia
New Delhi, India

**Osama Khan**
Department of Mechanical Engineering
Jamia Millia Islamia
New Delhi, India

**Aravind Kumar**
Department of Physics
Maulana Azad National Institute of Technology
Bhopal, India

**M.M. Malik**
Department of Physics
Maulana Azad National Institute of Technology
Bhopal, India

**Surabhi Mehrotra**
Maulana Azad National Institute of Technology
Bhopal, India

**Sachin Kumar Nikam**
LNCT University
Bhopal, India

**Anand Kumar Pandey**
Department of Electrical and Electronics Engineering
JSS Academy of Technical Education
Noida, India

**Mohd Parvez**
Department of Mechanical Engineering
Al Falah University
Faridabad, India

**Piyush K. Patel**
Department of Physics
Maulana Azad National Institute of Technology
Bhopal, India

**Bipasa Bimalendu Patra**
Department of Electrical Engineering
Padmashri Dr V B Kolte College of Engineering
Malkapur, India

**Ved Prakash**
Department of Mechanical Engineering
Galgotias College of Engineering & Technology
Greater Noida, India

**Rahul Kumar Rajak**
Energy Centre
Maulana Azad National Institute of Technology
Bhopal, India

**Jyoti Rani**
Novel Materials Laboratory
Department of Physics
Maulana Azad National Institute of Technology
Bhopal, India

**Tejaswi Ratre**
Novel Materials Laboratory
Department of Physics
Maulana Azad National Institute of Technology
Bhopal, India

**Sanju Saini**
Electrical Engineering Department
Deenbandhu Chhotu Ram University of Science and Technology
Murthal, India

**Olusegun David Samuel**
Federal University of Petroleum Resources
Effurun, Nigeria

**Vishal Saraswat**
Department of Mechanical Engineering
Galgotias College of Engineering & Technology
Greater Noida, India

**Sakshi Sarathe**
Energy Centre
Maulana Azad National Institute of Technology
Bhopal, India

**Mohammad Shajar**
Department of Mechanical Engineering
Jamia Millia Islamia
New Delhi, India

**Shivansh Sharma**
Department of Architecture and Planning
Maulana Azad National Institute of Technology
Bhopal, India

**Rajeev Shrivastava**
LNCT University
Bhopal, India

**Bishnu Kant Shukla**
Department of Civil Engineering
JSS Academy of Technical Education
Noida, India

**Ashish Singh**
Department of Civil Engineering
JSS Academy of Technical Education
Noida, India

**Devesh Singh**
Department of Electrical Engineering
JSS Academy of Technical Education
Noida, India

**Kaustubh Singh**
Energy Centre
Maulana Azad National Institute of Technology
Bhopal, India

**Rajat Soni**
Department of Architecture and Planning
Maulana Azad National Institute of Technology
Bhopal, India

**Raj Suman**
Energy Centre
Maulana Azad National Institute of Technology
Bhopal, India

**Deeptam Trivedi**
Energy Centre
Maulana Azad National Institute of Technology
Bhopal, India

**Anshul Tyagi**
Department of Environmental Engineering
Delhi Technological University
New Delhi, India

**Puneet Verma**
School of Population and Global Health
The University of Western Australia
Perth, Australia

**Arun Kumar Wamankar**
LNCT Bhopal
Bhopal, India

**Ashok Kumar Yadav**
Department of Mechanical Engineering
Raj Kumar Goel Institute of Technology
Ghaziabad, India

**Sandeep Yadav**
Electrical Engineering Department
Deenbandhu Chhotu Ram University of Science and Technology
Murthal, India

Chapter 1

# Fueling the future

## Sustainable biodiesel production strategies for a cleaner tomorrow

*Sujeet Kesharvani, Sakhi Katre, Sachin Kumar Banait, Gaurav Dwivedi, and Ashwani Kumar*

### 1.1 INTRODUCTION

The search for alternatives to petroleum derivatives was driven by the anticipated shortage of fossil fuels, leading to the discovery of "biodiesel" as a viable alternative. However, the concept of biodiesel remains a subject of debate. Some definitions encompass any combination of edible oil and diesel, while others restrict biodiesel to combinations of vegetable oil or animal fat alkyl esters and diesel. The Brazilian Biodiesel Programme defines biodiesel as "A fuel derived from mixes, in varying proportions, of fossil diesel and alkyl esters of vegetable oils or animal fats" (Murugesan et al., 2009).

Apart from hydroelectricity and nuclear energy, most of the world's energy needs are met by petrochemicals, coal, and natural gas. However, these resources are finite and will be rapidly depleted if current consumption rates continue. Diesel is critical for transportation, the production of industrial and agricultural goods, and other aspects of a developing nation's industrial economy. In India, the demand for diesel is six times greater than the demand for gasoline worldwide. Diesel consumption in India increased from 28.30 million tonnes in 1994–1995 to 40.34 million tonnes in 2000–2001, accounting for 43.2% of all petroleum products used during that period. Economic growth consistently leads to a corresponding increase in the transportation sector, which has driven research into alternative fuel sources to replace petroleum-based fuels. An alternate fuel should be readily accessible, environmentally responsible, and technologically competitive. Triglycerides and their derivatives are among the promising fuels. Vegetable oils pose a lesser environmental impact, with reduced greenhouse gas emissions compared to diesel, owing to their renewable nature, widespread availability from various sources, and minimal sulfur content, approaching zero. Chemically, these oils consist of three long-chain fatty acid triglyceride molecules that are bonded to a single glycerol molecule via ester bonds. The length, orientation, and placement of the double bonds in the carbon chains of these fatty acids vary. Thus, the term "biodiesel" refers to lower

DOI: 10.1201/9781003521341-1

alkyl esters of long-chain fatty acids produced either through esterification of fatty acids or through transesterification with lower alcohols (*Biodiesel Production Methods*, 2017).

### 1.1.1 Biodiesel

According to technical definitions, biodiesel is an alkyl ester derived from fatty acids produced through the transesterification of animal or plant-based oils or fats with short-chain alcohols such as ethanol and methanol. Glycerine is a byproduct of the biodiesel manufacturing process. Over the last 30 years, biodiesel production techniques have evolved, progressing from batch processes through quasi-flow systems and then to continuous flow systems.

As a fuel, biodiesel offers advantages over regular diesel. It is made from sustainable materials, emits fewer pollutants into the environment, and degrades quickly after use, and perhaps most importantly, its production may be decentralized, thereby supporting rural economies. However, biodiesel production also present several challenges that are worth highlighting. Among the most frequently discussed are the high cost of raw materials and the need to select efficient methods for producing biodiesel from various feedstock sources (Pinto et al., 2005).

FFAs, phospholipids, sterols, water, odorants, and other contaminants are typically present in plant oils, preventing their direct use as fuel. To address this issue, the oil needs to undergo key chemical modifications such as transesterification, pyrolysis, and emulsification. Among these, transesterification is a crucial step in converting vegetable oils into cleaner, safer fuel. Biodiesel, which is used in compression ignition engines, is defined as the mono-alkyl esters of long-chain fatty acids produced from renewable feedstocks, such as vegetable oil or animal fats. By transesterifying triglycerides with methanol or ethanol, biodiesel, which is being investigated as a potential replacement for traditional diesel, is typically composed of fatty acid methyl or ethyl esters. Traditional diesel and biodiesel can be mixed together (Singh et al., 2019).

### 1.1.2 Biodiesel production sources

The source of biodiesel synthesis is selected based on the availability if resources in each country or region. While fatty acids from various sources can be used to produce biodiesel, the majority of scientific literature uses soybean as the biodiesel feedstock. However, because edible vegetable oils, such soybean oil, are more expensive than conventional diesel, waste vegetable oils, and non-edible raw vegetable oils are considered potential low-cost biodiesel feedstock. Low-quality, underutilized sources have also been used to produce biodiesel. This approach is projected to be more cost-effective than producing biodiesel from processed soybean oil because it is similar in composition, equivalent in engine efficiency, and similar in emissions.

There are four primary methods to convert vegetable oil and/or animal fat into diesel engine fuel: direct usage or blending, transesterification, pyrolysis or thermal cracking, and micro-emulsion. Among these, transesterification reaction is most favored of technique. This process enables the production of fuel with properties that closely mimic conventional diesel while allowing the use a variety of feedstocks. Through transesterification, oils and fats (triglycerides) are converted into to their alkyl esters, with a viscosity similar to that of diesel.

All systems need some form of feedstock to start the reaction, an alcohol, and typically a catalyst, along with processes for separation and purification. Gravity or centrifugation is frequently used for separation. Contaminants are often removed via water washing, a dry wash process using absorbents, ion exchange resins, filtering, or a combination of these. Depending on the type of purification being utilized, the methanol removal might be done either before or after the wash. This section describes some of the various production methods in use today. Each method has distinct benefits and drawbacks, as well as a distinctively practical feedstock nature.

### 1.1.3 Potential feedstock

According to the European Academies Science Advisory Council report, the feedstock used for the synthesis of biodiesel can be categorized into four generations (Zulqarnain, Ayoub, et al., 2021). Environmental factors such as regional weather patterns, topography, agricultural practices, and soil conditions play a significant role in determining the availability and cost of feedstock for biodiesel production (*Biodiesel Production Methods*, 2017). These feedstock categories can be classified as first, second, third, and fourth generations (Pinto et al., 2005; Singh et al., 2019; Zulqarnain, Ayoub, et al., 2021), which include edible, non-edible, algae and waste, oleaginous biomasses (Kaya et al., 2018), and synthetic biological cells, respectively (Murugesan et al., 2009). A comprehensive classification of these feedstocks is illustrated in Figure 1.1.

#### *1.1.3.1 First-generation feedstocks*

These feedstocks represent readily accessible feedstock options for biodiesel production, encompassing edible oils such as palm, rapeseed, animal fats, soybean, sunflower, and coconut. However, using these oils impacts the global food supply chain, thereby influencing both availability and price. Due to their high demand for food, edible oils are expensive, underscoring the importance of avoiding their use for biodiesel production.

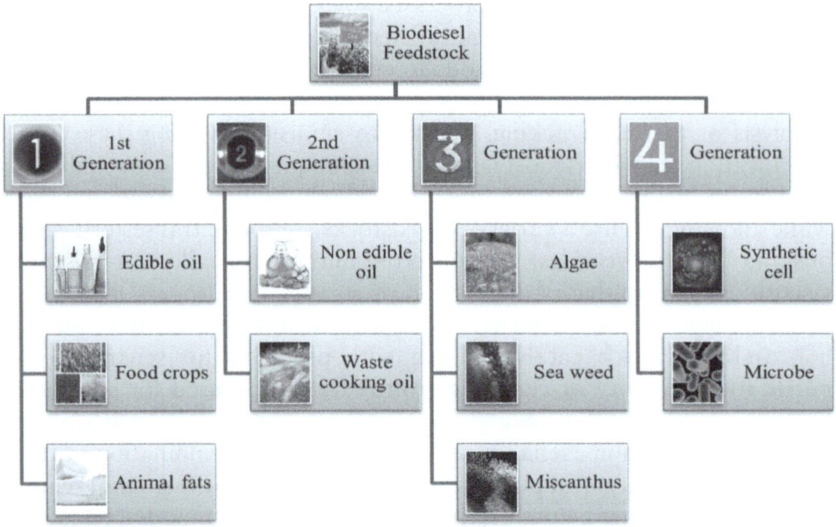


*Figure 1.1* Various generations of biodiesel feedstock (Zulqarnain, Ayoub, et al., 2021).

#### *1.1.3.2 Second-generation feedstocks*

Second-generation feedstocks were introduced to address the limitations associated with first-generation biofuels. These feedstocks primarily include food waste (Ayodele & Dawodu, 2014; Kaya et al., 2018), non-edible oils, and yellow grease (Lin & Chen, 2017). Non-edible oils have gained considerable traction in biodiesel production in recent years (Vasaki E et al., 2021). Compared to first-generation feedstocks, second-generation alternatives offer increased efficiency and greater ecological sustainability, potentially reducing reliance on first-generation sources (Karthikeyan et al., 2020; Wang et al., 2021). Algae, in particular, present several advantages, including lower costs, enhanced energy efficiency, higher energy yield per acre compared to conventional food crops, and carbon neutrality (Shah et al., 2018).

#### *1.1.3.3 Third-generation feedstocks*

Third-generation feedstocks, including miscanthus, seaweed, and algae biomass, have been developed to reduce dependence on previous alternatives. Algae, specifically, can be categorized into microalgae, macroalgae, and cyanobacteria based on their morphology and size (Mohiddin et al., 2021). Algae are environmentally friendly and require minimal space for cultivation, thriving on carbon dioxide, water, sunlight, and occasional nutrient supplementation. By consuming $CO_2$ during growth, they contribute to reducing greenhouse gas emissions (Ayodele & Dawodu, 2014; Bharathiraja et al., 2014).

#### 1.1.3.4 Fourth-generation feedstocks

Fourth-generation feedstocks encompass electric fuels, microbes, synthetic cells, and solar fuels. Solar biofuel production, a relatively novel area of research, involves converting raw materials into biodiesel using solar energy. These raw materials are inexhaustible, cost-effective, and readily accessible. Synthetic biology plays a crucial role in enabling these transformations. To achieve sustainability, innovative technologies such as synthetic living organisms and engineered microbes capable of directly converting solar energy into biodiesel are being explored (Holma et al., 2013; Razak et al., 2023). In future, fourth-generation biodiesel has the potential to address the limitations of the first three generations and offer a more efficient alternative to fossil fuels (Ahmad et al., 2011; Bharathiraja et al., 2014; Goh et al., 2019; Shah et al., 2018) (Table 1.1).

## 1.2 EDIBLE OILS

About 96% of the triacylglycerides in edible oils consist of various fatty acids. Other substances including phospholipids, FFAs, tocopherols, phytosterols, waxes, and additional antioxidants may also be present. When fatty acids undergo oxidative processes, whether they are free or bound to glycerol, they produce a range of volatile and non-volatile breakdown products. One of the biggest issues for the oil processing industry is maintaining the high quality of product from post-processing till end-user consumption. Additionally, the oxidative stability of edible oils is influenced by the origin of raw materials and the procedural steps taken during processing (Norjannah et al., 2016).

Vegetable oils constitute the majority of edible oils and undergo several processes to remove unwanted components. The majority of crude vegetable oils are refined through neutralization, bleaching, and deodorization to make them suitable for consumption. Olive oil is the only natural juice that can be consumed without any refinement.

### 1.2.1 Groundnut oil

Groundnut oil has been a primary dietary component in India since ancient times. Groundnut require sandy soil with good drainage or clay loam soil for optimal growth. The ideal pH range for soils is between 5.5 and 7. Groundnut cultivation thrives in tropical and subtropical climates. As an important oilseed crop, groundnut is also a cheap source of vegetable protein, containing 1.5 times the protein of meat, 2.5 times that of eggs, and 8 times that of fruits. Additionally, groundnuts are composed of 50% fat, making them a truly rich source of nutrients (Table 1.2).

*Table 1.1* Benefits and drawbacks of various generations of feedstock

| *Feedstock* | *Advantages* | *Disadvantage* | *References* |
|---|---|---|---|
| **First-generation feedstock** | | | |
| Animal fats<br>Rapeseed<br>Palm<br>Soybean<br>Sunflower<br>Canola | Easily available<br>Ensuring an adequate supply to meet both present and future biodiesel demands | Biodiversity loss<br>Deforestation<br>Water scarcity<br>Food insecurity<br>Soil degradation<br>Food insecurity High production cost | (Bharathiraja et al., 2014; Mohiddin et al., 2021; Razak et al., 2023) |
| **Second-generation feedstock** | | | |
| Neem<br>Karanja oil<br>Waste oil<br>Beef tallow<br>Restaurant grease<br>Jatropha<br>Rubber seed<br>Pork lard | Minimal cost of production<br>Food security | Low yield<br>Inadequate cold flow characteristics<br>Land for growth<br>Low-quality gasoline<br>Increased alcohol consumption | (Foteinis et al., 2020; Razak et al., 2023; Razzaq et al., 2020; Verma et al., 2021) |
| **Third-generation feedstock** | | | |
| Microalgae<br>Macroalga<br>Cyanobacteria<br>Algae weed | Carbon neutrality of algae biofuel<br>Elevated lipid content<br>No requirement for rich soil<br>Reduced greenhouse gas emissions<br>Reducing pressure on agricultural land<br>The rate of growth of biomass is high<br>Able to grow effluent<br>Wastewater may effectively be cleaned of N and P nutrients by microalgae<br>Elevated growth rate and productivity<br>Increased oil yield | Higher production cost<br>High need for sunlight<br>Need to be produced on large scale<br>Difficult for oil extraction<br>Used Cooking Oil (UCO) and Waste Animal Fat (WAF) are impure<br>Poor system for raw material collection<br>High concentration of FFAs in UFO | (Aro, 2016; Bharathiraja et al., 2014; Mohiddin et al., 2021; Pikula et al., 2020) |
| **Fourth-generation feedstock** | | | |
| Electro biofuels<br>Genetic and metabolic modified microorganisms and crops<br>Synthetic cell<br>Photobiological solar biodiesel | Elevated lipid concentration<br>Optimized nutrient utilization<br>Increased energy density<br>Enhanced $CO_2$ absorption | Decline in biodiversity<br>Toxic strains are introduced into the ecosystem<br>High initial expenditure<br>Ongoing research efforts | (Bharathiraja et al., 2014; Holma et al., 2013; Razzaq et al., 2020; Zulqarnain, Mohd Yusoff, et al., 2021) |

FFA, free fatty acid.

*Table* 1.2 Characteristics of notified varieties of groundnut oil (seednet.gov.in)

| *Variety* | *Released center* | *Year of release* | *Oil content* | *Recommended states* | *Yield potential* | *Safety features* |
|---|---|---|---|---|---|---|
| TG-39 | Raj AU, Bikaner | 2008 | | Arid and semi-arid regions of Rajasthan | 2,054–3,154 | Medium duration |
| TG 51 | BARC, Mumbai | 2008 | | Assam, West Bengal, Jharkhand, and Odisha | 2,675 | Resistant to root rot and stem rot, and well-suited for cultivation during *rabi*-summer season |
| ICGV 00348 | TNAU, Vridha chalam | 2008 | 47 | Tamil Nadu, Karnataka, southern Maharashtra, and Andhra Pradesh | 2,013 | Resilient to late leaf spot and rust, and advised for cultivation during the *kharif* season |
| Girnar 2 (PBS-24030) | NRCG, Junagadh | 2008 | 51 | Punjab, Uttar Pradesh, and northern Rajasthan | 2,907 | Virginia bunch type characterized by "stay green" leaves and bold seeds (Hundred Seed Mass (HSM) = 62 g); exhibits tolerance to rust, Late Leaf Spot (LLS), and Peanut Stem Necrosis Disease (PSND); and recommended for cultivation during the *kharif* season |
| Ajeya (R 2001–3) | UAS, Raichur | 2008 | 46–48 | Karnataka, southern Maharashtra, Andhra Pradesh, and Tamil Nadu | 2,440 | Tolerant to Peanut Bud Necrosis Disease (PBND), exhibits resilience to drought conditions, and recommended for the *kharif* season |

### 1.2.2 Sunflower oil

Sunflower oil is extracted from the seeds of the sunflower plant (*Helianthus annuus*). It has high levels of linoleic acid, an essential fatty acid. Some sunflower varieties are cultivated to have a high oleic acid content, producing a type of oil known as high-oleic sunflower oil. Sunflower oil serves as a dietary source of polyunsaturated fat and is used to lower cholesterol and the risk of heart disease. Additionally, it is used to treat eczema, high blood pressure, dry mouth, and dry skin (Khammee et al., 2021) (Table 1.3).

### 1.2.3 Safflower

Safflower (*Carthamus tinctorius*) is a perennial herbaceous plant in the Asteraceae family, which is characterized by multiple branches and resemblance to a thistle. In the early Spanish colonies along the Rio Grande, it was used as a substitute for saffron. It is currently commercially cultivated for the vegetable oil extracted from its seeds. Due to its high oil content and favorable properties, safflower is a viable option for biodiesel production. In-depth research has been conducted on the physical and chemical properties of safflower oil biodiesel. Without engine modifications, safflower oil biodiesel and its blends can be used in various engines (Khammee et al., 2021) (Table 1.4).

### 1.2.4 Sesame

Sesame, also referred to as "til" or "gingelly," is one of the oldest oilseed crops and a significant source of oil, with an oil content of 40%–50%. The oil come in a range of hues, including yellow, white, black, and red, depending on the type of seed. Sesame is cultivated in both large and small areas across almost all states in India. Due to its exceptional resistance to

*Table 1.3* Characteristics of notified varieties of sunflower oil (seednet.gov.in)

| *Variety* | *Released center* | *Year of release* | *Oil content* | *Recommended states* | *Yield potential* | *Safety features* |
|---|---|---|---|---|---|---|
| Phule Bhaskar (SS-0808) | ZARS Rahuri | 2016 | 37.9 | Maharashtra | 3,000 | |
| Phule Raviraj | MPKV, Rahuri | 2009 | 34 | Western Maharashtra | 1,795 | Tolerant to capitulum borer, *Alternaria*, and necrosis |
| RSFV-901 (Kanthi) | RARS-Raichur | 2012 | - | Karnataka | 1,200–1,400 | Tolerant to necrosis |

*Table 1.4* Characteristics of notified varieties of safflower oil (seednet.gov.in)

| *Variety* | *Released center* | *Year of release* | *Oil content* | *Recommended states* | *Yield potential* | *Safety features* |
|---|---|---|---|---|---|---|
| NARI-96 | AICRP, Nari, Phaltan | 2018 | 33 | Maharashtra, AP, Rajasthan, MP, Telangana, and Chhattisgarh | 2,020 | Recommended for irrigated area |
| DSH-185 | IIOR Hyderabad | 2018 | 28–29 | All India | 1,600–2,250 | Resistant to Fusarium wilt |
| SSF-708 | MPAU, Solapur | 2012 | 29 | Western Maharashtra | 1,300–2,200 | Exhibits moderate tolerance to aphids |
| PKV-Pink (AKS-311) | SRS, Oilseeds, PDKV, Akola | 2013 | 33 | Vidharbha region of Maharashtra | 2,200–2,500 | Displays moderate tolerance to wilt |
| NARI-57 | AICRP, Nari, Phaltan | 2015 | 29–30 | Maharashtra, WB, KNK, J&K, MP, Punjab, Raj, and UP | 1,500 | Shows resistance to wilt and is recommended for irrigated areas |
| SSF-658 | MPKV, Solapur | 2009 | 28 | All India | 1,430 | Demonstrates tolerance to aphids and wilt |

oxidation and rancidity, sesame is referred to as the "queen of oilseeds" and is also known as gingelly, til, or benne seed. In Andhra Pradesh, the Andhra Pradesh Southern Power Distribution Company Limited (APSSDCL) produced and distributed an average of 50–100 qtls of sesame per year from the 47 lakh hectares of land under sesame cultivation (Khammee et al., 2021) (Table 1.5).

### 1.2.5 Niger

Niger is grown primarily during the kharif season in India, covering an area of 2.61 lakh hectares. However, in Odisha, it is cultivated as winter crop. Assam, Gujarat, Karnataka, Maharashtra, West Bengal Andhra Pradesh, Chhattisgarh, Jharkhand, Madhya Pradesh, and Odisha are among the states where niger in grown. The average yield of niger in India is 3.21 q/ha (Khammee et al., 2021) (Tables 1.6 and 1.7).

## 1.3 BIODIESEL PRODUCTION TECHNOLOGIES

The prevalent technologies that enable the use of diverse fat and oil feedstocks as fuel in diesel engines include direct usage or blending of oils, transesterification, pyrolysis, and micro-emulsion. Among these, transesterification is the most preferred method as it produces fuel of higher quality, according to several scholars (Saengsawang et al., 2020) (Figure 1.2).

### 1.3.1 Pyrolysis

Pyrolysis is a chemical process in which thermal energy is applied in the absence of air or oxygen, or in the presence of a catalyst, to break molecular bonds, leading to the production of a variety of small molecules. The temperature range for pyrolysis typically falls between 400°C and 600°C. Depending on the rate of pyrolysis, the primary outputs are char, bio-oil, and gases.

The pyrolysis process can be categorized into three subdomains based on operational parameters: flash pyrolysis, fast pyrolysis, and traditional pyrolysis. The process used to produce bio-oil is rapid pyrolysis.

Diesel engines can run on biodiesel produced by the pyrolysis process, often referred as bio-oil. However, the elimination of oxygen during the process results in low-value components, leading to undesirable traits such as low heating value, incomplete volatility, and instability, which can limit the applicability of biodiesel produced through this method. Another drawback is the need for distillation equipment to separate various fractions. Moreover, the resulting product shares similarities with sulfur-containing gasoline, diminishing its environmental advantages.

*Table* 1.5 Characteristics of notified varieties of sesame oil (seednet.gov.in)

| *Variety* | *Released center* | *Year of release* | *Oil content* | *Recommended states* | *Yield potential* | *Safety features* |
|---|---|---|---|---|---|---|
| Jawahar Til–12 (PKDS-12) | ZARS, JNKVV, Powarkheda (MP) | 2008 | 700–750 | Summer-MP | 48–52 | White seed, matures in 82–85 days, and exhibits moderate resistance to *Macrophomina* stem/root rot |
| RT-346 | BAU, Bikaner | 2009 | 750–850 | Kharif, Maharashtra Rajasthan, Uttar Pradesh, Haryana, Gujarat, Himachal Pradesh, and Punjab | 49–51 | Features white seed, matures in 82–86 days, and exhibits resistance to leaf curl and moderate resistance to *Cercospora*, *Alternaria*, and *Macrophomina* |
| SWB-32-10-1 (Savitri) | ORS, Berhampore (WB) | 2008 | 1200–1500 | Rabi/summer-West Bengal | 48–52 | Featuring light brown seeds, ripens in 84–88 days, anddemonstrates tolerance to *Macrophomina* |
| Jawahar Til–14 (PKDS-8) | ZARS, JNKVV, Powarkheda (MP) | 2008 | 700–750 | Summer-MP | 50–53 | Duration 82–85 days |
| Gujarat Til-3 | ARS, GAU, Amreli (Gujarat) | 2009 | 750–800 | *Kharif*, Saurashtra region of Gujarat | 48–52 | White bold seed, matures in 84–88 days |

*Table* 1.6 Characteristics of notified varieties of niger (seednet.gov.in)

| *Variety* | *Released center* | *Year of release* | *Oil content* | *Recommended states* | *Yield potential* | *Safety features* |
|---|---|---|---|---|---|---|
| IGPN-2004-1 (Phule Karala-1) | ZARS, Igatpuri (MS) | 2010 | 650–700 | *Kharif* season in Karnataka and MS (high rainfall areas of Maharashtra) | 39–41 | Shining black seed, growth duration 95–100 days, and shows resistance to powdery mildew and *Alternaria* |
| Birsa Niger-3 (BNS-11) | BAU, Ranchi (Jharkhand) | 2010 | | Jharkhand, Kerala Chhattisgarh, WB, MP, Maharashtra, Odisha, and AP | | Early maturing variety, ready in just 85 days, shows strong tolerance to drought conditions |
| KBN-1 | UAS, Bangalore | 2009 | | Karnataka | | Shows moderate resistance to *Alternaria* leaf spot |
| BNS-10 (Pooja-1) | BAU, Ranchi | 2009 | 650–700 | All niger growing states | 36–38 | Shining black seed, duration 95–100 days, and displays resistance to diseases and pests |
| Utkal Niger-150 | RRTTS, OUA&T, Semiliguda (Odisha) | 2011 | 650–700 | Odisha | 38–40 | Black seed, duration 105–110 days, and tolerant to *Cercospora*, *Alternaria*, and *Cuscuta* |

*Table 1.7* Fuel properties of different straight vegetables oils

| *Type of Oil* | *Vegetable oils* | *Specific gravity @ 150* | *Kinematic viscosity cSt (38°C)* | *Flash point (°C)* | *Cetane number* | *Heating values (MJ/kg)* | *Cloud point (°C)* | *Pour point (°C)* |
|---|---|---|---|---|---|---|---|---|
| **Edible Oils** | Canola | 0.916 | 20.6 | 232 | 40 | 25 | −3.9 | −31.6 |
| | Cottonseed | 0.914 | 33.5 | 234 | 41.8 | 39.5 | 1.7 | −15.0 |
| | Rapeseed | 0.911 | 37.0 | 246 | 37.6 | 39.7 | −3.9 | −31.7 |
| | Soybean | 0.913 | 32.6 | 254 | 37.9 | 39.6 | −3.9 | −12.2 |
| | Sunflower | 0.914 | 33.2 | 314 | 37 | 39.5 | −6.5 | −8.7 |
| | Corn | 0.924 | 36.0 | 280 | 37 | 39.5 | 3.0 | −12.5 |

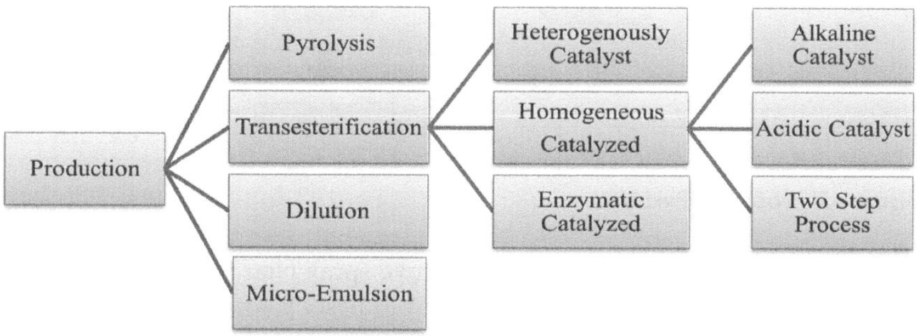


*Figure 1.2* Biodiesel production technologies.

The equipment required for thermal cracking and pyrolysis can be expensive, especially for operations with lower throughputs. Furthermore, while the chemical composition of the products resembles that of gasoline and diesel derived from petroleum, any environmental benefits of using an oxygenated fuel are nullified by the removal of oxygen during thermal processing. This process can yield some low-value materials and occasionally results in a surplus of gasoline compared to diesel (Mardoyan & Braun, 2015) (Figure 1.3).

## 1.3.2 Micro-emulsification

The high viscosity of raw vegetable oil poses a challenge to its use as a fuel, but one solution is creating a micro-emulsion. A micro-emulsion is a stable dispersion of oil, water, and surfactants, with droplet sizes ranging from 1 to 100 nm, typically between 10 and 50 nm. Ingredients for a biodiesel micro-emulsion include surfactants, vegetable oil, diesel, alcohol, and cetane improver. Alkyl nitrates enhance cetane, higher alcohols act as surfactants, and alcohols like methanol and ethanol reduce viscosity.

A micro-emulsion meeting diesel viscosity requirements can be created by combining vegetable oils, esters, dispersants, cetane improvers, surfactants,

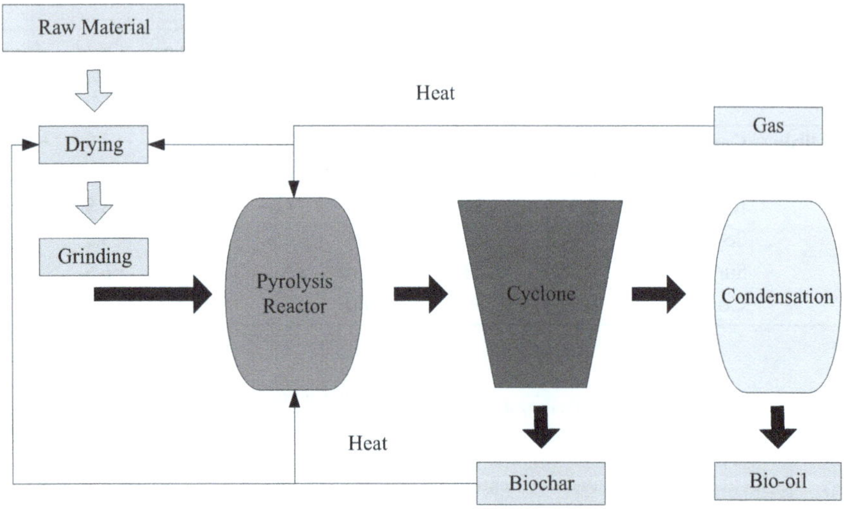


*Figure 1.3* Process of Pyrolysis

and alcohol. Micro-emulsions can improve spray characteristics by vaporizing low-boiling components, reducing biodiesel viscosity, and enhancing cetane number. Short-term performances of ethanol micro-emulsions in soybean oil are comparable to NO. 2 diesel, meeting ASTM specifications. However, the continuous use of micro-emulsified diesel may lead to issues such as carbon deposits, injector needle sticking, and incomplete combustion in engines (Ganesan et al., 2020).

### 1.3.3 Blending/dilution

Using edible oils directly in both direct and indirect diesel engines has generally been considered impractical and problematic due to several issues. These include high viscosity, gum formation, elevated FFA content, acid composition resulting from oxidation and polymerization during storage and combustion, carbon buildup, and thickening of lubricating oil. Additionally, two primary challenges associated with directly using edible oils as fuels are oil degradation and incomplete combustion.

To mitigate these issues, vegetable oils can be diluted with substances like diesel, gasoline, solvents, or ethanol. Dilution reduces the density and viscosity of vegetable oils. Adding 4% ethanol to diesel improves its thermal efficiency, brake torque, and brake power (García et al., 2020).

### 1.3.4 Transesterification

Transesterification is the most viable method for producing biodiesel from fat and oil feedstocks as it chemically approximates petroleum diesel.

This process reduces viscosity to levels close to diesel, transforming fats and oils (triglycerides) into their alkyl esters.

Therefore, the produced fuel shares characteristics with petroleum-based diesel, enabling it to be used in conventional diesel engines without modifications. Transesterification generally involves combining reactants, under controlled temperature and pressure, in a reversible reaction. The process can be expedited with the use of a catalyst.

In catalytic transesterification, triglycerides react with alcohol to produce esters and glycerol in the presence of a catalyst. Triglycerides consist of three fatty acids linked to glycerin, which determine the properties of oil or fat. The fatty acid composition, in turn, affects the quality of the biodiesel.

A key indicator of a successful transesterification reaction for biodiesel production is the prompt and efficient separation of the glycerol and ester layers after the reaction period. Glycerol, being a denser byproduct, can be subsequently refined for use in industries such as pharmaceuticals, cosmetics, and others (Chisti, 2007; Javed et al., 2019; Moreno-Garcia et al., 2017; Silva et al., 2021) (Figure 1.4).

### 1.3.5 Homogeneous acid-catalyzed transesterification

One of the earliest methods for producing biodiesel from palm oil involved the use of ethanol and sulfuric acid in a process called acid-catalyzed transesterification.

In this process, a triglyceride, a type of fat or oil, reacts with an alcohol in the presence of an acid catalyst. This reaction produces esters, commonly known as biodiesel, with glycerol as a byproduct. In particular, this approach makes it simple and practical to produce biodiesel from oil or fat sources with a high FFA content. However, compared to an alkali-catalyzed reaction, an acid-catalyzed reaction requires a longer reaction period and higher temperatures.

The process begins with a direct mixture of the oil and the acidified alcohol. Acid-catalyzed transesterification allows for both the segregation and transesterification to take place in a single step, with alcohol acting as both an esterification reagent and a solvent (Oren, 2005).

To prevent the formation of carboxylic acids, which can reduce alkyl ester yields, acid-catalyzed transesterification should be conducted in the absence of water. Researchers have studied the impact of water on the transesterification of oleic acid with methanol, using sulfuric acid as a catalyst. They examined fatty acid methyl ester yields at temperatures of 60°C and 80°C, with oil-to-methanol molar ratios of 1:3 and 1:6 (Xie et al., 2021). Their findings revealed that as the initial water content in the oil increased to 20%, the rate of oleic acid esterification significantly decreased.

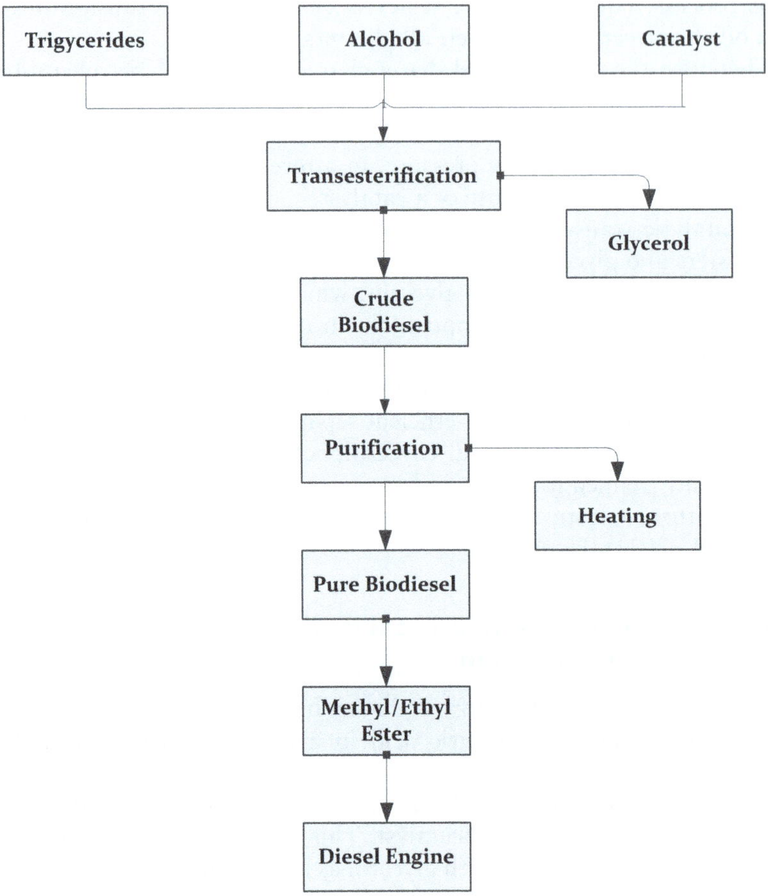


*Figure 1.4* Transesterification Process

### 1.3.6 Transesterification by homogeneous alkali catalysts

Alkaline-catalyzed transesterification converts triglycerides (fats or oils) with alcohol into biodiesel and glycerol. Metal alkoxides, hydroxides, or carbonates accelerate this reaction, making it preferred in industries due to its efficiency and compatibility with machinery compared to acid-catalyzed methods. However, the presence of water and a high concentration of FFAs in the feedstock can cause saponification, leading to an incomplete reaction during the alkaline transesterification process. This results in the formation of an emulsion, making the separation of glycerol problematic (Javed et al., 2019).

### 1.3.7 Heterogeneous catalyzed transesterification

The use of homogeneous catalysts, particularly base catalysts, is related to the demand for wastewater treatment and difficulties in purifying byproduct glycerol. Heterogeneous catalysts, typically solid base catalysts, are recommended as a solution to these issues. Solid base catalysts offer several advantages, including minimal impurity generation, easy separation, mild reaction conditions, and high efficiency. These catalysts can be categorized into three main types: acid solids, which catalyze the esterification of FFAs; base solids, which catalyze the transesterification of triglycerides; and bifunctional solids, which possess both acid and base properties, enabling them to catalyze both esterification and transesterification reactions simultaneously.

### 1.3.8 Transesterification catalyzed by lipase

An alternative method for transesterifying fats and oils for the synthesis of biodiesel involves the use of enzymes. This approach is preferable because it avoids issues such as saponification, purification, washing, and neutralization. Enzymatic catalysts are employed to enhance biodiesel production from feedstocks with high FFA content, enabling more efficient conversion of oil into biodiesel. However, enzyme catalysts have limitations, including longer reaction times and an higher costs. Due to these limitations, enzyme-catalyzed transesterification is infrequently employed (Xie et al., 2021).

In lipase-catalyzed transesterification, triglycerides (fats/oils) react with alcohol to produce esters (biodiesel) and glycerol. This process simplifies the removal of soap and catalysts and makes glycerol recovery easier, without the need for complex procedures. From an environmental perspective, this is more appealing compared to the conventional approaches.

### 1.3.9 Nano-catalyzed transesterification

Several recent advancements have been made in the catalytic transformation of fats and oils into biodiesel. Due to numerous advantages over traditional acid and base catalysts, the use of nanotechnology and ionic liquids for biodiesel production appears more promising.

Nanoparticles serve as catalysts in nano-catalysis for various applications in both heterogeneous and homogeneous catalysis. Their high catalytic activity is attributed to their large specific surface energy and surface area. Nano-catalysts enhance reaction efficiency by enabling reactions at lower temperatures, reducing side reactions, increasing recycling rates, and lowering energy consumption. Due to their high specific surface area and strong catalytic activity, nano-catalysts hold promise as potential substitutes for traditional catalysts, offering a more efficient way to produce biodiesel from fats and oils. Additionally, they solve the issue of mass transfer resistance.

### 1.3.10 Ionic liquid-catalyzed transesterification

Ionic liquids are organic compounds comprising cations and anions that remain liquid at room temperature. Cations determine the physical properties of ionic liquids, such as melting point, viscosity, and density, while anions govern their chemical properties and reactivity. One of the unique advantages of ionic liquids is in their controllability during synthesis, allowing them to meet specific reaction conditions. Another significant benefit of using ionic liquids in catalyzing transesterification for biodiesel production is the creation of a biphasic system at the end of the process. In this system, the catalyst, alcohol, ionic liquid, and glycerol produced during the reaction remain in the aqueous phase as the ionic liquid is insoluble in organic solvents.

### 1.3.11 Supercritical transesterification

The supercritical method is an effective approach to overcoming issues related to poor immiscibility between reactants and technical challenges associated with catalysts.

Supercritical alcohol transesterification occurs under extremely high pressure and temperature. When a gas or liquid is subjected to conditions beyond its critical point, it exhibits unique properties. In this process, the liquid and vapor phases are no longer distinct, resulting in a single supercritical fluid phase. Methanol and oil, which are typically immiscible liquids at normal temperatures, can combine in the supercritical transesterification process to form a homogeneous fluid. This occurs because methanol, under super-critical conditions, becomes non-polar due to its reduced solubility in water decreased dielectric constant it (Davoodbasha et al., 2021; Kalaimurugan et al., 2021; Patel et al., 2023; Xu et al., 2021). Without mass transfer restrictions, the reaction is significantly accelerated (Subramani et al., 2021) (Table 1.8).

*Table 1.8* Properties of edible oil biodiesel

| *Feedstock* | *Molar Ratio* | *Catalyst* | *Time (min)* | *Temperature (°C)* | *Yield* |
|---|---|---|---|---|---|
| Soybean oil | 12:1 | 6% CaFeA | 60 | 60 | 90 |
| Mustard oil | | KOH | 30 | 40–60 | 91 |
| Sunflower oil | 6:1–24:1 | 1% CaO | 90-330 | 23–60 | 99 |
| Canola oil | 6:1 | 0.5% KOH | - | 45 | 95 |
| Rapeseed oil | 3.5:1–42:1 | - | 120 | 200–500 | 95 |
| Corn oil | 6:1 | KOH | 60 | 40–65 | 97 |

## 1.4 IMPACT OF BIODIESEL ON ENGINE

Biodiesel impacts diesel engine performance, emissions, and combustion parameters significantly. Biodiesel typically has a slightly lower energy content than petroleum diesel, which can result in a minor decrease in engine power and fuel economy. However, its better lubricity can reduce engine wear and potentially extend engine life. Biodiesel produces lower emissions of particulate matter (PM), carbon monoxide (CO), and unburned hydrocarbons (HC) compared to petroleum diesel, contributing to cleaner air. Additionally, since biodiesel contains virtually no sulfur, it significantly reduces sulfur oxide ($SO_x$) emissions, which are harmful to the environment. Biodiesel generally has a higher cetane number than petroleum diesel, which can improve ignition quality and result in a smoother combustion process. However, it can also lead to the formation of deposits in the combustion chamber and fuel injectors if not properly managed. Biodiesel's higher viscosity can impact fuel atomization, leading to potential issues with incomplete combustion in some engine types (Figures 1.5–1.7).

## 1.5 CONCLUSION

The search for alternative energy sources for diesel engines has become more significant due to the depletion of petroleum reserves and the harmful effects of exhaust emissions from petroleum-powered engines.

While soybean is the most widely used feedstock for biodiesel, other sources are also used depending on regional availability. Ideal seed sources ought to have high oil content, high yields per hectare, and low production costs. Recycled oil is more frequently used as a feedstock for biodiesel because it is more cost-effective than raw or refined oils. From a chemical perspective, suitable oils for the synthesis of biodiesel needs to be rich in long-chain and low unsaturated fatty acids.

Among the four methods discussed in this chapter, namely, pyrolysis, micro-emulsification, dilution/blending, and transesterification, transesterification stands out as the optimal method for biodiesel production. Although new catalysts, particularly for heterogeneous processes, are continuously being developed, the performance of heterogenous catalysts remains less favorable than alkaline homogeneous catalysts.

Biodiesel is a clean-burning, renewable, and biodegradable fuel. However, a recent United Nations report urges governments to consider the impact of transitioning to plant-based energy sources on people and the environment. Significant technical obstacles remain, including the development of more efficient and effective catalysts, advancements in technology for producing high-quality biodiesel, the use of non-fossil solvents, the conversion of waste products like glycerol into useful products like ethanol and methanol, and the creation of low-cost photobioreactors.

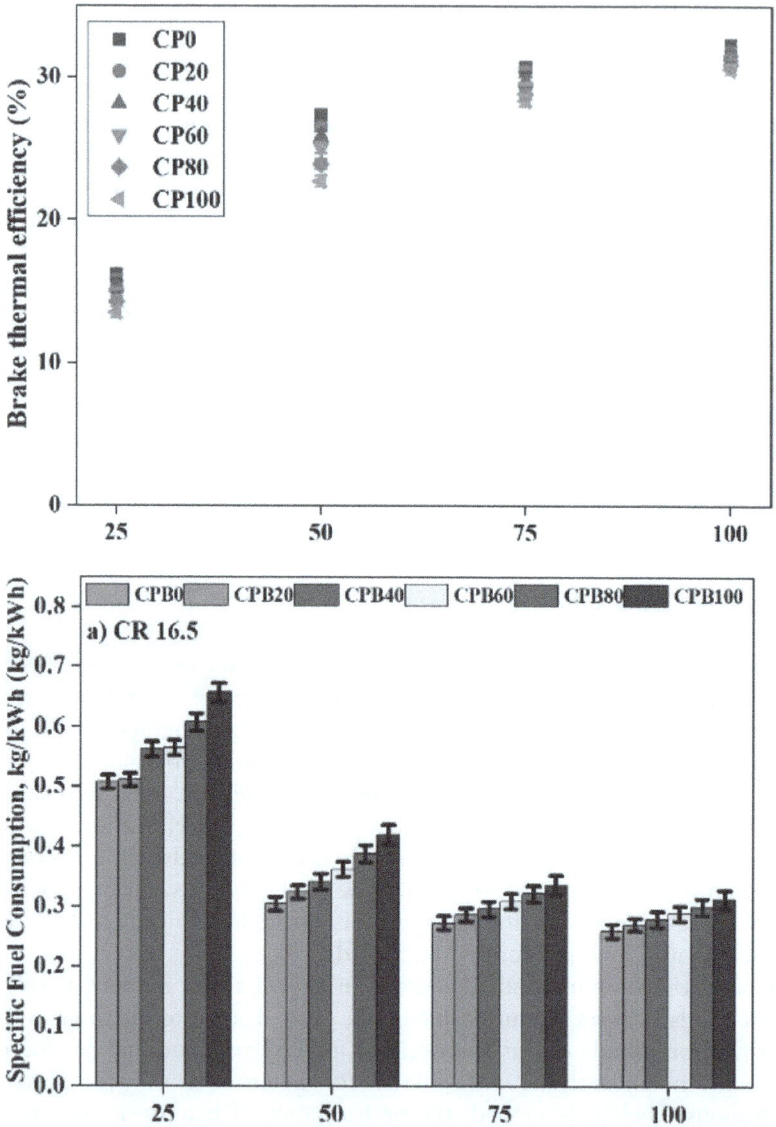


Figure 1.5 (a) Brake thermal efficiency vs load (b) Specific fuel consumption vs load

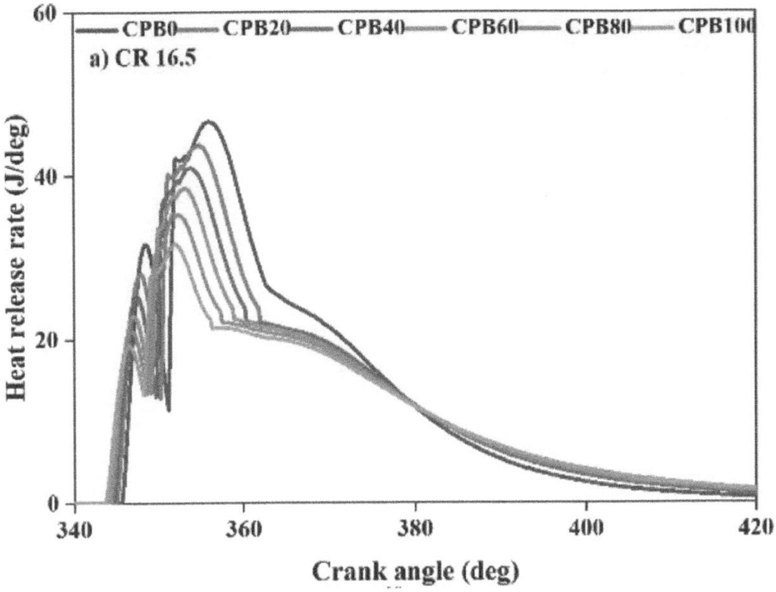


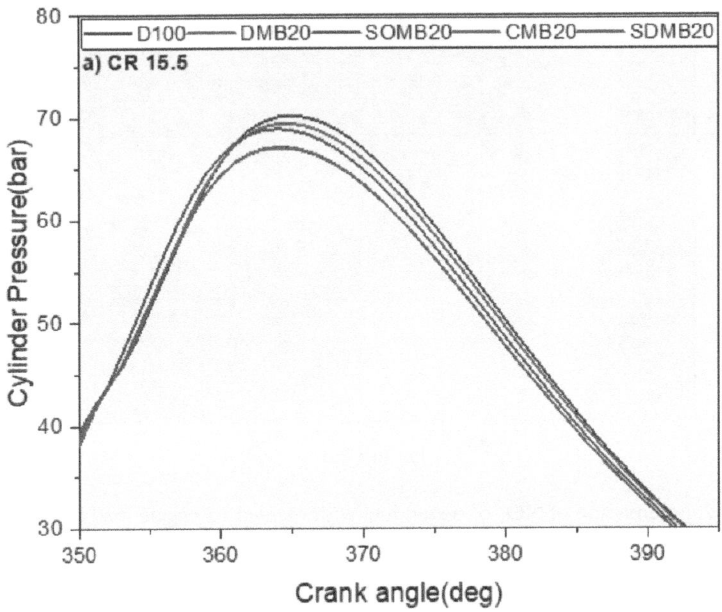


*Figure 1.6* (a) Cylinder heat release rate of tested fuel with respect to crank angle (b) Cylinder pressure wrt crank angle

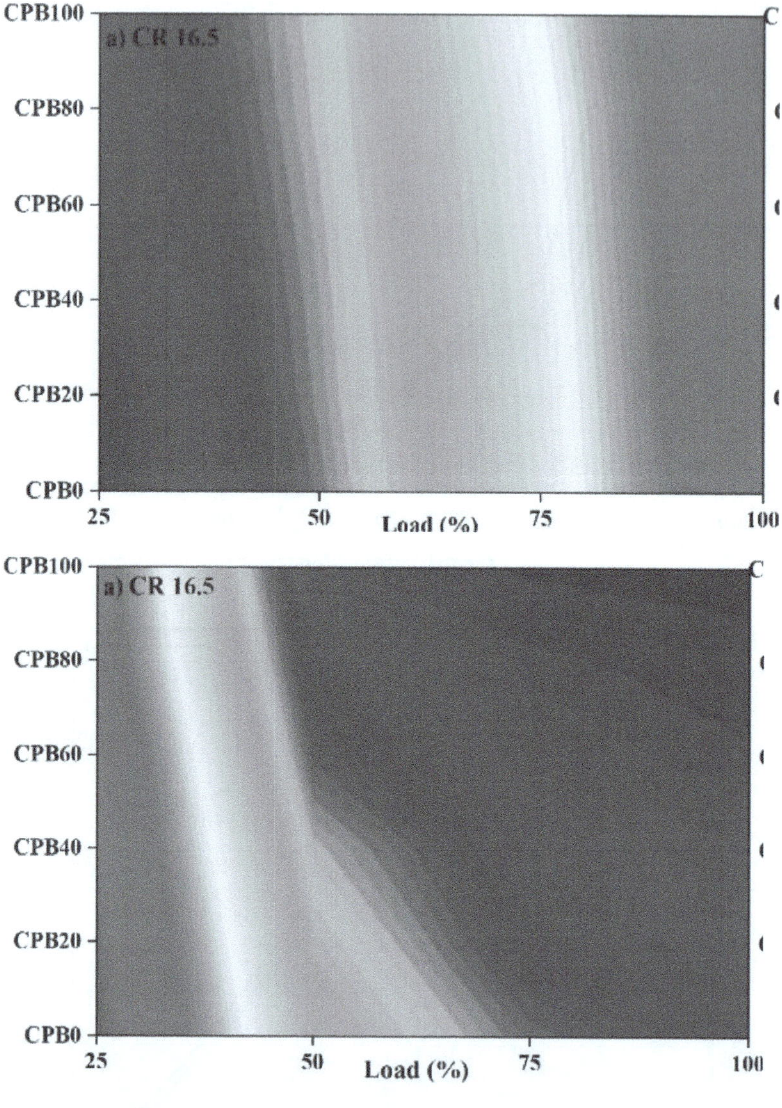


*Figure 1.7* (a) Emission of $NO_x$ of tested fuel with respect to engine load (b) Particulate matter emission of tested fuel with respect to engine load

## REFERENCES


Ahmad, A. L., Mat Yasin, N. H., Derek, C. J. C., & Lim, J. K. (2011). Optimization of microalgae coagulation process using chitosan. *Chemical Engineering Journal*, *173*(3), 879–882. https://doi.org/10.1016/J.CEJ.2011.07.070

Aro, E. M. (2016). From first generation biofuels to advanced solar biofuels. *Ambio*, *45*(1), 24–31. https://doi.org/10.1007/S13280-015-0730-0/FIGURES/1

Ayodele, O. O., & Dawodu, F. A. (2014). Production of biodiesel from Calophyllum inophyllum oil using a cellulose-derived catalyst. *Biomass and Bioenergy*, *70*, 239–248. https://doi.org/10.1016/J.BIOMBIOE.2014.08.028

Bharathiraja, B., Chakravarthy, M., Kumar, R. R., Yuvaraj, D., Jayamuthunagai, J., Kumar, R. P., & Palani, S. (2014). Biodiesel production using chemical and biological methods – A review of process, catalyst, acyl acceptor, source and process variables. *Renewable and Sustainable Energy Reviews*, *38*, 368–382. https://doi.org/10.1016/J.RSER.2014.05.084

*Biodiesel Production Methods*. (2017).

Chisti, Y. (2007). Biodiesel from microalgae. *Biotechnology Advances*, *25*(3), 294–306. https://doi.org/10.1016/J.BIOTECHADV.2007.02.001

Davoodbasha, M. A., Pugazhendhi, A., Kim, J. W., Lee, S. Y., & Nooruddin, T. (2021). Biodiesel production through transesterification of Chlorella vulgaris: Synthesis and characterization of CaO nanocatalyst. *Fuel*, *300*, 121018. https://doi.org/10.1016/J.FUEL.2021.121018

Foteinis, S., Chatzisymeon, E., Litinas, A., & Tsoutsos, T. (2020). Used-cooking-oil biodiesel: Life cycle assessment and comparison with first- and third-generation biofuel. *Renewable Energy*, *153*, 588–600. https://doi.org/10.1016/J.RENENE.2020.02.022

Ganesan, R., Manigandan, S., Samuel, M. S., Shanmuganathan, R., Brindhadevi, K., Lan Chi, N. T., Duc, P. A., & Pugazhendhi, A. (2020). A review on prospective production of biofuel from microalgae. *Biotechnology Reports*, *27*, e00509. https://doi.org/10.1016/J.BTRE.2020.E00509

García, L. M., Gariépy, Y., Barnabé, S., & Raghavan, V. (2020). Biorefinery of microalgae biomass cultivated in wastewaters. *Refining Biomass Residues for Sustainable Energy and Bioproducts: Technology, Advances, Life Cycle Assessment, and Economics*, 149–180. https://doi.org/10.1016/B978-0-12-818996-2.00007-7

Goh, B. H. H., Ong, H. C., Cheah, M. Y., Chen, W. H., Yu, K. L., & Mahlia, T. M. I. (2019). Sustainability of direct biodiesel synthesis from microalgae biomass: A critical review. *Renewable and Sustainable Energy Reviews*, *107*, 59–74. https://doi.org/10.1016/J.RSER.2019.02.012

Holma, A., Koponen, K., Antikainen, R., Lardon, L., Leskinen, P., & Roux, P. (2013). Current limits of life cycle assessment framework in evaluating environmental sustainability – case of two evolving biofuel technologies. *Journal of Cleaner Production*, *54*, 215–228. https://doi.org/10.1016/J.JCLEPRO.2013.04.032

Javed, F., Aslam, M., Rashid, N., Shamair, Z., Khan, A. L., Yasin, M., Fazal, T., Hafeez, A., Rehman, F., Rehman, M. S. U., Khan, Z., Iqbal, J., & Bazmi, A. A. (2019). Microalgae-based biofuels, resource recovery and wastewater treatment: A pathway towards sustainable biorefinery. *Fuel*, *255*, 115826. https://doi.org/10.1016/J.FUEL.2019.115826

Kalaimurugan, K., Somasundaram, D., Thirugnanam, C., & Dharmaprabhakaran, T. (2021). WITHDRAWN: Combustion behaviour of ZnO nanoparticles added algae biodiesel on CI engine. *Materials Today: Proceedings*. https://doi.org/10.1016/J.MATPR.2021.02.508

Karthikeyan, S., Periyasamy, M., Prathima, A., & Sabariswaran, K. (2020). Combustion analysis of single-cylinder CI engine fueled with S. Marginatum Macro algae biofuel - diesel blends. *Materials Today: Proceedings*, *33*, 3475–3480. https://doi.org/10.1016/J.MATPR.2020.05.360

Kaya, T., Kutlar, O. A., & Taskiran, O. O. (2018). Evaluation of the Effects of Biodiesel on Emissions and Performance by Comparing the Results of the New European Drive Cycle and Worldwide Harmonized Light Vehicles Test Cycle. *Energies 2018, Vol. 11, Page 2814*, *11*(10), 2814. https://doi.org/10.3390/EN11102814

Khammee, P., Ramaraj, R., Whangchai, N., Bhuyar, P., & Unpaprom, Y. (2021). The immobilization of yeast for fermentation of macroalgae Rhizoclonium sp. for efficient conversion into bioethanol. *Biomass Conversion and Biorefinery*, *11*(3), 827–835. https://doi.org/10.1007/S13399-020-00786-Y/TABLES/3

Lin, J. J., & Chen, Y. W. (2017). Production of biodiesel by transesterification of Jatropha oil with microwave heating. *Journal of the Taiwan Institute of Chemical Engineers*, *75*, 43–50. https://doi.org/10.1016/J.JTICE.2017.03.034

Mardoyan, A., & Braun, P. (2015). Analysis of Czech Subsidies for Solid Biofuels. *International Journal of Green Energy*, *12*(4), 405–408. https://doi.org/10.1080/15435075.2013.841163

Mohiddin, M. N. Bin, Tan, Y. H., Seow, Y. X., Kansedo, J., Mubarak, N. M., Abdullah, M. O., Chan, Y. S., & Khalid, M. (2021). Evaluation on feedstock, technologies, catalyst and reactor for sustainable biodiesel production: A review. *Journal of Industrial and Engineering Chemistry*, *98*, 60–81. https://doi.org/10.1016/J.JIEC.2021.03.036

Moreno-Garcia, L., Adjallé, K., Barnabé, S., & Raghavan, G. S. V. (2017). Microalgae biomass production for a biorefinery system: Recent advances and the way towards sustainability. *Renewable and Sustainable Energy Reviews*, *76*, 493–506. https://doi.org/10.1016/J.RSER.2017.03.024

Murugesan, A., Umarani, C., Chinnusamy, T. R., Krishnan, M., Subramanian, R., & Neduzchezhain, N. (2009). Production and analysis of bio-diesel from non-edible oils—A review. *Renewable and Sustainable Energy Reviews*, *13*(4), 825–834. https://doi.org/10.1016/J.RSER.2008.02.003

Norjannah, B., Ong, H. C., Masjuki, H. H., Juan, J. C., & Chong, W. T. (2016). Enzymatic transesterification for biodiesel production: a comprehensive review. *RSC Advances*, *6*(65), 60034–60055. https://doi.org/10.1039/C6RA08062F

Oren, A. (2005). A hundred years of Dunaliella research: 1905–2005. *Saline Systems 2005 1:1*, *1*(1), 1–14. https://doi.org/10.1186/1746-1448-1-2

Patel, A., Gami, B., Patel, P., & Patel, B. (2023). Biodiesel production from microalgae Dunaliella tertiolecta: a study on economic feasibility on large-scale cultivation systems. *Biomass Conversion and Biorefinery*, *13*(2), 1071–1085. https://doi.org/10.1007/S13399-020-01191-1/FIGURES/4

Pikula, K., Zakharenko, A., Stratidakis, A., Razgonova, M., Nosyrev, A., Mezhuev, Y., Tsatsakis, A., & Golokhvast, K. (2020). The advances and limitations in biodiesel production: feedstocks, oil extraction methods, production, and environmental life cycle assessment. *Green Chemistry Letters and Reviews*, *13*(4), 11–30. https://doi.org/10.1080/17518253.2020.1829099

Pinto, A. C., Guarieiro, L. L. N., Rezende, M. J. C., Ribeiro, N. M., Torres, E. A., Lopes, W. A., De Pereira, P. A. P., & De Andrade, J. B. (2005). Biodiesel: An overview. *Journal of the Brazilian Chemical Society*, *16*(6 B), 1313–1330. https://doi.org/10.1590/S0103-50532005000800003

Razak, A. A., Mohamed, M. A., & Derawi, D. (2023). Patents on Biodiesel. *Biodiesel Technology and Applications*, 361–375. https://doi.org/10.1002/9781119724957.CH13

Razzaq, L., Farooq, M., Mujtaba, M. A., Sher, F., Farhan, M., Hassan, M. T., Soudagar, M. E. M., Atabani, A. E., Kalam, M. A., & Imran, M. (2020). Modeling Viscosity and Density of Ethanol-Diesel-Biodiesel Ternary Blends for Sustainable Environment. *Sustainability 2020, Vol. 12, Page 5186*, *12*(12), 5186. https://doi.org/10.3390/SU12125186

Saengsawang, B., Bhuyar, P., Manmai, N., Ponnusamy, V. K., Ramaraj, R., & Unpaprom, Y. (2020). The optimization of oil extraction from macroalgae, Rhizoclonium sp. by chemical methods for efficient conversion into biodiesel. *Fuel*, *274*, 117841. https://doi.org/10.1016/J.FUEL.2020.117841

Shah, S. H., Raja, I. A., Rizwan, M., Rashid, N., Mahmood, Q., Shah, F. A., & Pervez, A. (2018). Potential of microalgal biodiesel production and its sustainability perspectives in Pakistan. *Renewable and Sustainable Energy Reviews*, *81*, 76–92. https://doi.org/10.1016/J.RSER.2017.07.044

Silva, M. R. O. B. da, Moura, Y. A. S., Converti, A., Porto, A. L. F., Viana Marques, D. de A., & Bezerra, R. P. (2021). Assessment of the potential of Dunaliella microalgae for different biotechnological applications: A systematic review. *Algal Research*, *58*, 102396. https://doi.org/10.1016/J.ALGAL.2021.102396

Singh, D., Sharma, D., Soni, S. L., Sharma, S., & Kumari, D. (2019). Chemical compositions, properties, and standards for different generation biodiesels: A review. *Fuel*, *253*, 60–71. https://doi.org/10.1016/J.FUEL.2019.04.174

Subramani, S., Natarajan, K., & Lakshmi Narayana Rao, G. (2021). Optimization of injection timing and anti-oxidants for multiple responses of CI engine fuelled with algae biodiesel blend. *Fuel*, *287*, 119438. https://doi.org/10.1016/J.FUEL.2020.119438

Vasaki E, M., Karri, R. R., Ravindran, G., & Paramasivan, B. (2021). Predictive capability evaluation and optimization of sustainable biodiesel production from oleaginous biomass grown on pulp and paper industrial wastewater. *Renewable Energy*, *168*, 204–215. https://doi.org/10.1016/J.RENENE.2020.12.038

Verma, T. N., Shrivastava, P., Rajak, U., Dwivedi, G., Jain, S., Zare, A., Shukla, A. K., & Verma, P. (2021). A comprehensive review of the influence of physicochemical properties of biodiesel on combustion characteristics, engine performance and emissions. *Journal of Traffic and Transportation Engineering (English Edition)*, *8*(4), 510–533. https://doi.org/10.1016/J.JTTE.2021.04.006

Wang, C., Kong, Y., Hu, R., & Zhou, G. (2021). Miscanthus: A fast-growing crop for environmental remediation and biofuel production. *GCB Bioenergy*, *13*(1), 58–69. https://doi.org/10.1111/GCBB.12761

Xie, S. R., Li, Y., Liang, M. H., Yan, B., & Jiang, J. G. (2021). Creatinine combined with light increases the contents of lutein and β-carotene, the main carotenoids of Dunaliella bardawil. *Enzyme and Microbial Technology*, *151*, 109913. https://doi.org/10.1016/J.ENZMICTEC.2021.109913

Xu, L., Pan, W., Yang, G., Tang, X., Martin, R. M., Liu, G., & Zhong, C. (2021). Impact of light quality on freshwater phytoplankton community in outdoor mesocosms. *Environmental Science and Pollution Research*, *28*(41), 58536–58548. https://doi.org/10.1007/S11356-021-14812-7/TABLES/2

Zulqarnain, Ayoub, M., Yusoff, M. H. M., Nazir, M. H., Zahid, I., Ameen, M., Sher, F., Floresyona, D., & Budi Nursanto, E. (2021). A Comprehensive Review on Oil Extraction and Biodiesel Production Technologies. *Sustainability 2021, Vol. 13, Page 788*, *13*(2), 788. https://doi.org/10.3390/SU13020788

Zulqarnain, Mohd Yusoff, M. H., Ayoub, M., Ramzan, N., Nazir, M. H., Zahid, I., Abbas, N., Elboughdiri, N., Mirza, C. R., & Butt, T. A. (2021). Overview of Feedstocks for Sustainable Biodiesel Production and Implementation of the Biodiesel Program in Pakistan. *ACS Omega*, *6*(29), 19099–19114. https://doi.org/10.1021/ACSOMEGA.1C02402/ASSET/IMAGES/LARGE/AO1C02402_0009.JPEG

Chapter 2

# An overview of artificial neural networks for photovoltaic system optimization and performance enhancement

*Sourabh Kaithwas, Bharat Girdhani, and Meena Agrawal*

## 2.1 INTRODUCTION

Energy consumption is rapidly rising across all sectors, including agriculture, residential, and commercial, due to population growth and industrialization. Currently, most electrical energy production relies on non-renewable resources such as coal, natural gas, and oil, all of which have adverse environmental effects (Ahmed et al., 2022b; Yilmaz & Corapsiz, 2022). However, the use of solar and wind energy is expanding more rapidly today. Among these, PV energy is preferred over wind energy due to its more consistent daily production. Wind turbine output, in contrast, is highly intermittent due to wind activity. Additionally, PV systems have lower ongoing operating and maintenance costs because they does not have any moving parts.

There is a need to develop Maximum Power Point Tracking (MPPT) algorithms that ensure that PV generators operate at its maximum capacity, many of which are discussed in the literature (Ko et al., 2022). A PV cell is a semiconductor material that can generate direct current when exposed to direct sunlight. Because it produces energy from a free source, is an established technology, requires less maintenance, and is less expensive than other methods of generating power, the usage of PV systems has recently gained popularity.

Additionally, a PV system is a durable way to produce electricity and does not need mechanical parts (Alkhalaf et al., 2022). However, PV systems do have some drawbacks: PV systems are sensitive to weather conditions, have a finite scope, have low system efficiency, and involve high setup costs. The PV output feature typically includes an MPP, where the PV module produces its peak output power and energy. This point is erratic and reliant on the load impedance and local weather conditions (Ahmed et al., 2022a; Raj & Samuel, 2022b; Wu et al., 2022; Awan, 2022). MPPT techniques are classified as classical, intelligent, optimization, and hybrid. The advantages of classical methods include their ease of use and superior performance in environments with constant radiation and temperature conditions.

DOI: 10.1201/9781003521341-2

However, these techniques struggle to locate the MPP under various environmental conditions. This chapter aims to analyze the cutting-edge methods used in MPPT devices that employ Artificial Neural Networks (ANNs) or a hybrid approach involving ANN-based models. Hybrid approaches combine conventional methods, improved conventional methods, intelligent algorithms, and metaheuristic approaches, resulting in significant improvements over other MPPT techniques. Integrating renewable energy systems into the grid or load results in power quality issues. As a result, ANNs are utilized to address these power quality issues and may also be employed in combination with other techniques, as will be covered in the following sections of this chapter. There are subsections within this section that provide a succinct overview of the PV system and power quality.

### 2.1.1 PV system

Solar energy is widely used and can be converted into a usable form using PV technology. When light falls on PV cells, they produce direct current through the photo-electric effect. A single PV cell cannot generate high power output. To achieve high power output, PV cells are interconnected to form solar modules. These modules are arranged in series, parallel, or series and parallel combinations to form PV panels, which are arranged in an array field to form PV arrays. The output power produced from a PV array varies with the variation in solar irradiance and temperature.

The effectiveness of a solar cell is influenced by temperature, the fundamental characteristics of the cell material, and the intensity of sunlight. To estimate this, the load must be converted into an analogous circuit using a real load. Figure 2.1 depicts corresponding electrical circuit of a PV cell. Developing a PV cell simulation model is required for enabling the solar panel's MPP to optimize PV conversion efficiency, which is influenced by the light intensity and ambient temperature.

To better understand the electronic behavior of a solar cell, it is helpful to develop an electrically analogous model. Since no solar cell is perfect in reality, the equivalent circuit, as shown in Figure 2.1, includes series resistance and shunt resistance connected to a diode. The series resistance

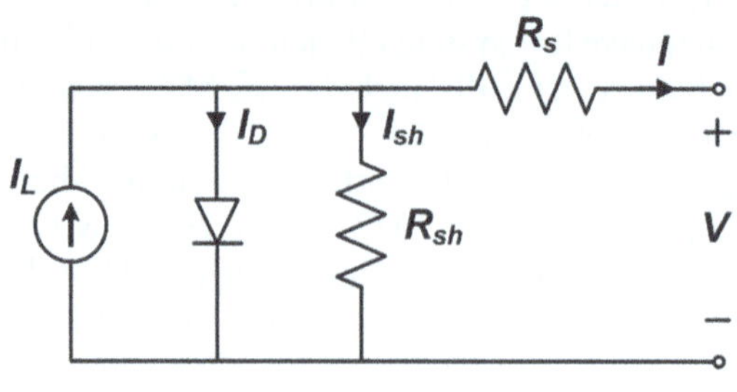


*Figure 2.1* Solar PV electrical equivalent circuit model.

represents the sum of various structural resistances and internal electrical losses. Shunt resistance is connected to denote the leakage current losses.

Let $I_{sc}$ and $I_o$ be the short-circuit and saturation currents, respectively, and let $N_s$ be the series-connected cell, $T$ the cell's temperature, $K$ the Boltzmann constant which is equal to $1.38\times10^{-23}$ J/K, $q$ the charge of an electron, which is equal to $1.6\times10^{-19}$ C, $a$ is the quality factor of diode, and $R_s$ and $R_{sh}$ be the array's series and shunt resistances, respectively. Therefore, Eqs. (2.1) and (2.2) are appropriate to express the solar cell current (Roy et al., 2021):

$$I = I_{sc} - I_o\left[\exp\left(\frac{V+R_sI}{\frac{N_sKT}{q}a}\right)-1\right]-\left(\frac{V+R_sI}{R_{sh}}\right) \tag{2.1}$$

$$I = I_{sc} - I_o\left[\exp\left(\frac{V+R_sI}{V_ta}\right)-1\right]-\left(\frac{V+R_sI}{R_sh}\right) \tag{2.2}$$

If the thermal voltage of the array is substituted as $V_t$=$N_sKT/q$, then Eq. (2.1) may be written as Eq. (2.2). The value of PV current changes with the value of solar irradiance; it increases with an increase in the irradiance value and decreases with a decrease in the irradiance value. As PV output power depends on the values of current and voltage, it is directly proportional to irradiance, as shown in Figure 2.2a (Aihua et al., 2022).

The value of PV voltage decreases with an increase in temperature, and vice versa. A decrease in voltage due to an increase in temperature reduces the output power, as power depends on the current and voltage. Therefore, a decrease in temperature increases the output power, as shown in Figure 2.2b (Raj & Samuel, 2022a; Babes et al., 2022).

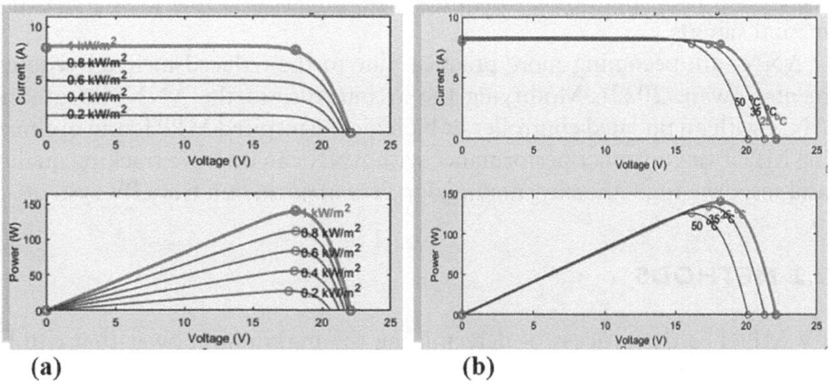


(a) (b)

*Figure 2.2* (a) Effect of varying irradiance and (b) effect of varying temperature.

The performance of the PV array is further impacted by variations in irradiance from panel to panel, which can be caused by the shadows cast by moving clouds, plants, surrounding structures, seasonal conditions, dust accumulation, the decaying effect of the panel, etc.

If all PV panels receive the same amount of irradiation, the P-V curve will only have a single peak. However, if one or more PV panels are exposed to the shadow effect, the P-V curve will have many peaks. To maximize the power output of PV systems, a control technique known as an MPPT mechanism is necessary, as it can continuously monitor the MPP while operating (Raj & Samuel, 2022b).

### 2.1.2 ANN systems

ANNs, commonly known as neural networks (NNs) or neural nets, are computer architectures that work analogously to the human brain, making decisions according to the situation. NNs require only input and output data. For PV systems, the input data typically include solar irradiance and temperature. After the ANN is trained by using a suitable algorithm, for example, Levenberg–Marquardt algorithm, it adjusts the duty cycle of the DC-to-DC converter in response to changes in input values, providing the desired output (Yilmaz and Corapsiz, 2022; Alkhalaf et al., 2022).

A trained ANN can analyze the provided data, predict outcomes, and offer solutions to new problems. Similar to human brains, ANNs can handle ambiguous data and adapt to situations lacking clear algorithmic solutions. They are used in many applications involving unused data, including financial applications, signal refining, robotics, etc. This advantage allows ANNs to succeed in areas where conventional computers frequently struggle.

The architecture of an ANN refers to the arrangement of neurons, which may be organized into structures called layers. These layers together form an ANN.

The fundamental models include three distinct layers: the input layer, hidden layer(s), and the output layer, as shown in Figure 2.3. Data are received at the input layer, which may contain sensors that gather environmental signals.

ANNs are becoming more practical due to the reduced memory requirements (Awan, 2022). Modifying the architecture of the ANN, integrating ANN with an updated controller or biologically inspired MPPT, and optimizing MPPT or controller performance with ANN can improve tracking quality and response time, thereby maximizing power extraction from PV systems.

## 2.2 METHODS

PV MPPT is the process of determining the maximum power that can be drawn from or generated by the PV system. MPPT is an electrical system that adjusts the duty cycle of a DC-to-DC converter to calculate the MPP.

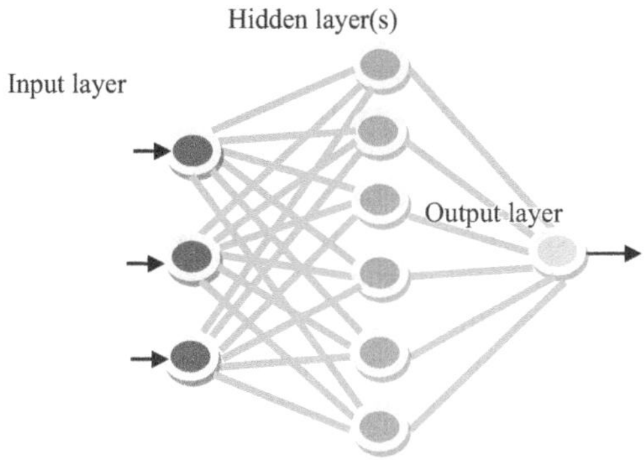


*Figure 2.3* Simple neural network.

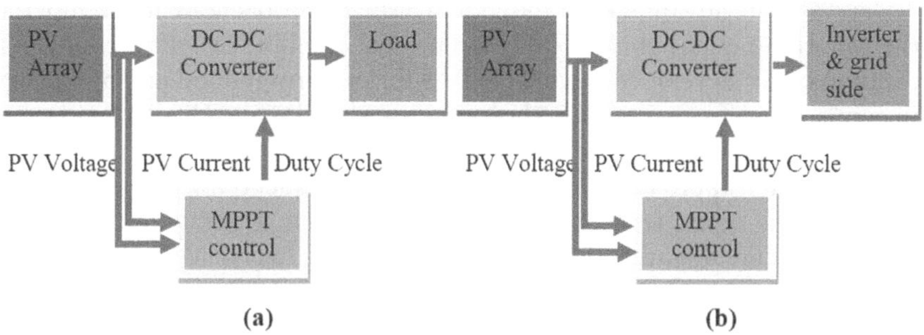


*Figure 2.4* (a) Standalone systems block diagram and (b) grid-connected systems block diagram.

As previously mentioned, power varies with changes in solar irradiance and temperature, therefore, this MPP is variable, and the system must be handled at this point to ensure the PV system's reliability and efficiency (Badar et al., 2021; Talaat et al., 2022). Figure 2.4a represents standalone mode, and Figure 2.4b represents the grid mode of PV systems.

### 2.2.1 Classical methods

Because of their straightforward structure and inexpensive implementation costs, classical MPPT controllers are preferred by the industry over soft computing techniques. However, classical MPPT approaches have certain drawbacks, such as slow tracking speed, excessive steady-state oscillation, inability to operate under the Partial Shading Condition (PSC), drifting problems, etc. To enhance the performance of the classical MPPT methods, modified conventional MPPT methods focus on addressing the aforementioned flaws (Ahmed et al., 2022a). The most commonly used classical techniques are Perturb and Observe (P&O) and Incremental Conductance (INC) (Raj & Samuel, 2022b).

### 2.2.2 Soft computing methods

Soft computing methods are a good option for resolving non-linear issues. These algorithms provide quick convergence and are reliable. MPPT soft computing methods offer enormous potential for dealing with non-linear optimization issues. In contrast to traditional MPPT methods, these soft computing systems can operate under any weather conditions and can effectively track global MPP in the PSC. These soft computing techniques involve intelligent algorithms such as ANNs, Fuzzy Logic Controller (FLC), and optimization techniques such as Particle Swarm Optimization (PSO), Ant Colony Optimization (ACO), Genetic Algorithm (GA), Artificial Bee Colony (ABC), and Grey Wolf Optimizer (GWO).

### 2.2.3 Hybrid methods

To leverage the advantages of individual MPPT approaches while removing their shortcomings, hybrid MPPT systems that combine two different MPPT techniques are developed. Hybrid MPPT outperform individual approaches in terms of performance, whether tracking the global MPP under partial shading without added complexity or combining accuracy and speed. Hybrid MPPT approaches are categorized into sequential (two-staged) and simultaneous types, with the latter being extremely accurate but more difficult to execute. Figure 2.5 shows an example of a hybrid MPPT controller.

## 2.3 LITERATURE SURVEY

ANNs serve different purposes in optimizing and improving the performance of renewable energy systems, which are outlined as follows.

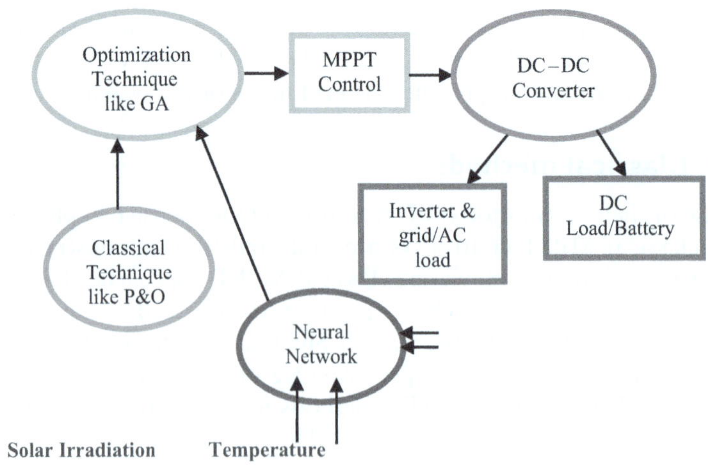


*Figure 2.5* Block diagram of hybrid MPPT controller.

### 2.3.1 Role of conventional ANN technique in MPPT control

Table 2.1 shows the improvement of PV MPPT control using ANNs compared to conventional MPPT techniques under different environmental conditions using MATLAB/Simulink.

### 2.3.2 Role of ANNs in hybrid configurations for MPPT control

ANNs in hybrid configurations offer high power tracking efficiency, reduce the ripples in capacitor voltage and inductor currents, and decrease Total Harmonic Distortion (THD) values for stator current and torque ripple in Brushless Direct Current (BLDC) motors compared to P&O and INC techniques (Tables 2.2 and 2.3).

*Table 2.1* ANN's performance over conventional MPPT techniques

| *MPPT technique* | *DC-DC converter type* | *Conventional technique involved* | *Improvement over conventional technique* | *References* |
|---|---|---|---|---|
| ANN | Boost | P&O and INC | Good settling time and reduced overshoot | Yilmaz and Corapsiz (2022) |
| ANN | Boost | P&O | Fewer oscillations around MPP and accurate MPP tracking with high efficiency | |

*Table 2.2* Improvement of PV MPPT control using the ANN-based hybrid techniques over the other MPPT techniques under different environmental conditions

| *Proposed hybrid approach* | *Other technique(s) involved* | *Improvement over other technique(s) involved* | *Tracking under PSC* | *References* |
|---|---|---|---|---|
| Hybrid GS-PS ANN-P&O | P&O, β method, FOCV, INC, FLC, and Adaptive Neuro-Fuzzy Inference System (ANFIS) | More output power, high convergence speed, less power fluctuations around the MPP, and high accuracy | Yes | Alkhalaf et al. (2022) |
| Hybrid SFL-PS-ANN-P&O | P&O, INC, FLC, and ANFIS | High response speed, more tracking efficiency, and less oscillation around MPP | No | Jiang et al. (2021) |

*(Continued)*

*Table 2.2 (Continued)* Improvement of PV MPPT control using the ANN-based hybrid techniques over the other MPPT techniques under different environmental conditions

| *Proposed hybrid approach* | *Other technique(s) involved* | *Improvement over other technique(s) involved* | *Tracking under PSC* | *References* |
|---|---|---|---|---|
| BOSS-D-RBFN | RBFN, FLC, PSO, P&O, and INC | More tracking speed, more rapid convergence, stable functioning, and reduced oscillation | Yes | Raj & Samuel (2022a) |
| ACO-ANN | INC and PSO-ANN | More tracking efficiency, more tracking speed, and reduced oscillation | No | Babes et al. (2022) |

*Table 2.3* Improvement of PV MPPT control using ANFIS and ANFIS-based hybrid techniques over the other MPPT techniques under different environmental conditions

| *Proposed hybrid approach* | *Other technique(s) involved* | *Improvement over other technique(s) involved* | *Tracking under PSC* | *References* |
|---|---|---|---|---|
| Hybrid CS-PS ANFIS-INC | P&O, INC, FLC, and ANFIS | More output power, reduced power oscillations, a quick transient response, and minimal power loss in the steady state | No | Ahmed et al. (2022b) |
| Hybrid CS-PS ANFIS-INC | HC, P&O, INC, FLC, and HCS-PS-ANN-INC | Better tracking efficiency, less response time, low oscillation around MPP, and more output power | Yes | Wu et al. (2022) |
| GOA-ANFIS-INC | INC, FLC, HC, P&O, β method, and ANN | More efficient, less oscillation around MPP, and more response speed | Yes | Aihua et al. (2022) |
| ANFIS | FL | More efficiency and more rapid convergence | Yes | Revathy et al. (2022) |

### 2.3.3 Role of ANNs in power quality

Mishra et al. (2021) combined a down-sampling empirical mode decomposition with an optimized random forest machine learning technique to identify non-islanding power quality events and detect islanding situations in a distribution generating system with wind energy high penetration.

The proposed approach outperformed other machine learning approaches including ANNs in terms of accuracy and detection speed.

Yarar et al. (2022) used hourly solar irradiance and temperature data to perform harmonic estimation for a 250 kW PV system employing an instantaneous ANN and non-linear autoregressive exogenous feedback model. The prediction accuracy of the ANN was 0.98, whereas that of the non-linear autoregressive exogenous model was 0.96.

Vanam et al. (2022) evaluated a dynamic voltage restorer with a solar source using an ANN and the hysteresis voltage control technique. The ANN controller consistently adjusted the voltage under difficult circumstances, improving system voltage quality. When comparing the performance of ANNs and hysteresis controllers, ANN control was preferable for modifying sag and swell circumstances, while hysteresis voltage control was more effective for managing defective scenarios in a three-phase system.

### 2.3.4 Role of ANNs in solar power forecasting

Samy et al. (2022) proposed a novel statistical and ANN-based decision-making mechanism for optimizing hybrid models. The results of the ANN-based "PSO" study were superior and more reliable than those of statistical analysis, providing highly accurate predictions.

Alghamdi et al. (2022) estimated the true harmonic distortions of a solar PV converter using location-specific data and ANN technology. A basic power system was developed and simulated to enhance the prediction capabilities of the ANN system. In the IEEE 34-bus test feeder, a maximum error of <10% and a median error of 5.4% were observed in the calculation of harmonic components.

Jahan et al. (2022) used seven models to forecast four power quality indicators and compared the accuracy and computation time of the models. The study found that bagging decision trees provided the most accurate predictions, while the ANN had the highest processing time and linear regression required the least.

## 2.4 DISCUSSION

The role of ANN techniques, both as an individual or in hybrid combinations, has been discussed in the Literature section. Tables 2.4 and 2.5 present the major outcomes of these techniques in improving the performance of PV or renewable energy systems, and these strategies can be successfully implemented under various operating situations.

With a special focus on PV systems, ANNs continue to have a significant influence on rapidly emerging sectors such as renewable energy. This chapter presents a comprehensive evaluation of recent studies, with a special emphasis on application of ANNs in PV systems.

*Table 2.4* Review of features for hybrid MPPT algorithms with ANN as a one technique employed for the PV system

| *References* | *Proposed technique* | *Tracking efficiency (%)* | *Response time (seconds)* | *Oscillation around MPP (watts)* | *Conditions* | *Remarks* |
|---|---|---|---|---|---|---|
| Alkahlaf et al. (2022) | Hybrid GS-PS-ANN-P&O | 99.31 | 0.16 | 4.36 | Variable *T* and constant G | Better than P&O, β method, FOCV, INC, FLC, and ANFIS techniques |
| Wu et al. (2022) | Hybrid CS-PS-ANFIS-INC | 99.19 | 0.17 | 5.1 | PSCs | Better than HC, P&O, INC, FLC, and HCS-PS-ANN-INC under PSCs |
| (Alhua et al, 2022) | GOA-ANFIS-INC | 99.32 | 0.17 | 6.45 | PSCs | Better as compared to INC, FLC, HC, P&O, β method, ANN techniques |
| (Babes et al, 2022) | ACO-ANN | 99.20 | 0.18 | - | Variable G and constant T | Better as compared to PSO-ANN and INC techniques |

*Table 2.5* ANN's role as an individual or in a hybrid configuration for improvement in power quality issues caused by integration of renewable/hybrid energy systems to grid/load

| *References* | *Proposed technique(s)* | *System(s) involved* | *Remarks* |
|---|---|---|---|
| Irfan et al. (2021) | DSTATCOM with ANFIS controller | PV system with battery for electric vehicle charging | THD values of source current and load current are 3.54% and 19.74%, respectively, by using DSTATCOM with ANFIS controller in MATLAB environment, which is less than those achieved by traditional controllers |
| Dheeban & Muthu Selvan (2021) | ANFIS | PV-UPQC system | Using an ANFIS controller, the load current's THD values for harmonics of the third, fifth, seventh, and ninth orders are 2.74%, 1.13%, 2.41%, and 3.41%, respectively. For harmonics of the 15th, 21st, and 33rd orders, the values have been reduced to 1.44%, 0.73%, and 0.59%, respectively. These are less than those of a PI controller |
| Albert Alexander et al. (2022) | ANN | (PV)-fed cascaded multilevel inverter | THD value of load side voltage using the proposed technique is 4.52%, whereas using pulse width modulation, digital logic, and FLC techniques, it is 10.79%, 6.79%, and 5.49%, respectively |
| Zanib et al. (2022) | ANN with Unified Power Quality Connditioner | PV and WT hybrid system | The average THD of the load voltage is about 1.4% with an ANN controller and around 1.35% with a PI controller. |
| R et al. (2022) | ANN-GA | PV,Wind System and Grid | The value of Grid current THD is 1.93% (Simulation) and 2.40 % (Hardware) whereas using PI controller, it is 3.03 %(Simulation) and 4.43%(Hardware) |

DSTATCOM, distribution static synchronous compensator.

## 2.5 CONCLUSION

This chapter presents a review of different authors' work on optimizing and improving the performance of PV systems using ANNs. The literature from recent years was thoroughly examined, with a primary focus on ANNs and PV systems. Some studies use standard ANN algorithms that work well under various environmental conditions, achieving an average tracking efficiency of 98% compared to classical MPPT techniques. However, these studies do not address MPPT under partially shaded conditions, and the use of ANNs introduces memory and sensor requirements. In some studies, the THD value is reduced to <5% using the ANN approach, which also plays a vital role in minimizing errors in solar PV energy forecasting. However, techniques like ANFIS and FL outperform ANNs in resolving power quality issues under all loading conditions. After analyzing the work of many authors, we found stability and power quality problems in ANN-based methods, leading to the idea of a hybrid technique to optimize results based on PV. Many authors have worked on hybrid MPPT techniques that combine ANN with other methods, showing significant improvements such as better time response (0.11–0.17 seconds), reduced oscillations (4–7 W), and high tracking efficiency (99.10%–99.68%) compared to conventional ANN and other MPPT techniques under both uniform and varying environmental conditions, including PSCs in some papers. Furthermore, hybrid techniques such as ANFIS outperform ANN and conventional controllers in reducing THD values below 5% under all loading conditions, resulting in improved power quality. This chapter may help readers and researchers in making the ANN-based MPPT technique simpler, less complicated, and less time-consuming for training the ANN by adopting hybrid techniques and improving power quality using the ANN or hybrid techniques. In addition, it is recommended that future studies evaluate some of these hybrid techniques using real hardware.

## REFERENCES

Ahmed, E. M., Norouzi, H., Alkhalaf, S., Ali, Z. M., Dadfar, S., & Furukawa, N. (2022a) Enhancement of MPPT controller in PV-BES system using incremental conductance along with hybrid crow-pattern search approach based ANFIS under different environmental conditions. *Sustainable Energy Technologies and Assessments* 50, 101812.

Ahmed, S., Mekhilef, S., Mubin, M. B., & Tey, K. S. (2022b) Performances of the adaptive conventional maximum power point tracking algorithms for solar photovoltaic system. *Sustainable Energy Technologies and Assessments* 53, 102390.

Aihua, G., Yihan, X., & Rezvan, A. (2022) A performance improvement of maximum power point tracking for photovoltaic system using grasshopper optimization algorithm based ANFIS under different conditions. *Optik* 270, 169965.

Albert Alexander, S., Harish, R., Srinivasan, M., & Sarathkumar, D. (2022) Power quality improvement in a solar PV assisted microgrid using upgraded ANN-based controller. *Mathematical Problems in Engineering* 2022, 2441534.

Alghamdi, T. A. H., Abdusalam, O. T. E., Anayi, F., & Packianather, M. (2022) An artificial neural network based harmonic distortions estimator for grid- connected power converter-based applications. *Ain Shams Engineering Journal* 14, 101916.

Alkhalaf, S., Ali, Z. M., & Oikawa, H. (2022) A novel hybrid gravitational and pattern search algorithm based MPPT controller with ANN and perturb and observe for photovoltaic system. *Soft Computing* 26, 7293–7315.

Awan, M. M. A. (2022) Technical review of MPPT algorithms for solar photovoltaic system: SWOT analysis of MPPT. *Sir Syed University Research Journal of Engineering & Technology* 12(1), 98–106.

Babes, B., Boutaghane, A., & Hamouda, N. (2022) A novel nature-inspired maximum power point tracking (MPPT) controller based on ACO-ANN algorithm for photovoltaic (PV) system fed arc welding machines. *Neural Computing and Applications* 34(1), 299–317.

Badar, M., Ahmad, I., Mir, A. A., Ahmed, S., & Waqas, A. (2021) An autonomous hybrid DC microgrid with ANN-fuzzy and adaptive terminal sliding mode multi-level control structure. *Control Engineering Practice* 121, 105036.

Dheeban, S. S., & Muthu Selvan, N. B. (2021) ANFIS-based power quality improvement by photovoltaic integrated UPQC at distribution system. *IETE Journal of Research* 69, 2353–2371.

Irfan, M. M., Rangarajan, S. S., Collins, E. R., & Senjyu, T. (2021) Enhancing the power quality of the grid interactive solar photovoltaic-electric vehicle system. *World Electric Vehicle Journal* 12(3), 98.

Jahan, I. S., Blazek, V., Misak, S., Snasel, V., & Prokop, L. (2022) Forecasting of power quality parameters based on meteorological data in small-scale household off-grid systems. *Energies* 15, 5251.

Jiang, M., Ghahremani, M., Dadfar, S., Chi, H., Abdallah, Y. N., & Furukawa, N. (2021) A novel combinatorial hybrid SFL–PS algorithm based neural network with perturb and observe for the MPPT controller of a hybrid PV-storage system. *Control Engineering Practice* 114, 104880.

Ko, J. S., Huh, J. H., & Kim, J. C. (2022) Overview of maximum power point tracking methods for PV system micro grid. *Electronics (Switzerland) MDPI AG* 9(5), 1–22.

Mishra, S., Mallick, R. K., Gadanayak, D. A., & Nayak, P. (2021) A novel hybrid downsampling and optimized random forest approach for islanding detection and non-islanding power quality events classification in distributed generation integrated system. *IET Renewable Power Generation* 15(8), 1662–1677.

Raj, S. A., & Samuel, G. G. (2022a) BOSS-D-RBFN: BOosted Salp Swarm optimization based Deep RBFN for MPPT under partial shading condition in photovoltaic systems. *Optik* 259, 168876.

Raj, S. A., & Samuel, G. G. (2022b) Survey of AI based MPPT algorithms in PV systems. *Proceedings of the 4th International Conference on Smart Systems and Inventive Technology, ICSSIT,* pp. 597–604.

Revathy, S. R., Kirubakaran, V., Rajeshwaran, M., Balasundaram, T., Sekar, V. S. C., Alghamdi, S., Rajab, B. S., Babalghith, A. O., & Anbese, E. M. (2022) Design and analysis of ANFIS: Based MPPT method for solar photovoltaic applications. *International Journal of Photoenergy* 2022, 1–9.

Roy, R. B., Rokonuzzaman, M., Amin, N., Mishu, M. K., Alahakoon, S., Rahman, S., Mithulananthan, N., Rahman, K. S., Shakeri, M., & Pasupuleti, J. (2021) A Comparative Performance Analysis of ANN Algorithms for MPPT Energy Harvesting in Solar PV System. *IEEE Access* Issue- 9, 102137–102152

Samy, M. M., Almamlook, R. E., Elkhouly, H. I., & Barakat, S. (2022) Decision-making and optimal design of green energy system based on statistical methods and artificial neural network approaches. *Sustainable Cities and Society* 84, 104015.

Talaat, M., Alblawi, A., Tayseer, M., & Elkholy, M. H. (2022) FPGA control system technology for integrating the PV/wave/FC hybrid system using ANN optimized by MFO techniques. *Sustainable Cities and Society* 80, 103825.

Vanam, S., & Bharath, B. S. P. (2022) Power Quality Improvement in Grid System with PV Based SVPWM-DVR for Sag & Swell Mitigation with a Novel ANN Controller. *Proceedings of the 2022 International Conference on Electronic Systems and Intelligent Computing, ICESIC 2022*, 200–205.

Wu, X., Furukawa, N., Tao, H., & Farajian, H. (2022) An efficient tracking of MPP in PV systems using hybrid HCS-PS algorithm based ANFIS under partially shaded conditions. *Soft Computing* 26(12), 5699–5717.

Yarar, N., Yagci, M., Bahceci, S., Onen, A., & Ustun, T. S. (2022) Artificial neural networks based harmonics estimation for real university microgrids using hourly solar irradiation and temperature data. *Energy Nexus* 9, 100172.

Yilmaz, M., & Corapsiz, M. (2022) Artificial neural network based MPPT algorithm with boost converter topology for stand-alone PV system. *Erzincan Üniversitesi Fen Bilimleri Enstitüsü Dergisi* 15(1), 242–257.

Chapter 3

# Advances and prospects of tin-based perovskites for efficient and stable photovoltaic devices

*Tejaswi Ratre and Jyoti Rani*

## 3.1 INTRODUCTION

Urbanization and the increasing use of technological devices are driving a significant rise in the global energy demand. Nevertheless, most of the required electricity is entirely generated from non-renewable fossil fuels, which are harmful for the environment and human health. Hence, renewable energy sources such as wind, water, and solar energy have been identifies as the best alternatives for electrical energy generation. Among these, solar energy stands out due to the abundant availability of sunlight. The development of solar cells has increased significantly over the past few decades because PV films can easily convert solar energy into electrical energy, thus more effectively addressing both the energy crisis and climate threat posed by carbon emissions. Si-based PV cells currently dominate the solar cell market because of their well-established technology, greater stability (over two decades), and high power conversion efficiency (PCE) (26.7%) [1]. Despite this, several studies on the production of affordable and flexible thin-film PV cells using low-temperature printing or solution-based methods.

Over the past 10 years, organic–inorganic hybrid perovskites have received much attention for their potential in cost-effective and simple fabrication, as well as their excellent intrinsic properties. These properties include tunable bandgap, long carrier diffusion length, high carrier mobility, and strong absorption coefficient. The general chemical formula of organic–inorganic perovskite material is $ABX_3$, where A is a monovalent organic cation (e.g., ($CH_3NH^{3+}$ ($MA^+$), $HC(NH_2)_2)^+$ ($FA^+$), or $Cs^+$), B is a divalent metal cation (e.g., $Pb^{2+}$ and $Sn^{2+}$), and X is a halide anion (e.g., I, Cl, Br, or a mixture of these). The most extensively researched perovskite is methylammonium lead iodide ($MAPbI_3$); it is known for its 1.5 eV bandgap [2], which allows the intake of a broad range of solar wavelengths and exhibits ambipolar properties crucial for PV cells [3–5]. Although methylammonium and lead-based hybrid halide perovskites show potential as inexpensive and industrially scalable PV materials that are capable of achieving more than 23% solar energy conversion efficiency [6], they also have significant

DOI: 10.1201/9781003521341-3

drawbacks. These include instability due to high reactivity with external moisture and oxygen when exposed to sunlight over an extended period, as well as toxicity concerns associated with the use of lead. To address these issues, researchers are exploring non-toxic, eco-friendly alternatives. The candidate materials $MASnX_3$ and $CsSnX_3$ appear to be promising for PV applications.

## 3.2 CHARACTERISTICS OF SN-BASED PEROVSKITES

$Sn^{2+}$ differs from its homolog $Pb^{2+}$ in several ways, including a higher energy band for Sn active electrons and higher Lewis acidity (as shown in Figure 3.1a, e, and c). These characteristics of Sn-based films present several challenges for the performance of their solar devices (as shown in Figure 3.1b, d, and f). In this section, we will discuss the properties and issues of Sn-based perovskites and propose some solutions from perspective of PV solar cell (PSC) engineering.

### 3.2.1 High band energy of $Sn^{2+}$

The divalent ion radii of $Sn^{2+}$ (114 pm) [7–10] and $Pb^{2+}$ (120 pm) and valence structures of Sn and Pb result in matched electronic structures of Pb-based and Sn-based perovskites, both of which are members of the IVA group of the periodic table. However, their energy band topologies are very different due to $Sn^{2+}$'s higher band energy. The conduction band minimum (CBM) of $AMX_3$ perovskite is determined by the hybridization between Sn-p states

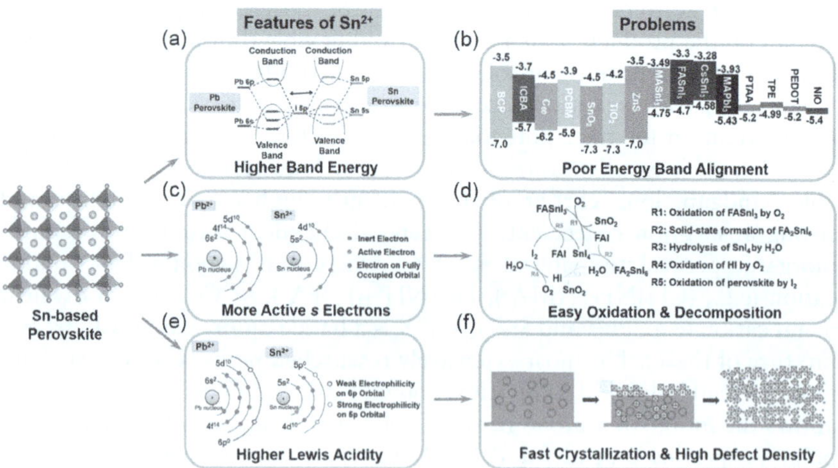


*Figure 3.1* (a–f) Problems associated with Sn-based PSCs that are derived from the features of $Sn^{2+}$. (Reprinted with the permission of Ref. [15]. Copyright 2021 American Chemical Society.)

and X-s levels, where Sn-p states play a significant role. In contrast, valence band maximum (VBM) results from the hybridization between X-p and Sn/Pb states [11]. The Sn-based perovskites exhibit higher VBM and CBM levels than their lead counterparts due to the higher energy state of tin-5p and -5s atomic shells (Figure 3.1a). Furthermore, sharp CBM and VBM edges and low effective masses of holes and electrons of the Sn-based perovskites are caused by the elevated energy levels of the tin-5p and -5s atomic orbitals [12]. The lower $Sn^{2+}$ can also cause lattice deformation and unwinding of the perovskite cage, leading to altered energy band levels [13].

For Sn-based PV cells to achieve excellent performance, the charge carrier transport layers must be specially designed due to the differences in band structures between Sn-based perovskites and their Pb analogs. In the current examples of Sn-based PSCs, regardless of whether they have a p-i-n and an n-i-p device structure, conventional hole transport layer (HTLs) and electron transport layer (ETLs) from lead-based PSCs are often used directly. This results in poor band alignment within the device (Figure 3.1b), which worsens open circuit voltage ($V_{oc}$) and fill factor (FF) loss. Therefore, to develop effective and stable Sn-based PSCs, it is crucial to design innovative HTLs and ETLs that are appropriate for tin-based perovskites.

### 3.2.2 $Sn^{2+}$ active electrons

The simple oxidation of $Sn^{2+}$ is a major obstacle to the production of stable tin-based PV cells. The electrons in $Pb^{+}$'s 6s orbital exhibit the inert pair effect because they have a powerful shielding effect on the fully filled electrons in 4f and 5d orbitals, a result of lanthanide shrinkage (Figure 3.1c). In contrast, the facile oxidation of $Sn^{2+}$ to $Sn^{4+}$ is caused by the tendency of $Sn^{2+}$ to lose its two 5s electrons due to the electrons' decreased shielding effect on the acquired 4d orbital (Figure 3.1c).This oxidation is further exacerbated by the p-type nature of Sn-based perovskites, in which 5s electrons of $Sn^{2+}$ are easily depleted by p-type dopants. As a result, the $Sn^{2+}$ oxidation process can occur even when the film strip is created in an environment devoid of water and oxygen. This oxidation problem is the primary cause of poor Sn-based film quality and reduced PV efficiency in Sn-based PSCs. Although the defects in tin-based perovskites are superficial, their high density leads to poor morphology, which negatively impacts crystal quality of tin-based perovskite films. This, in turn, causes significant carrier recombination and energy dissipation in tin-based PSCs [14].

### 3.2.3 High Lewis acidity of $Sn^{2+}$

$Sn^{2+}$ exhibits stronger Lewis acidity than $Pb^{2+}$ due to the higher orbital energy of the 5p atom in $Sn^{2+}$ (Figure 3.1e), which results in a rapid reaction with MAI/FAI to form the perovskite structure. According to the classical LaMer theory for crystal growth, this rapid crystallization can result in the formation of small crystal grains. Because of this quick crystallization, the solution-processed films often exhibit irregular orientation and has several flaws (Figure 3.1f). Even worse, rapid crystallization favors the less forming energy of Sn and the ease with which faulty perovskite crystal formations with $Sn^{4+}$ as the octahedral coordination core may be formed. These defects, which are formed when $Sn^{2+}$ is easily ionized and when the tin-iodine bond in the crystal lattice is broken (e.g., $FA_2SnI_6$ structure), function as p-dopants inside Sn-based perovskite compounds. This increases the concentration of hole carriers and creates high recombination centers.

## 3.3 LEAD-FREE SN-BASED PEROVSKITES

Tin is a promising ion replacement for lead. Lead-free tin-based perovskites have the potential to become leaders in the energy market. Sn is considered a viable candidate for replacing lead in the lattice, offering significant advantages to researchers. However, to achieve this, efficiency must be the primary focus.

The highest PCE of 35% is predicted by the Shockley–Queisser limit [7] for a single-junction cell with a bandgap of 1.34 eV. In addition to efficiency, stability is another factor that must be considered in the global PV market, which is essential for the long-term survival of perovskites [8].

Similar to Pb-based perovskites, lead-free Sn-based perovskites offer improved light absorption characteristics [10] and a reduced optical energy bandgap [9]. The lead 6p orbital is critical in the band structure of conventional organic–inorganic hybrid lead-based perovskites, whereas lead-free tin-based perovskites exhibit better charge carriers than Pb-based perovskites [11]. Furthermore, when exposed to the atmosphere, they oxidize to $Sn^{4+}$, forming environmentally friendly $SnO_2$ [6]. According to experimental reports, the efficiency of Sn-based perovskites can reach 12.96% [12].

The PV properties of several tin-based lead-free perovskites utilized as absorbers in solar cells with diverse device topologies are summarized in Table 3.1.

### 3.3.1 $MASnX_3$

The first fully lead-free solar cell based on methylammonium tin iodide perovskite was reported by Noel et al. [16]. Equimolar amounts of $CH_3NH_3$ and $SnI_2$ were mixed in N-dimethylformamide (DMF) solution, forming $CH_3NH_3SNI_3$, a perovskite with a tetragonal crystal structure. The absorption spectra showed that the bandgap was 1.23 eV. The constructed device has an efficiency of more than 6% under one sun's light.

A method for constructing $MASnI_3$-based solar devices was developed by Song et al. [17,18], in which the solvent used was a Dimethyl sulfoxide (DMSO) solution. Then, 20 mol% $SnF_2$ was added to this solution and

*Table 3.1* The PV characteristics of several Sn-based Pb-free perovskites used as absorbers in the solar cell

| *Absorber* | *Device* | $V_{oc}$ *(V)* | $J_{sc}$ *(mA/cm²)* | *FF (%)* | *PCE (%)* | *References* |
|---|---|---|---|---|---|---|
| $MASnI_3$ | FTO/c-$TiO_2$/m-$TiO_2$/absorber/spiro-OMeTAD/Au | 0.88 | 16.8 | 42 | 6.4 | [16] |
| $MASnI_3$+20% $SnF_2$/hydrazine | FTO/c-$TiO_2$/m-$TiO_2$/absorber/PTAA/Au | 0.378 | 19.92 | 51.73 | 3.89 | [17] |
| $MASnIBr_2$+30% $SnF_2$ | FTO/c-$TiO_2$/m-$TiO_2$/absorber/Spiro-OMeTAD/Au | 0.45 | 13.77 | 59.58 | 3.7 | [22] |
| $MASn(I_{0.33}Br_{0.67})_3$ | FTO/c-$TiO_2$/m-$TiO_2$/absorber/Spiro-OMeTAD/Au | 0.58 | 11.09 | 49.50 | 3.20 | [19] |
| $MASnI_3$ | FTO/c-$TiO_2$/m-$TiO_2$/absorber/PTAA/Au | 0.486 | 22.91 | 64 | 7.13 | [20] |
| $MASnI_3$ | ITO/PEDOT:PSS/absorber/$PC_{60}$BM/Ag | 0.57 | 20.68 | 66 | 7.78 | [21] |
| $MASnI_3$ | FTO/c-$TiO_2$/m-$TiO_2$/absorber/Spiro-OMeTAD/Au | 0.68 | 16.30 | 48 | 5.23 | [23] |
| $MASnI_2Br$ | FTO/c-$TiO_2$/m-$TiO_2$/absorber/Spiro-OMeTAD/Au | 0.77 | 14.38 | 50 | 5.48 | [23] |
| $MASnIBr_2$ | FTO/c-$TiO_2$/m-$TiO_2$/absorber/Spiro-OMeTAD/Au | 0.82 | 12.30 | 57 | 5.73 | [23] |
| $MASnBr_3$ | FTO/c-$TiO_2$/m-$TiO_2$/absorber/Spiro-OMeTAD/Au | 0.88 | 8.26 | 59 | 4.27 | [23] |
| $MASnI_3$ | FTO/c-$TiO_2$/m-$TiO_2$/absorber/Spiro-OMeTAD/Au | 0.716 | 15.18 | 50.07 | 5.44 | [24] |
| $CsSnI_3$ | ITO/c-$TiO_2$/absorber/Spiro-OMeTAD/Au | 0.86 | 23.2 | 65 | 12.96 | [25] |
| $CsSnBr_3$ | ITO/c-$TiO_2$/absorber/Spiro-OMeTAD/Au | 0.85 | 21.23 | 58 | 10.46 | [25] |
| $CsSnCl_3$ | ITO/c-$TiO_2$/absorber/Spiro-OMeTAD/Au | 0.87 | 19.82 | 56 | 9.66 | [25] |
| $CsSnI_3$+10% $SnF_2$ | FTO/c-$TiO_2$/absorber/PTAA/Au | 0.41 | 18 | 46.3 | 3.4 | [26] |
| $CsSnI_3$+10% $SnCl_2$ | FTO/c-$TiO_2$/absorber/PTAA/Au | 0.43 | 17.4 | 52.3 | 3.9 | [26] |
| $CsSnI_3$+10% $SnBr_2$ | FTO/c-$TiO_2$/absorber/PTAA/Au | 0.44 | 18.5 | 52.9 | 4.3 | [26] |
| $CsSnI_3$+Hydrazine | FTO/c-$TiO_2$/m-$TiO_2$/absorber/PTAA:TPFB/Au | 0.381 | 25.71 | 49.05 | 4.81 | [27] |
| $CsSnBr_3$+20% $SnF_2$ | FTO/c-$TiO_2$/m-$TiO_2$/absorber/PTAA/Au | 0.367 | 13.96 | 59.36 | 3.04 | [17] |
| $CsSnIBr_2$+60% $SnF_2$+$H_3PO_2$ | FTO/c-$TiO_2$/m-$TiO_2$/m-$Al_2O_3$/absorber/C | 0.31 | 17.4 | 56 | 3.2 | [28] |

PV, photovoltaic.

was heated to 150 C. The ratio of $Sn^{2+}$ to $Sn^{4+}$ decreased by more than 20% due to the introduction of a reducing vapor atmosphere. This caused severely filtered charge carrier recombination, leading to a PCE of 3.89% for $MASnI_3$-based solar cells as a final outcome.

Nguyen et al. [19] fabricated several mixed tin-based $MASn(II\text{-}xBrx)_3$ solar devices, demonstrating that the concentration of DMSO in the DMF solution could control the rapid nucleation and slow growth of $MASnI_3$ crystals.

Li et al. [20] introduced a novel two-step chemical technique for producing dense, uniform $MASnI_3$ films with high reproducibility and high crystallinity. In the first step, a hydrazinium tin iodide ($HASnI_3$) perovskite layer was originally deposited using $SnI_2$ and hydrazinium iodide (HAI). The second stage involved converting $HASnI_3$ into $MASnI_3$ using a cation displacement method. Using a cation displacement technique in two steps, a premium $MASnI_3$ film was created, with a maximum PCE of 7.13%.

Perovskite film nucleation and crystallization must be controlled to produce highly efficient solar cell systems. To this end, Wang et al. [21] recently developed a synthetic method that uses ion exchange/insertion processes to create $MASnI_3$ thin films. By combining $SnF_2$ with MAI, the method produced consistent, pinhole-free $MASnI_3$ films. The resulting film had successful suppression of the unfavorable oxidation due to the substantial amount of remaining $SnF_2$. In addition, the SEM of both the unaltered $SnF_2$/PEDOT: PSS solar film and the one that had been subjected to an ion exchange/insertion reaction for 60 minutes. The final solar cell had a PCE of 7.78% with excellent reproducibility and stability. This strategy presents a novel method for producing low-cost, steady tin-based lead-free PVs. Table 3.1 depicts the important progress that has been made in $MASnX_3$-based PV cells up to date.

### 3.3.2 $CsSnX_3$

Kumar et al. [29] developed a low-pressure and low-temperature solution procedure for generating $CsSnI_3$ film wherein $SnI_2$ and CsI were stoichiometrically combined in suitable solvents. By using $CsSnI_3$ as an absorber in a solar cell, they demonstrated the viability of employing solution-processed Pb-free perovskites for extremely effective PVs with more photocurrent densities (>22 $mAcm^2$). However, the PCEs were hampered by their low Voc.

Wang et al. [30] used a simple solution method to deposit B-c-$CsSnI_3$ as an absorber. They used NiOx, which had thermodynamically favorable band orientations, to solve the organic HTL incompatibility problem. This approach resulted in a PCE of 3.31% due to the stable management of the PV cell architecture.

For the first time, Chen et al. [25] demonstrated how to make $CsSnX_3$ quantum rods using a straightforward solvothermal technique. They showed that by modifying the halides, $CsSnX_3$'s photoluminescence and optical properties could be tuned. The findings revealed that $CsSnI_3$ had a record PCE of 12.96%.

Heo et al. [26] used density functional theory and related experiments to investigate the impacts of $SnX_2$ additives on the stability of Pb-free Sn-based perovskites. Their study demonstrated that $SnX_2$ additives could improve the durability of $CsSnI_3$ PV devices. Although it is widely believed that additives eliminate flaws, this study found that additives also effectively passivate surfaces and stabilize perovskites, contributing to the stability of $CsSnI_3$. The overall findings indicated that $SnBr_2$ was the most potential candidate for sustained stability, yielding a high PCE of 4.3%.

Song et al. [27] described $CsSnI_3$ thin-film perovskites in which excessive $SnI_2$ was uniformly diffused into the perovskite film. This approach served as a suppressor and compensator for $Sn^{2+}$ vacancies, effectively reducing hole conductivity and improving the performance of $CsSnI_3$ devices on mesoporous $TiO_2$. Several devices were created using various $CsI/SnI_2$ molar ratios. The best-performing device, with a maximum PCE of 4.81%, had a molar ratio of 0.4. Table 3.2 summarizes the advancements in the performance of $CsSnX_3$-based PV cells to date.

Table 3.1 represents the FF, current density, open circuit voltage, and efficiency of different materials [20]. Solar Energy Materials and Solar Cells 204 (2020) 110212. B. Parida. Coordination chemistry review, H. Moghe, L. Wang, P. R. Brown, et al. 2020 [34].

## 3.4 CHALLENGES IN THE FABRICATION OF PB-FREE SN-BASED PEROVSKITE THIN FILMS

Tin has been identified as a viable alternative to lead in the perovskite lattice. However, to date, PV cells based on tin have not achieved performance levels comparable to lead-based equivalents. This "under per" performance can be attributed to several factors, including poor film quality, electrical characteristics, and most crucially, poor stability caused by the spontaneous oxidation of $Sn^{2+}$ to $Sn^{4+}$. The fabrication of solar cells using lead-free tin-based perovskite thin films as absorbers typically requires an inert atmosphere, raising concerns about the approach's viability in real-world applications. Furthermore, manufacturing such devices in a natural environment remains impractical. Several studies have proposed various methods for improving film quality, including gas pumping treatment and hybrid deposition [28,35–37]. However, these strategies have yielded inadequate PCEs, highlighting their limited ineffectiveness. Lead analogs have superior electrical properties, whereas lead-free and tin-based perovskites have inferior electronic properties due to the poor film quality and chemical composition. The presence of broad grain boundaries and numerous extremely tiny crystallite domains in tin perovskite films, which act as trap states, increase recombination, and reduce charge separation and transport, is prevalent [38]. Moreover, it is difficult to manage the material's conductivity effectively due to the rapid oxidation of $Sn^{2+}$ to $Sn^{4+}$, which leads to gaps and poor stability in lead-free tin-based perovskite solar cells under all

*Table 3.2* The stability of different $MASnX_3$ and $CsSnX_3$ materials with time

| *Absorber* | *Device structure* | *Method used* | *Stability* | *Year* | *References* |
|---|---|---|---|---|---|
| $MASnI_3$ + 20% $SnF_2$/hydrazine | FTO/c-$TiO_2$/m-$TiO_2$/absorber/PTAA/Au | Reducing vapor method | Unstable | 2017 | [17] |
| $MASn(I_{0.33}Br_{0.67})_3$ | FTO/c-$TiO_2$/m-$TiO_2$/absorber/ spiro-OMeTAD/Au | Solvent optimization | Poor stability | 2019 | [19] |
| $MASnI_3$ | ITO/PEDOT: PSS/absorber/$PC_{60}$BM/Ag | Ion exchange reactions | Better stability | 2020 | [21] |
| $MASnIBr_{1.8}Cl_{0.2}$ + 20% $SnF_2$ | FTO/c-$TiO_2$/m-$TiO_2$/m-$Al_2O_3$/ absorber/C | Co-crystallization of halides | Better long-term stability | 2017 | [31] |
| $CsSnBr_3$ + 20% $SnF_2$ | FTO/c-$TiO_2$/m-$TiO_2$/absorber/ spiro-OMeTAD/Au | – | Unstable in atmosphere | 2016 | [32] |
| $CsSnI_3$ + 10% $SnCl_2$ | ITO/absorber/$PC_{61}$BM/BCP/Al | HTL-free protocol | Excellent stability | 2016 | [33] |
| $CsSnI_3$ + 10% $SnCl_2$ | ITO/absorber/$PC_{61}$BM/BCP/Al | Solvothermal synthesis | Moderate stability | 2016 | [25] |
| $CsSnI_3$ + 10% $SnF_2$ | FTO/c-$TiO_2$/absorber/PTAA/Au | – | Stability degraded quickly over time | 2018 | [26] |
| $CsSnI_3$ + 10% $SnCl_2$ | FTO/c-$TiO_2$/absorber/PTAA/Au | – | Stability degraded quickly over time | 2018 | [26] |
| $CsSnI_3$ + 10% $SnBr_2$ | FTO/c-$TiO_2$/absorber/PTAA/Au | – | Excellent stability | 2018 | [26] |

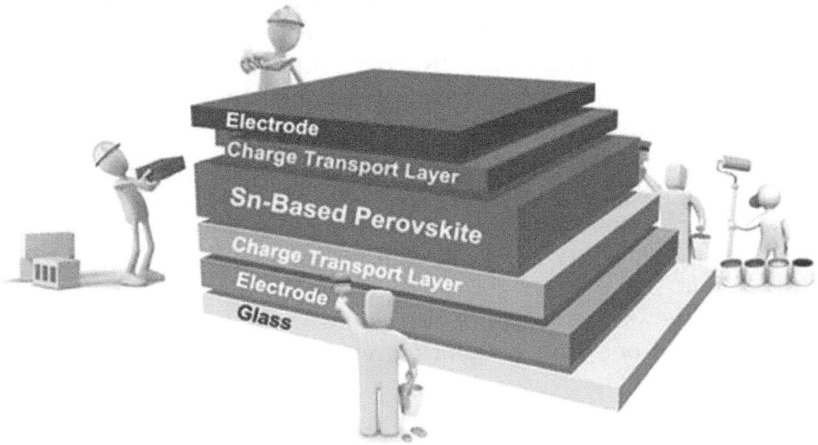


*Figure 3.2* Architecture engineering of Sn-based perovskite solar cells. (Reprinted with the permission of Ref. [15]. Copyright 2021 American Chemical Society (ACS).)

circumstances [16]. There have been numerous attempts to stop the oxidation of $Sn^{2+}$ to $Sn^{4+}$, including the use of $SnF_2$ and other additives and the synthesis of mixed lead and tin perovskites [39,32]. However, the resulting thin films and PV devices have consistently been found to be inferior to lead analogs (Figure 3.2).

## 3.5 MATERIAL STABILITY

Lead-free perovskite PV cells based on Sn show promise for achieving high performance. Unfortunately, the stability of these materials is significantly poorer than that of lead-based perovskite solar cells. Yet, several strategies have been developed in recent years to concurrently increase the PCE and stability of tin-based perovskite solar cells. For example, $MASnI_3$-based solar cells demonstrated a PCE of 1.94% and showed no loss in conversion efficiency over 146 days [27]. This cell was constructed with 20% $SnF_2$. After 60 days of storage in a nitrogen-filled glove box, the PV cell retained over 80% of its initial efficiency [40]. Ion exchange/insertion reactions have also been used to increase the stability of $MASnI_3$-based PV cells. This technique resulted in a high PCE of 7.78%, strong repeatability, and good stability [41]. Additional recent methods to improve the stability of methylammonium $SnX_2$-based perovskites include co-crystallization of three halides [42] and a low-temperature vapor-assisted solution procedure [25].

Additionally, the decreasing vapor atmosphere approach and the vapor deposition method have been found to increase the stability of $CsSnBr_3$. In contrast to the MA, Cs cation-based PV cells were tested and showed excellent stability and efficiency. The cation showed outstanding air stabilities when used in perovskite solar cells along with FA and MA cations [38]. Moreover, combining FA and MA cations can enhance the efficiency and stability of tin-based lead-free perovskite solar cells. An anti-solvent dripping technique was used to create $FA_{0.75}MA_{0.25}Sn_3$-based solar cells, which achieved a high PCE and retained 75% of its initial form after 30 days in an encased state, demonstrating long-term stability. Table 3.2 summarizes the development of stability in several reported tin-based PV cells from 2016 to the present.

Table 3.2 shows the stability of different $MASnX_3$ and $CsSnX_3$ materials over time [34]. Coordination chemistry review, "H. Moghe, L. Wang, P. R. Brown et al 2020.

## 3.6 CONCLUSION

Significant progress has been made in the field of PVs, especially in perovskites over the past few decades. Efforts are ongoing to improve efficiencies and optimize each factor to make PV panels more viable on a larger scale. While early developments in this field primarily focused on lead-based perovskites, recent reports indicate that tin-based perovskites can be competent. For instance, $CsSnI_3$ has achieved a PCE of 12.96%, the highest PCE for tin-based photocells to date. This illustrates the potential for tin-based materials to narrow the bandgap and offer favorable outcomes for PVs. However, a lot of challenges remain for lead-free tin-based perovskites in PV cells to reach the level of pre-established PV technology. Due to the change of Sn from its +2 to +4 oxidation state in the ambient environment, the stability of tin-based perovskite is a more significant challenge than their lead counterparts. While the $SnX_2$ addition has been widely employed to avoid this procedure, further research is needed to explore alternative materials and strategies that can protect Sn in its +2 state and minimize the conversion to the +4 state. Other than these strategies, more effective and practically feasible strategies are required to enhance both stability and efficiency of tin-based perovskites in real-life experiments. These might include improvements in film quality, advances fabrication techniques, and the use of more advanced additives.

## ACKNOWLEDGMENT

The author would like to thank the Department of Physics, MANIT Bhopal, for providing a platform to carry out this research.

## REFERENCES


1. M.A. Green, Y. Hishikawa, E.D. Dunlop, D.H. Levi, J. Hohl-Ebinger, M. Yoshita, A. W.Y. Ho-Baillie, Solar cell efficiency tables (Version 53). *Prog. Photovolt. Res. Appl.* 27 (2019) 3–12.
2. S.D. Stranks, G.E. Eperon, G. Grancini, C. Menelaou, M.J.P. Alcocer, T. Leijtens, L. M. Herz, A. Petrozza, H.J. Snaith, Electron-hole diffusion lengths exceeding 1 micrometer in an organometal trihalide perovskite absorber. *Science* 342 (2013) 341.
3. J.H. Noh, S.H. Im, J.H. Heo, T.N. Mandal, S.I. Seok, Chemical management for colorful, efficient, and stable inorganic–organic hybrid nanostructured solar cells. *Nano Lett.* 13 (2013) 1764–1769.
4. Q. Dong, Y. Fang, Y. Shao, P. Mulligan, J. Qiu, L. Cao, J. Huang, Electron-hole diffusion lengths > 175 μm in solution-grown $CH_3NH_3PbI_3$ single crystals. *Science* 347 (2015) 967.
5. S. De Wolf, J. Holovsky, S.-J. Moon, P. Loper, B. Niesen, M. Ledinsky, F.-J. Haug, J.-H. Yum, C. Ballif, Organometallic halide perovskites: Sharp optical absorption edge and its relation to photovoltaic performance. *J. Phys. Chem. Lett.* 5 (2014) 1035–1039.
6. J.J. Yoo, G. Seo, M.R. Chua, T.G. Park, Y. Lu, F. Rotermund, Y.K. Kim, C.S. Moon, N.J. Jeon, J.P.C. Baena, V. Bulovic, S.S. Shin, M.G. Bawendi, J. Seo, Efficient perovskite solar cells via improved carrier management. *Nature* 590 (2021) 587.
7. C. Gai, J. Wang, Y. Wang, J. Li, The low-dimensional three-dimensional tin halide perovskite: Film characterization and device performance. *Energies* 13 (2020) 2.
8. G. Nasti, A. Abate, Tin halide perovskite ($ASnX_3$) solar cells: A comprehensive guide toward the highest power conversion efficiency. *Adv. Energy Mater.* 10 (2020) 1902467.
9. Y. Jiang, X. Zang, Z. Zhou, Y. Li, H. Wei, Q. Ning, Tin halide perovskite solar cells: An emerging thin-film photovoltaic technology. *Acc. Mater. Res.* 2 (2021) 210–219.
10. Y. Yan, T. Pullerits, T. Zheng, K. Liang, Advancing tin halide perovskites: Strategies toward the $ASnX_3$ paradigm for efficient and durable optoelectronics. *ACS Energy Lett.* 5 (2020) 2052–2086.
11. S. Schmidt, I. Brocks, G. Jiang, J. Tranca, I. Meerholz, K. Olthof, Absolute energy level positions in tin- and lead-based halide perovskites. *Nat. Commun.* (2019) 2560.
12. M.K. Jeon, T. Park, H. I. Lee, J. M. Nam, S. A. Kim, Effective control of crystal grain size in $CH_3NH_3PbI_3$ perovskite solar cells with a pseudohalide $Pb(SCN)_2$ additive. *CrystEngComm* 18 (2016) 6090–6095.
13. A. McKechnie, S. Pashov, D. Tumas, W., van Schilfgaarde, M., V. Stevanovic, Origin of pronounced nonlinear band gap behavior in lead–tin hybrid perovskite alloys. *Chem. Mater.* 30 (2018) 3920–3928.
14. B. Li, B. Chang, L. Pan, Z. Li, L. Fu, Z. He, L. Yin, Tin based defects and passivation strategies in tin-related perovskite solar cells. *ACS Energy Lett.* 5 (2020) 3752–3772.

15. N. Sun, W. Gao, H. Dong, Y. Liu, X. Liu, Z. Wu, L. Song, C. Ran, Y. Chen, Architecture engineering of p-i-n Sn-based perovskite solar cells: Characteristics, advances, and perspectives. *ACS Energy Lett.* 6 (2021) 2863–2875.
16. N.K. Noel, S.D. Stranks, A. Abate, C. Wehrenfennig, S. Guarnera, A.A. Haghighirad, A. Sadhanala, G.E. Eperon, S.K. Pathak, M.B. Johnston, A.M. Petrozza, L.M. Herz, H.J. Snaith, Lead-free organic–inorganic tin halide perovskites for photovoltaic applications. *Energy Environ. Sci.* 7(9) (2014) 3061–3068.
17. T.B. Song, T. Yokoyama, C.C. Stoumpos, J. Logsdon, D.H. Cao, M.R. Wasielewski, S. Aramaki, M.G. Kanatzidis, Importance of reducing vapor atmosphere in the fabrication of tin-based perovskite solar cells. *J. Am. Chem. Soc.* 139(2) (2017) 836–842.
18. T. Handa, T. Yamada, H. Kubota, S. Ise, Y. Miyamoto, Y. Kanemitsu, Photocarrier recombination and injection dynamics in long-term stable lead-free $CH_3NH_3SnI_3$ perovskite thin films and solar cells. *J. Phys. Chem. C.* 121(30) (2017) 16158–16165.
19. B.P. Nguyen, D. Shin, H.R. Jung, J. Kim, T.T. Nguyen, S. Yoon, Y. Yi, W. Jo, Phase formation and local charge transport of lead-free $CH_3NH_3Sn(I1\text{-}xBrx)_3$ (0 < x < 1) perovskite solar cells fabricated by solvent optimization. *Sol. Energy* 186 (2019) 136–144.
20. F. Li, C. Zhang, J.H. Huang, H. Fan, H. Wang, P. Wang, C. Zhan, C.M. Liu, X. Li, L. M. Yang, Y. Song, K.J. Jiang, A cation-exchange approach for the fabrication of efficient methylammonium tin iodide perovskite solar cells. *Angew. Chem. Int. Ed.* 58(20) (2019) 6688–6692.
21. P. Wang, F. Li, K.J. Jiang, Y. Zhang, H. Fan, Y. Zhang, Y. Miao, J.H. Huang, C. Gao, X. Zhou, F. Wang, L.M. Yang, C. Zhan, Y.L. Song, Ion exchange/insertion reactions for fabrication of efficient methylammonium tin iodide perovskite solar cells. *Adv. Sci.* 1903047 (2020) 1058–1067.
22. M. Xiao, S. Gu, P. Zhu, M. Tang, W. Zhu, R. Lin, C. Chen, T. Yu, J. Zhu, Tin-based perovskite with improved coverage and crystallinity through tin-fluoride-assisted heterogeneous nucleation. *Adv. Optical Mater.* 6(1) (2018).
23. T.F. Hao, C.C. Stoumpos, D.H. Cao, R.P. Chang, M.G. Kanatzidis, Lead-free solid-state organic–inorganic halide perovskite solar cells. *Nat. Photonics.* 8(6) (2014) 489.
24. M. Hao, C.C. Stoumpos, R.P. Chang, M.G. Kanatzidis, Anomalous band gap behavior in mixed Sn and Pb perovskites enable broadening of absorption spectrum in solar cells. *J. Am. Chem. Soc.* 136(22) (2014) 8094–8099.
25. L.J. Chen, C.R. Lee, Y.J. Chuang, Z.H. Wu, C. Chen, Synthesis and optical properties of lead-free cesium tin halide perovskite quantum rods with the high-performance solar cell application. *J. Phys. Chem. Lett.* 7 (24) (2016) 5028–5035.
26. J.H. Heo, J. Kim, H. Kim, S.H. Moon, S.H. Im, K.H. Hong, Roles of $SnX_2$ (X= F, Cl, Br) additives in tin-based halide perovskites toward highly efficient and stable lead-free perovskite solar cells. *J. Phys. Chem. Lett.* 9(20) (2018) 6024–6031.

27. T.B. Song, T. Yokoyama, S. Aramaki, M.G. Kanatzidis, Performance enhancement of lead-free tin-based perovskite solar cells with reducing atmosphere-assisted dispersible additive. *ACS Energy Lett.* 2(4) (2017) 897–903.
28. T. Handa, T. Yamada, H. Kubota, S. Ise, Y. Miyamoto, Y. Kanemitsu, Photocarrier recombination and injection dynamics in long-term stable lead-free $CH_3NH_3SnI_3$ perovskite thin films and solar cells. *J. Phys. Chem. C.* 121(30) (2017) 16158–16165.
29. M.H. Kumar, S. Dharani, W.L. Leong, P.P. Boix, R.R. Prabhakar, T. Baikie, C. Shi, H. Ding, R. Ramesh, M. Asta, M. Graetzel, Lead-free halide perovskite solar cells with high photocurrents realized through vacancy modulation. *Adv. Mater.* 26(41) (2014) 7122–7127.
30. N. Wang, Y. Zhou, M.G. Ju, H.F. Garces, T. Ding, S. Pang, X.C. Zeng, N.P. Padture, X.W. Sun, Heterojunction-depleted lead-free perovskite solar cells with coarse-grained B-c-$CsSnI_3$ thin films. *Adv. Energy Mater.* 6(24) (2016) 1601130.
31. M. Tsai, N. Mohanta, C.Y. Wang, Y.P. Lin, Y.W. Yang, C.L. Wang, C.H. Hung, E. W. Diau, Formation of stable tin perovskites co-crystallized with three halides for carbon-based mesoscopic lead-free perovskite solar cells. *Angew. Chem.* 129(44) (2017) 14007–14011.
32. S. Tong, Z. Song, D.H. Kim, X. Chen, C. Chen, A.F. Palmstrom, P.F. Ndione, M.O. Reese, S.P. Dunfield, O.G. Reid, J. Liu, Carrier lifetimes of > 1 ls in Sn-Pb perovskites enable efficient all-perovskite tandem solar cells. *Science* 364(6439) (2019) 475–479.
33. K.P. Marshall, M. Walker, R.I. Walton, R.A. Hatton, Enhanced stability and efficiency in hole-transport-layer-free $CsSnI_3$ perovskite photovoltaics. *Nat. Energy* 1(12) (2016) 1–9.
34. H. Moghe, L. Wang, C.J. Traverse, A. Redoute, M. Sponseller, P.R. Brown, V. Bulovic, R.R. Lunt, All vapor-deposited lead-free doped $CsSnBr_3$ planar solar cells. *Nano Energy* 28 (2016) 469–470.
35. B. Lee, B. Shin, B. Park, Uniform $Cs_2SnI_6$ thin films for lead-free and stable perovskite optoelectronics via hybrid deposition approaches. *Electron. Mater. Lett.* 15(2) (2019) 192–200.
36. X.L. Li, L.L. Gao, Q.Q. Chu, Y. Li, B. Ding, G.J. Yang, Green solution-processed tin-based perovskite films for lead-free planar photovoltaic devices. *ACS Appl. Mater. Interfaces* 11(3) (2018) 3053–3060.
37. X.W. Ke, C.C. Stoumpos, I. Spanopoulos, M. Chen, M.R. Wasielewski, M.G. Kanatzidis, Diammonium cations in the $FASnI_3$ perovskite structure lead to lower dark currents and more efficient solar cells. *ACS Energy Lett.* 3(7) (2018) 1470–1476.
38. X.F. Yang, F. Igbari, Y.H. Lou, Z.K. Wang, L.S. Liao, Tin halide perovskites: Progress and challenges. *Adv. Energy Mater.* 10(13) (2019) 1902584.
39. D. Mosconi, P. Umari, F. De Angelis, Electronic and optical properties of mixed Sn–Pb organohalide perovskites: A first-principles investigation. *J. Mater. Chem. A* 3(17) (2015) 9208–9217.
40. Z. Zhao, F. Gu, Y. Li, W. Sun, S. Ye, H. Rao, Z. Liu, Z. Bian, C. Huang, Mixedorganic-cation tin iodide for lead-free perovskite solar cells with an efficiency of 8.12%. *Adv. Sci.* 4(11) (2017) 1700204.

41. J. Jiang, C.K. Onwudinanti, R.A. Hatton, P.A. Bobbert, S. Tao, Stabilizing lead-free all-inorganic tin halide Perovskites by ion exchange. *J. Phys. Chem. C.* 122(31) (2018) 17660–17667.
42. E. Jokar, C.H. Chien, C.M. Tsai, A. Fathi, E.W. Diau, Robust tin-based perovskite solar cells with hybrid organic cations to attain efficiency approaching 10%. *Adv. Mater.* 31(2) (2019) 1804835.

Chapter 4

# Study of super-hydrophobic surfaces for solar cells

*Aravind Kumar, Sonal Santosh Bagade, Sudhansu B. Barik, Jyoti Rani, M.M. Malik, and Piyush K. Patel*

## 4.1 INTRODUCTION

We explored various synthesis methods for creating super-hydrophobic surfaces, with a focus on CVD, solution immersion, layer-by-layer deposition, anodization, and chemical etching. These methods are cost-effective. The industrial applications of materials such as ceramics, metals, and polymers depend on their long-term properties. As compare to ceramics, metals, and polymers. metals and alloys are particularly relevant because of their excellent electrical and thermal conductivity, high strength, efficiency, unique mechanical properties, and accessibility. However, the physical and mechanical properties of metals are influenced by their surrounding environment due to high atmospheric reactivity and weak surface properties [1,2].

Metals can degrade through corrosion and wear, resulting in material loss from the surface [3]. Degradation of metals usually occurs in the presence of moisture and ions. Therefore, limiting surface reactions and preventing destruction of metals are crucial for their industrial applications. The need for corrosion protection is critical not only for safety but also for financial reasons. While metals and alloys are corrosion-resistant, the development of specialized surface coatings that offer wet corrosion protection has attracted much attention [4–6]. Surface wetting is the skill of the surface to be wet or damp in its interaction with water. The wetting properties of surfaces are influenced by their morphology and microstructure and morphology [7,8]. Super-hydrophobic coating and traditional chromate-based coatings have been used to protect metals from corrosion [9–13]. Different chromate-free anti-corrosion coatings, are used in metals. Aluminum, magnesium and stainless steel [14–16]. The introduction of super-hydrophobicity in coatings enhances corrosion protection achieved through surface modification with low-phase energy components.

The main objective of this research is to develop an easy and inexpensive method to design super-hydrophobic aluminum alloy surfaces with thermal and physical stability. The 6061 aluminum alloy is used in many applications including aircraft construction, yard construction, automotive parts,

DOI: 10.1201/9781003521341-4

and aluminum bins, new contributions presented in this chapter are as follows: Hydrochloric and acetic acids can be used as a straightforward and inexpensive method for developing super-hydrophobic surfaces on hard aluminum alloy surfaces. This work describes cost-effective, practical, and easy method for creating a super-hydrophobic aluminum surface that is both physically and thermally stable, suitable for many industrial applications. Direct application of HCl and $CH_3COOH$ is more synergistic and less costly for achieving the desired rough surface texture than the commonly used anodizing process. Super-hydrophobic aluminum surfaces with a $H_2O$ contact angle of 165°±2° and a sliding angle of less than 3° were attained using the simple and inexpensive method described above, which is superior to other super-hydrophobic aluminum surfaces described in previous works that mostly use expensive methods and modifiers. The resulting super-hydrophobic aluminum surfaces have shown short-term stability underwater and over 6 months. Super-hydrophobic aluminum surfaces have thermal stability up to biomimetic surfaces based on simulating biological surfaces, leading to major developments in creating synthetic surfaces with unique properties. Some natural surfaces, such as the water strider's leg, shark skin, mosquito eyes, and lotus leaves, exhibit unique and unusual interactions with surrounding fluids, especially water [17–20]. Among these, the lotus leaves and Nelumbo luciferase have received much attention because of their wet and self-cleaning properties, as described by author. They revealed the interdependence of surface roughness, water retention, and self-cleaning properties by characterizing the microstructure of the lotus leaf surface. Air pockets accumulate on the leaf surface, which is an important cause of the abnormal water dissipation and self-purification of the lotus leaf [21–24].

Recently, super-hydrophobic surfaces have received much attention due to a variety of applications (e.g., corrosion barrier improvement [25], friction drag reduction [26], heat transfer enhancement [27], anti-icing [28], self-cleaning [29,30]). In the following sections, we first introduce theoretical models of liquid–solid interaction and then briefly discuss the basics and recent improvements in the creation of artificially designed super-hydrophobic surfaces.

Wetness refers to the tendency of a fluid to stick or spread on a fluid surface, which is defined by the contact angle of the fluid. In case of liquid drop, the tangent line of the liquid–vapor interface interacts with the solid surface angle ($\theta$), which measures the wetting of the solid surface [31,32]. On the other hand, any contact with the liquid phase and angle measured at the point of contact with the moving solid surface is referred to as the dynamic contact angle. Different methods have been reported for studying dynamics interaction of a fluid with a rigid solid surface, such as rolling angle (RA) and contact angle hysteresis (CAH). The metastable drop forms a moving drop on the oblique surface. The downward movement and the upward decrease in the contact angle is called CAH, which is characterized by surface chemistry and topography [33,34]. The rolling angle is the smallest tilting angle at which a droplet can roll over a surface. Water, as a liquid phase, exhibits a water contact angle (WCA) below 90° and below 10°, which is called hydrophilic and super-hydrophilic, respectively. On the other hand, the term "hydrophobic surface" refers to a surface with a WCA greater than 90°. Surfaces with a WCA≥150° and a CAH<10 are called super-hydrophobic, in which a droplet forms a nearly perfect sphere on the surface, and the contact angle of a fluid at is almost smooth. Super-hydrophobic coatings have been proposed in recent years as anti-icing coatings. As described above, the presence of air pockets on super-hydrophobic surfaces reduces the time available for water droplets to freeze, thus minimizing ice formation. Blizzards damages equipment such as electrical transmission equipment, communication systems, aerospace installations, and highways every year. Numerous methods have been developed to mitigate this type of damage, such as local heating, and to prevent ice formation by chemical reactions [35], although these methods have some drawbacks in practical applications. In contrast, super-hydrophobic surfaces can prevent the ice formation without requiring special equipment. An important application is in transmission line insulators, which are needed to prevent ice formation in cold areas [36,37]. A comparative study was performed to evaluate the efficiency of two types of coatings, showing that increasing the static contact angle increases hydrophobicity. The results imply that KOH coating is more effective than NaOH coating for achieving an efficient hydrophobic layer.

Super-hydrophobic surfaces have been successfully applied to rigid substrates. However, in applications involving materials like silk fabrics and nylon, where the substrate (factory) is not rigid anymore, the effectiveness of super-hydrophobicity diminishes. Therefore, producing super-hydrophobic surfaces that are soft and flexible is essential. To withstand scratching, abrasion, tensile stress, or compressive force, super-hydrophobic fabrics must be both durable and flexible [38]. Several researches have been conducted on oil pollution in seawater and rivers due to industrial accidents and environmental disasters such as deep water horizon spill. Removing oil contamination from water is challenging and costly which is prompting the scientists to develop various methods for its elimination. These methods can be divided into three main groups: removal of water, removal of oil, and intelligent, controllable partitions [39,40]. Water removal filters are super-hydrophilic and super-lyophobic; these filters operate underwater, where the presence of water on the filter surface prevents oil from passing through the filter pores if they are saturated with water. The oil removal method is more effective since the quantity of oil is often less than the quantity of water, making it more practical to extract oil from water. To remove oil from water, the material must be super-hydrophobic and super-lyophilic. Super-hydrophobic oil removal filters play a critical role in the oil extraction process.

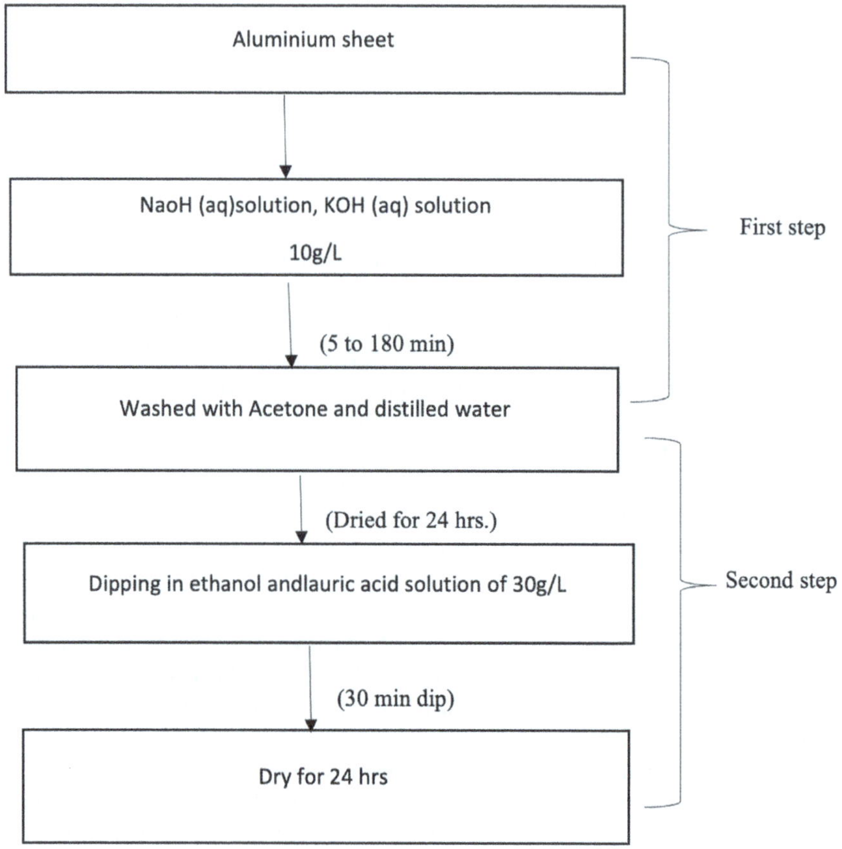


*Figure 4.1* Etching technique for creating the super-hydrophobic surface.

## 4.2 EXPERIMENTAL METHOD

The super-hydrophobic surface was created using a two-step etching technique (Figure 4.1).

### 4.2.1 Synthesis of super-hydrophobic surface

The formation of super-hydrophobic surfaces is influenced by two factors: surface roughness and chemical composition· The super-hydrophobic surface is developed through the combination of two processes: (i) preparing a rough surface and (ii) reducing surface energy of prepared rough surface.

Super-hydrophobic surfaces can be synthesized using the following methods:

### 4.2.2 Layer-by-layer method

The layer-by-layer method used to create super-hydrophobic surfaces is independent of the substrates. This technique (layer by layer technique) involves plunging the substrate into solutions containing oppositely charged materials. After each dip, the coated film is washed with water and then dried.

### 4.2.3 Sol–gel method

The sol–gel synthesis is based on solution reactions. Hydrophobic additives are incorporated to achieve super-hydrophobicity in the sol–gel coating. In the presence of a solvent, a sol is prepared through a hydrolysis reaction, and a gel is formed by impregnating the solvent into the solution [41–44]. Figure 4.2 illustrates the sol–gel method. The sol is applied to the substrate, and the resulting surface properties depend on the sol processing and functional sol groups. A chemical precursor is transformed into a glassy material as it undergoes hydrolysis and polycondensation reactions, forming a layer on the substrate. The surface roughness can be controlled by adjusting the conditions and the mixture of reactions [45].

### 4.2.4 Anodization

The anodization method is an established technique for producing nanoscale and micro-roughness using an electrochemical cell to form an oxide layer on the metal surface. The process of is illustrated in Figure 4.3.

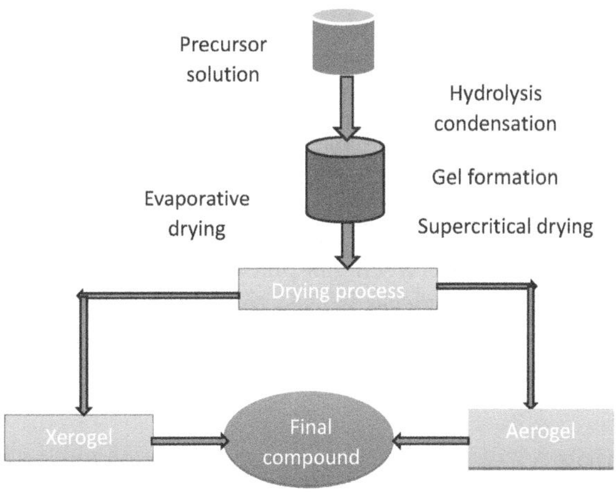


*Figure 4.2* Sol–gel process.

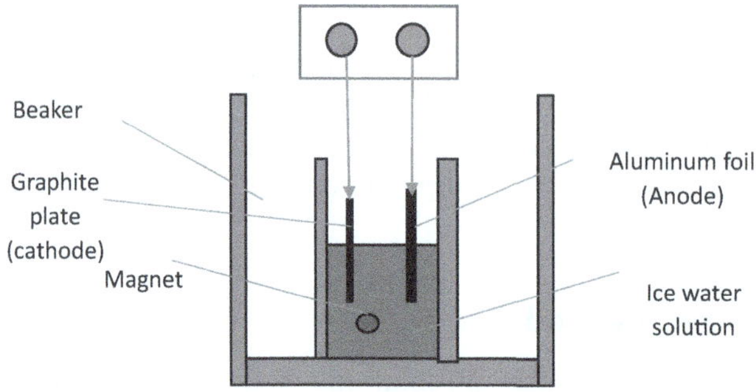


*Figure 4.3* Atomization process.

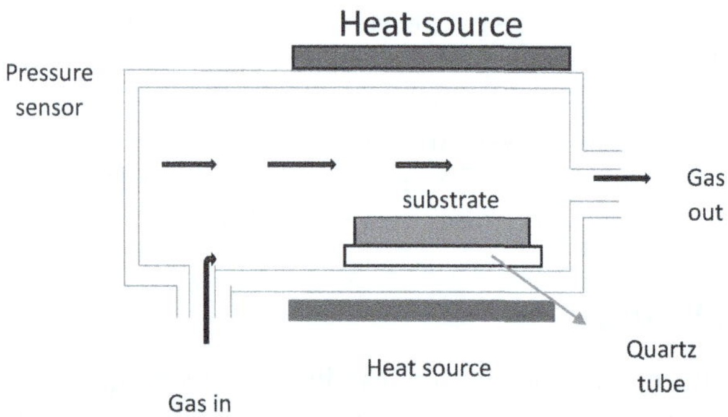


*Figure 4.4* Chemical vapor deposition.

Anodization is a type of classical electrochemical conversion that encompasses electrode reactions in combination with an electric field-driven diffusion of metal and oxygen ions, resulting in the formation of an oxide film on the anode surface.

## 4.2.5 CVD

Chemical vapour deposition (CVD) is a vacuum deposition method that uses a chemical reaction between gas-phase precursors to create thin films on a heated surface. The process is used to produce high-quality, high-performance solid materials, such as metallic, ceramic, and semiconducting thin films (Figure 4.4).

## 4.2.6 Etching

Surface roughness method can be achieved using a simple etching method to form super-hydrophobic surfaces on various materials. The roughness of the surface is due to the differential etching rates between the crystal matrix and/or planes compared to the crystalline regions [46]. Chemical etching can be used to create superhydrophobic surfaces on materials like stainless steel, aluminum, and copper. While the chemical etching method has its advantages, it also comes with certain disadvantages.

## 4.3 RESULTS AND DISCUSSION

### 4.3.1 Effect of etching concentration versus static contact angle

The effects of two etchant concentrations, NaOH and KOH, both at 10 g/L, were compared under rapid etching. High roughness was observed with NaOH, resulting in a static contact angle of 133.7° after 20 minutes of etching. As the etching time increased, the scale of microstructural pits also increased, with static contact angles measured as 148.6°, 159.9°, 163.8°, and 163.8°, respectively [47,48]. For KOH, the static contact angle reached 144° after 20 minutes, and as the size and depth of the microstructure increased, the static contact angle also increased. Overall, the NaOH etchant concentration showed better super-hydrophobic behavior than the KOH etchant concentration (Figure 4.5 and Table 4.1).

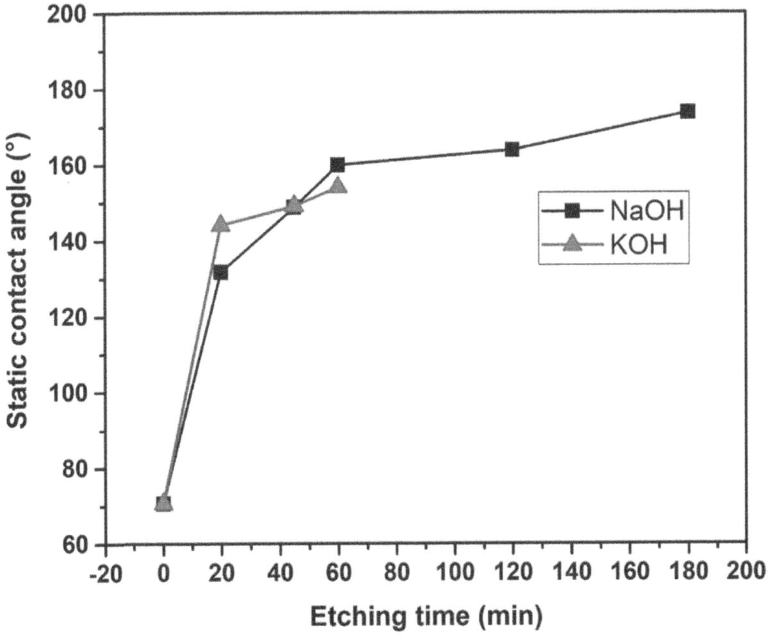


*Figure 4.5* Variation of static contact angle as a function of etching time.

*Table 4.1* Comparing etching time vs contact angle

| *Using NaOH (aq) solution* | | *Using KOH (aq) solution* | |
|---|---|---|---|
| *Etching time (minutes)* | *Static contact angle (°)* | *Etching time (minutes)* | *Static contact angle (°)* |
| 0 | 70.6 | 0 | 70.6 |
| 20 | 131.7 | 20 | 144 |
| 45 | 148.7 | 45 | 149 |
| 60 | 159.9 | 60 | 154 |
| 120 | 163.8 | | |
| 180 | 173.6 | | |

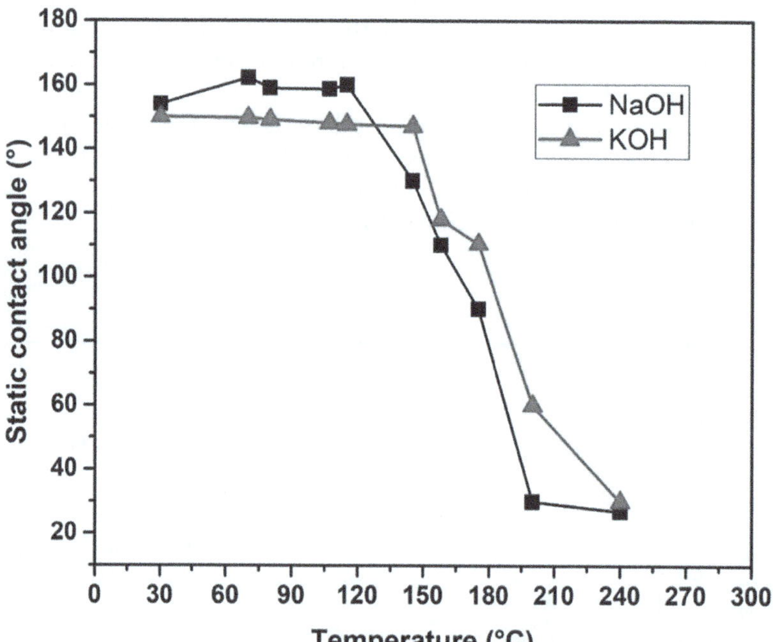


*Figure 4.6* Graph of temperature versus static contact angle.

## 4.3.2 Effect of temperature versus static contact angle

Figure 4.6 shows the variation of temperature versus static contact angle, highlighting the effect of annealing on the wettability of the coating. For this comparison, samples were annealed at different high temperatures (75°C–275°C) for up to 1 hour in a warm oven and the contact angles were measured following heat treatment. Within the 80°C–140°C temperature

*Table 4.2* Temperature and static contact angle for using NaOH and KOH

| *Using NaOH* | | *Using KOH* | |
|---|---|---|---|
| *Temperature (°C)* | *Static contact angle (°)* | *Temperature (°C)* | *Static contact angle (°)* |
| 30 | 154 | 30 | 150 |
| 70 | 162.2 | 70 | 149.6 |
| 80 | 159 | 80 | 149 |
| 107 | 158.7 | 107 | 148 |
| 115 | 160 | 115 | 147.6 |
| 145 | 130 | 145 | 147 |
| 158 | 110 | 158 | 118 |
| 175 | 90 | 175 | 110.4 |
| 200 | 30 | 200 | 60 |
| 240 | 27 | 240 | 30 |

range [47], it was observed that the static and sliding contact angles of the samples remained constant, indicating that the surface was super-hydrophobic, stabilized, and unaffected by this temperature range [49]. However, when the annealing temperature exceeded 170°C, the WCA began to decrease. Because of static angle of contact, the droplets spread rapidly on the annealed surfaces at 250°C and 260°C (Table 4.2).

### 4.3.3 Effect of UV exposure versus static contact angle

The ultraviolet (UV) durability of the synthesized super-hydrophobic aluminum surface was tested by exposing the sample to UV light for up to 60 hours. Figure 4.7 shows the variation of the static contact angle versus UV. It was observed that the coating did not lose its super-hydrophobicity until 53 hours of UV exposure, but the surface began to degrade after 48 hours. After 54 hours of UV exposure, a static contact angle of 133.2° was recorded. Additionally, it was observed that the super-hydrophobic aluminum surfaces remained stale in open air for six months, resisting exposure without losing their super-hydrophobic properties.

## 4.4 CONCLUSION

This chapter compared super-hydrophobic coatings on aluminum surfaces synthesized using a two-step etching chemical process. The aluminum surfaces were first etched by immersing them in NaOH and KOH solutions, followed by reducing the surface energy by dipping the etched aluminum in a lauric acid solution. These coatings were created using different types of materials at the nanoscale level. The successful fabrication

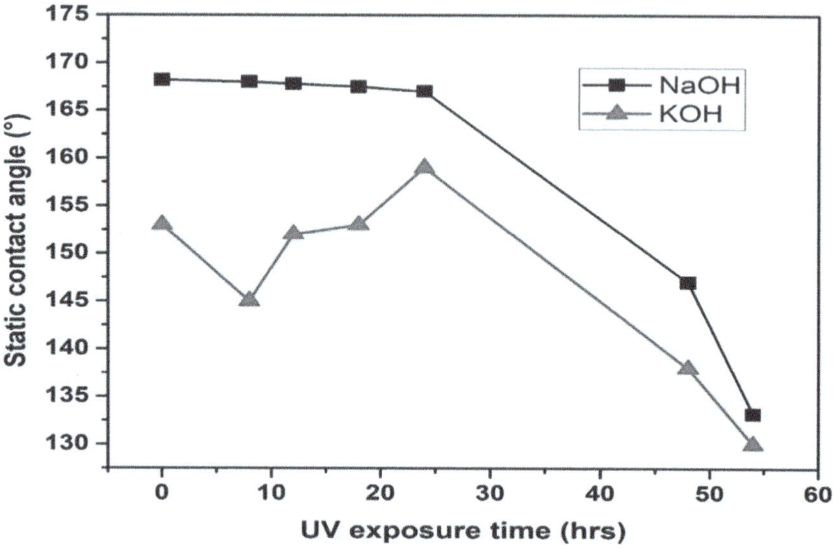


*Figure 4.7* Graph of UV exposure time vs static contact angle.

of super-hydrophobic surfaces requires controlled generation of hierarchical rough morphology and the selection of hydrophobic polymers, such as fluoropolymers or perfluorooctanoic acids, to further reduce the surface energy. First, the comparison focused on etching time and static contact angle, in which aluminum was immersed in NaOH and KOH solutions. As the etching time increased, the surface energy of the etched aluminum decreased, and the static contact angle increased after treatment with lauric acid. Second, the comparison involved temperature and static contact angle. After continuous heating at 290°C for six hours, the super-hydrophobic coatings lost their properties and became super-hydrophilic. However, the super-hydrophobicity of the coatings was restored by simply immersing the aluminum surface in lauric acid for 30 minutes. Third, the comparison was made concerning UV exposure.

## REFERENCES


1. Barthlott, Wilhelm, and Christoph Neinhuis. "Purity of the sacred lotus, or escape from contamination in biological surfaces." *Planta* 202 (1997): 1–8.
2. Pourbaix, Marcel. "Applications of electrochemistry in corrosion science and in practice." *Corrosion Science* 14(1) (1974): 25–82.
3. Dennis III, Robert Vincent. Towards lightweight nanocomposite coatings for corrosion inhibition: Graphene, carbon nanotubes, and nanostructured magnesium as case studies. Diss. State University of New York at Buffalo (2015).

4. Tang, Junlei, et al. "A corrosion resistant cerium oxide based coating on aluminum alloy 2024 prepared by brush plating." *Applied Surface Science* 257(7) (2011): 2806–2812.
5. Pourbaix, Marcel. "Applications of electrochemistry in corrosion science and in practice." *Corrosion Science* 14(1) (1974): 25–82.
6. He, Tian, et al. "Super-hydrophobic surface treatment as corrosion protection for aluminum inseawater." *Corrosion science* 51(8) (2009): 1757–1761.
7. Mallik, Abul K., et al. "Antibacterial surface modification to prevent biofilm formation on polymeric biomaterials." In Pandey, Lalit M., and Abshar Hasan (eds.), *Nanoscale Engineering of Biomaterials: Properties and Applications*. Singapore: Springer Nature Singapore (2022), pp. 425–455.
8. Tang, J., et al. "A corrosion resistant cerium oxide based coating on aluminum alloy 2024 prepared by brush plating." *Applied Surface Science A* 257(2011): 2806–2812.
9. Zhang, Chi, et al. "Probing nanoscopic cell surface areas for rapid and label-free plasmon enhanced Raman detection." In *Optics and the Brain*. Optica Publishing Group (2016), pp. JM3A-53.
10. Pathankar, Neelesh A. "Mimicking the locus effect: The effect of twin rough structures and thin poles." *Langmuir* 20 (2004): 8209–8213.
11. Li, Wei, and Zhixin Kang. "Fabrication of corrosion resistant superhydrophobic surface with self-cleaning property on magnesium alloy and its mechanical stability." *Surface and Coatings Technology* 253 (2014): 205–213.
12. Rasouli, S. A. (2021). *Fabrication of superhydrophobic-superoleophilic membranes for oil-water separation applications* (Doctoral dissertation, Memorial University of Newfoundland).
13. Kendig, M. W., A. J. Davenport, and H. S. Isaacs. "The mechanism of corrosion inhibition by chromate conversion coatings from X-ray absorption near edge spectroscopy (XANES)." *Corrosion Science* 34(1) (1993): 41–49.
14. Li, Wei, and Zhixin Kang. "Fabrication of corrosion resistant superhydrophobic surface with self-cleaning property on magnesium alloy and its mechanical stability." *Surface and Coatings Technology* 253 (2014): 205–213.
15. Conde, A., et al. "Effective corrosion protection of 8090 alloy by cerium conversion coatings." *Electrochimica Acta* 53(26) (2008): 7760–7768.
16. Antonini, Carlo, et al. "Understanding the effect of superhydrophobic coatings on energy reduction in anti-icing systems." *Cold Regions Science and Technology* 67(1–2) (2011): 58–67.
17. Conde, A., et al. "Effective corrosion protection of 8090 alloy by cerium conversion coatings." *Electrochimica Acta* 53(26) (2008): 7760–7768.
18. Nazeri, A., P. P. Trzaskoma-Paulette, and D. Bauer. "Synthesis and properties of cerium and titanium oxide thin coatings for corrosion protection of 304 stainless steel." *Journal of Sol-Gel Science and Technology* 10 (1997): 317–331.
19. Rudd, Amy L., Carmel B. Breslin, and Florian Mansfeld. "The corrosion protection afforded by rare earth conversion coatings applied to magnesium." *Corrosion Science* 42(2) (2000): 275–288.
20. Cui, Xiufang, et al. "Corrosion behaviors in physiological solution of cerium conversion coatings on AZ31 magnesium alloy." *Applied Surface Science* 257(23) (2011): 9703–9709.

21. Zheng, Shunli, et al. "Fabrication of self-cleaning superhydrophobic surface on aluminum alloys with excellent corrosion resistance." *Surface and Coatings Technology* 276 (2015): 341–348.
22. Esmailard, A., M.V. Rukosuyev, M. B. G. June, and F. Van Weigel. "A cost-effective method for creating super-hydrophobic aluminum alloy surfaces that are physically, thermally stable and stable." *Surface and Quotings Technology* 285(2016): 227–234.
23. Castano, Carlos E., Matthew J. O'Keefe, and William G. Fahrenholtz. "Cerium-based oxide coatings." *Current Opinion in Solid State and Materials Science* 19(2) (2015): 69–76.
24. Zheng, Shunli, et al. "Fabrication of self-cleaning superhydrophobic surface on aluminum alloys with excellent corrosion resistance." *Surface and Coatings Technology* 276 (2015): 341–348.
25. Wenzel, Robert N. "Resistance of solid surfaces to wetting by water." *Industrial & Engineering Chemistry* 28(8) (1936): 988–994.
26. Koishi, Takahiro, et al. "Coexistence and transition between Cassie and Wenzel state on pillared hydrophobic surface." *Proceedings of the National Academy of Sciences* 106(21) (2009): 8435–8440.
27. De Oliveira, R. R. L., et al. "Measurement of the nanoscale roughness by atomic force microscopy: Basic principles and applications." In Victor Bellitto (ed.), *Atomic Force Microscopy-Imaging, Measuring and Manipulating Surfaces at the Atomic Scale*, vol. 3. BoD – Books on Demand: Norderstedt, Germany, (2012), pp. 147–174.
28. Yu, Ranbo, et al. "Controlled synthesis of $CeO_2$ flower-like and well-aligned nanorod hierarchical architectures by a phosphate-assisted hydrothermal route." *The Journal of Physical Chemistry* C 112(50) (2008): 19896–19900.
29. Hirano, Masanori, and Etsuro Kato. "Hydrothermal synthesis of cerium (IV) oxide." *Journal of the American Ceramic Society* 79(3) (1996): 777–780.
30. Cho, Young Jun, et al. "Direct growth of cerium oxide nanorods on diverse substrates for superhydrophobicity and corrosion resistance." *Applied Surface Science* 340 (2015): 96–101.
31. Li, Lingjie, et al. "Robust biomimetic-structural superhydrophobic surface on aluminum alloy." *ACS Applied Materials & Interfaces* 7(3) (2015): 1449–1457.
32. Lee, S. H., et al. "Microstructures and mechanical properties of 6061 aluminum alloy processed by accumulative roll-bonding." *Materials Science and Engineering*: A 325(1–2) (2002): 228–235.
33. Ouyang, Jiahu, Eswar Yarrapareddy, and Radovan Kovacevic. "Microstructural evolution in the friction stir welded 6061 aluminum alloy (T6-temper condition) to copper." *Journal of Materials Processing Technology* 172(1) (2006): 110–122.
34. Gao, Xuefeng, and Lei Jiang. "Water-repellent legs of water striders." *Nature* 432(7013) (2004): 36–36.
35. Hu, David L., and John W.M. Bush. "Meniscus- climbing insects." *Nature* 437(7059) (2005): 733–736.
36. Feng, Lin, et al. "Super-hydrophobic surfaces: from natural to artificial." *Advanced Materials* 14(24) (2002): 1857–1860.

37. Li, Lingjie, et al. "Robust biomimetic-structural superhydrophobic surface on aluminum alloy." *ACS Applied Materials & Interfaces* 7(3) (2015): 1449–1457.
38. Liu, Hongqin, et al. "Preparation of superhydrophobic coatings on zinc as effective corrosion barriers." *ACS Applied Materials & Interfaces* 1(6) (2009): 1150–1153.
39. Aljallis, Elias, et al. "Experimental study of skin friction drag reduction on superhydrophobic flat plates in high Reynolds number boundary layer flow." *Physics of Fluids* 25(2) (2013): 025103.
40. Cho, Young Jun, et al. "Direct growth of cerium oxide nanorods on diverse substrates for superhydrophobicity and corrosion resistance." *Applied Surface Science* 340 (2015): 96–101.
41. Ruan, Min, et al. "Preparation and anti-icing behavior of superhydrophobic surfaces on aluminum alloy substrates." *Langmuir* 29(27) (2013): 8482–8491.
42. Nakajima, Akira, et al. "Transparent superhydrophobic thin films with self-cleaning properties." *Langmuir* 16(17) (2000): 7044–7047.
43. Hu, David L., and John W.M. Bush. "Meniscus- climbing insects." *Nature* 437(7059) (2005): 733–736.
44. Anderson, William G. "Wettability literature survey-part 1: Rock/oil/brine interactions and the effects of core handling on wettability." *Journal of Petroleum Technology* 38(10) (1986): 1125–1144.
45. Kwok, Daniel Y., and August W. Neumann. "Contact angle measurement and contact angle interpretation." *Advances in Colloid and Interface Science* 81(3) (1999): 167–249.
46. Joanny, J. F., and Pierre-Gilles De Gennes. "A model for contact angle hysteresis." *The Journal of Chemical Physics* 81(1) (1984): 552–562.
47. Cho, Y. J., H. Zhang, K.-S. Lee, and D.M.R. Kim. "Direct growth of cerium oxide nanorods on diverse substrates for superhydrophobicity and coronal resistance." *Applied Surface Science* 340(2015): 96–101.
48. Feng, Lin, et al. "Super-hydrophobic surfaces: From natural to artificial." *Advanced Materials* 14(24) (2002): 1857–1860.
49. Zhu, Shitong, Wenyi Deng, and Yaxin Su. "Recent advances in preparation of metallic superhydrophobic surface by chemical etching and its applications." *Chinese Journal of Chemical Engineering* 61 (2023): 221–236.

Chapter 5

# A holistic review on energy-efficient green buildings considering environmental, economical, and technical factors

*Bishnu Kant Shukla, Anand Kumar Pandey, Ashish Singh, Devesh Singh, and Mohd Anas*

## 5.1 INTRODUCTION

The economy, society, and environment are all significantly impacted by the construction industry. Buildings, one of the primary outputs of the construction industry, strongly reflect these effects throughout the course of their lives. Some advantages of construction operations include the creation of structures and facilities that meet human needs, as well as the direct and indirect employment opportunities in sectors linked to the building industry (Shukla and Gupta, 2020; Bharti et al., 2022a,b; Varadharajan et al., 2022; Hurukadli et al., 2022).

Certifying agencies report that building materials account for 40% of all energy use (Nelson and Grayson, 2017). In addition to consuming energy, construction activities also release greenhouse gases (GHGs), which contribute to global warming. Building-related carbon emissions are projected to increase by 43% from 2007 levels, reaching 42.4 billion tons globally by 2035 (Outlook, 2010). Furthermore, building repairs, renovations, and retrofitting consume energy and natural resources, leading to GHG emissions, noise pollution, and other pollutants (Figure 5.1). After a building's useful life ends, its disposal is also associated with energy use and waste production.

Green building principles provide a technique to increase a building's energy efficiency. For a building to be labeled green, it must meet specific criteria that minimize its environmental impact by reducing energy, water, and waste. The most widely applied standard for green building construction globally is the U.S. Green Building Council's (USGBC) Leadership in Energy and Environmental Design standard (LEED) standard. The USGBC introduced LEED in 2000. As of 2016, more than 90,000 commercial buildings in 165 countries, predominantly in the United States, had earned LEED certification (MacNaughton et al., 2018).


DOI: 10.1201/9781003521341-5

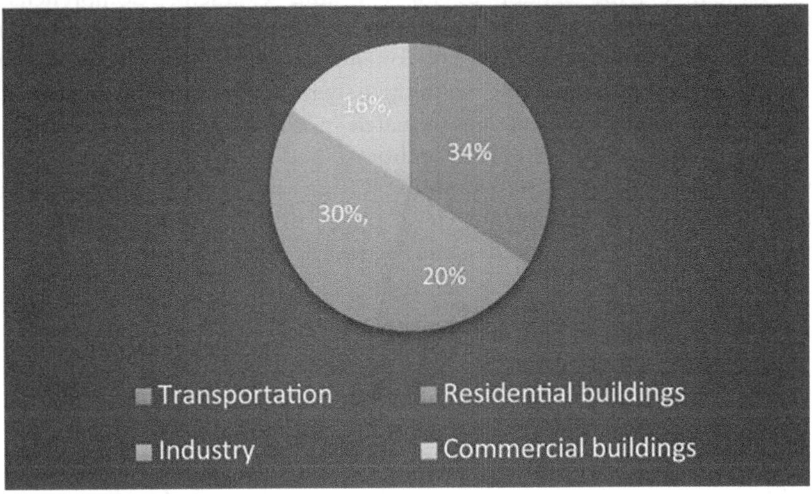


*Figure 5.1* Energy used by various sectors. (Reproduced with the permission of Shukla et al. (2020).)

Analyses of numerous studies examining the impacts of green buildings on human health (Allen et al., 2016; Colton et al., 2015; Thiel et al., 2014) have identified several recurrent benefits for residents of green buildings, including fewer symptoms of illness within the building, reduced hospitalizations for asthma, and lower mortality rates in hospital settings. Residents in environmentally designed green buildings were also found to perform better on cognitive tests due to improvements in ventilation, lighting, and chemical concentrations (Allen et al., 2016; MacNaughton et al., 2017; Wargocki et al., 2000). Increased productivity has also been connected to green buildings. As a result of the efficiency of energy attained by green construction approaches, emissions of harmful air pollutants and GHG, such as nitrogen oxides, particulate matter and sulfur dioxide are decreased (Buonocore et al., 2016).

## 5.2 CONCEPT OF GREEN BUILDING

The general population is becoming increasingly aware of green construction. However, there has been considerable debate about what creates a green building construction and what it should include. The ambiguity surrounding the term "green building" makes it more challenging to promote and implement these green buildings.

Prominent green building assessment tools include BRE Environmental Assessment Method (BREEAM, United Kingdom), LEED (United States), DGNB (Germany), Green Mark Scheme (Singapore), Pearl Rating System for Estidama (Abu Dhabi Urban Planning Council), Comprehensive Assessment System for Built Environment Efficiency (Japan), and Hong Kong Building Environmental Assessment Method (Malaysia). Numerous studies are also dedicated to developing new rating systems for green buildings or customizing existing ones to match specific local framework, such as economic development, climatic conditions, and geographic circumstances. For instance, the SABA Green Building Rating System for Jordan was developed using in-depth expert interviews and by drawing on resources including BREEAM, LEED, and GBTool (Ali and Al Nsairat, 2009).

## 5.3 ENVIRONMENTAL AND TECHNICAL CONSIDERATIONS

The environmental aspects of sustainability have historically been the focus of studies on green buildings. Significant research on the ecological sustainability of buildings supports this, including the reduction of GHG emissions and efficient use of energy, water, and resource (Golić et al., 2011; Hughes et al., 2011; Pacheco et al., 2012; Rahman and Khondaker, 2012; Sadineni et al., 2011; Zabaneh, 2011). For instance, fly ash can be used as a component in green building designs, which not only contributes to energy savings but also reduces the amount of waste dumped in landfills (Drochytka et al., 2013; Danatzko et al., 2013). Similarly, the use of precast or prefabrication technologies considerably reduces the amount of waste generated during construction and deconstruction (Jaillon et al., 2009).

Prefabrication is also considered as one of the most effective strategies for preventing accidents, especially those associated with sustainable elements like "elevated, high-voltage, and confined-space construction" (Dewlaney and Hallowell, 2012).

## 5.4 SOCIAL FACTORS

Rising concerns about sustainability, particularly social sustainability in buildings, have emerged over the past few decades due to the social nature of construction (Abowitz and Toole, 2010). The three key components of social sustainability are living standards, opportunities for future professional growth, and workplace safety and health (Petrovic-Lazarevic, 2008; Zuo et al., 2012a). Social responsibility of corporates is a common theme in sustainability reports released by construction businesses (Zuo et al., 2012b).

In the context of building construction, social sustainability refers to ensuring a safe and healthy environment for all stakeholders, including users, operators, and construction workers, during the sustainable design process (Wong and Fan, 2013). Zuo et al (2012a) emphasized that social sustainability in construction should extend beyond individual buildings to include the surrounding community. Social sustainability is another crucial element in green building and its evaluation (Sarkis et al., 2012). By adding social sustainability indicators to the LEED framework, Sarkis et al. (2012) established an innovative paradigm for sustainability-based contractor selection processes (Berardi, 2013). This framework also offers a mechanism for simulating and developing teams.

## 5.5 SOCIOECONOMIC VARIABLES

The social and economic prerequisites for green buildings, according to Berardi, include factors like accessibility, inclusiveness, education, affordability, cohesiveness, impacts on the local economy, economic value, societal awareness, interior well-being, and motivation (Alwaer and Clements-Croome, 2010). Therefore, a green building can be defined broadly (e.g., by adopting a triple bottom line approach) or narrowly (e.g., by focusing solely on environmental sustainability) (Shi et al., 2013; Mateus and Bragança, 2011). However, the significance of economic, social, and cultural factors in the sustainability of green building initiatives is rarely discussed (Zhang et al., 2017).

## 5.6 BASELINE ENERGY CONSUMPTION

Brazil, Germany, China, Turkey, and India were chosen as the five nations representing the largest LEED markets across various regions (Taskgroup, 2015). The International Energy Agency's (IEA) energy performance indicators were used to determine the Energy Use Intensity baseline (EUIbaseline) for institutional and commercial buildings in these nations during the period of 2000–2012 (Owen et al., 2017). The predicted EUIbaseline for Turkey was the average for the EU. The IEA's World Energy Balances and Statistics database and the energy statistics from the institutional and commercial sectors were used to compute the energy mix (natural gas, electricity, coal, fuel oil, heat, and biomass) during the period of 2000–2014 (Wang and Wu, 2012). The IEA energy price database contained information on residential energy prices in Germany and Turkey from 2000 to 2015 (OVO Energy, 2018). We gathered the most current average national costs for natural gas (Zhang and Paltsev, 2016; Sargent, 2017), electricity (International Gas Union, 2013), coal and fuel oil (Aden et al., 2009; Henry and Frascaria-Lacoste, 2012) in order to estimate the raw energy costs in these nations. Since the most recent data was available, it was conservatively assumed that the EUIbaseline and the energy mix/prices would remain constant.

## 5.7 GREEN BUILDING CONSTRUCTION ADVANTAGES

There is an abundance of research inspecting the expenses and advantages of green construction initiatives. These primary goal of these studies is to support the advantages of adopting green practices, thereby aiding in the decision-making process. It is particularly relevant in the context of the current global financial situation, where client have relatively limited financial competence and financial institutions are making more cautious lending decisions. A common approach in existing studies is to compare the characteristics of green buildings with those of conventional buildings, focusing on aspects such as water and energy efficiency, thermal comfort, indoor environmental quality, productivity, and health.

## 5.8 ENVIRONMENTAL ADVANTAGES

It is extensively acknowledged that green buildings provide a range of benefits. From an environmental perspective, green buildings contribute to enhanced ecosystem protection and urban biodiversity through sustainable land use (Bianchini and Hewage et al., 2012; Akadiri and Olomolaiye, 2012). Sustainable building design must include waste reduction strategies during construction and demolition (C&D) (Yeheyis et al., 2013; Coelho and De Brito, 2012). To reduce the environmental effects of demolition and construction waste, including the use of recycled and reused materials in new structures, the recycling rate needs to exceed 90% (Jo et al., 2009; Marszal et al., 2011). Generally, green buildings outperform conventional ones in terms of water efficiency, energy efficiency, and reduction of carbon emissions. For example, if LEED rating methods were used in all new construction projects in Seoul, a significant reduction in $CO_2$ emissions (created from efficiency of energy) could be achieved (Jo et al., 2009; Ross et al., 2007). According to Turner and Frankel's research (Turner et al., 2008), LEED-certified buildings can save an additional 28% energy compared to the national average (Figure 5.2, prepared from data extracted from The Economist, 2004).

## 5.9 ECONOMIC ADVANTAGES

Enhanced building performance is also linked to cost savings, especially when observed from a life cycle perspective. Operation costs are optimized as a result. Compared to conventional structures, green buildings can reduce energy usage by 30% (Ross et al., 2007). The largest share of

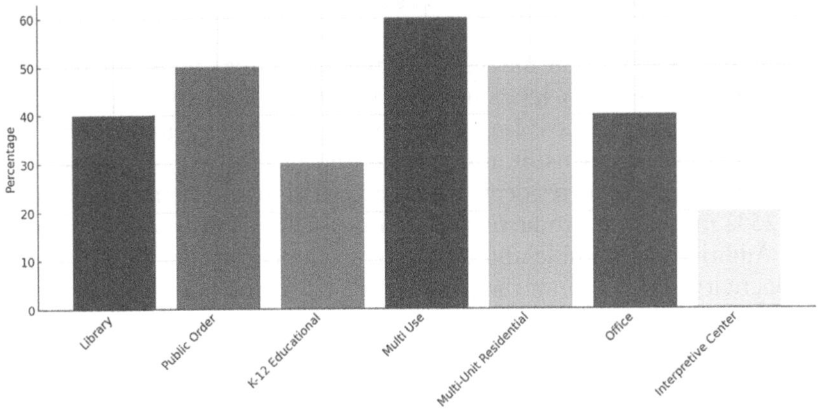


*Figure 5.2* Energy preservation in LEED-certified structures.

the cost of green buildings is attributed to the construction component (including labor and materials) (Fowler et al., 2010). Additionally, Ross et al. (2007) financial modeling demonstrated that LEED-certified buildings incur an additional cost of 10%. According to research commissioned by the Administration of General Services, green buildings outperform their conventional equivalents in terms of water, energy, and cost efficiency regarding maintenance (Zhang and Altan, 2011).

## 5.10 HUMAN FACTORS

Some academics have argued that there are other advantages to green building that are not directly related to costs. This research gives special attention to the positive effects of green buildings on people. This occurs as a result of the amount of time individuals spend inside buildings (Mekhilef et al., 2012).

## 5.11 THERMAL COMFORT

Thermal comfort, which involves the interplay of humidity and temperature, is directly related to user satisfaction in buildings (Bisoniya et al., 2013; Sicurella et al., 2012). Researchers have carefully considered this aspect while calculating and simulating the thermal comfort levels in green buildings in comparison to traditional buildings. Recommended room temperature ranges can thus be determined (Aminuddin et al., 2012; Djongyang et al., 2010). Adaptive thermal comfort may also be influenced by cultural, psychological, and behavioral factors (Singh et al., 2011; Ter Mors et al., 2011; Pan et al., 2008).

## 5.12 HEALTH AND PRODUCTIVITY

Studies have shown that moving into green buildings improves residents' health and productivity levels (Gou et al., 2012). For example, Ries et al. (2006) suggested that it is important to consider the economic benefits of green buildings in terms of absenteeism and productivity. Their research showed that relocating residents from traditional to green buildings resulted in a 25% increase in productivity and a considerable decrease in absenteeism. Additionally, Deuble and de Dear recommended including occupant productivity and health in the post-occupancy evaluation process (Deuble and de Dear, 2012).

Energy conservation should not come at the cost of occupants' satisfaction, productivity, or health (Korkmaz et al., 2010; Omer, 2008; US EIA, 2014). Maintaining a delicate balance between indoor environmental quality and the efficient use of energy and resources by heating, ventilation, and air conditioning systems remains a challenging task (IEA Statistics, 2011).

## 5.13 ENERGY SAVINGS FROM LEED-CERTIFIED PROJECTS

Authors collated data from Green Building Information Gateway (GBIG) (Owen et al., 2017) regarding the energy efficiency and floor space of the LEED-certified projects worldwide from 2000 to 2016. For LEED Building Design and Construction projects, energy efficiency was defined as the intended average percentage reduction from the baseline energy code. Eq. (5.1) can be used to determine the annual energy saved by current LEED-certified developments in year $n$.

$$\text{Annual energy saving}_n = \sum_{i=2,000}^{n} \left(\text{EUI}_{\text{baseline},i} \times \text{Floor space} \times \text{Energy efficiency}_i\right) \tag{5.1}$$

The majority of regions and countries found that LEED-certified buildings have an energy efficiency rating between 20% and 40%.

## 5.14 EMISSION REDUCTION

The IEA's emissions of $CO_2$ from the fuel combustion database were used to extract $CO_2$ emission factors for each nation's power production during the period of 2000–2014. Depending on $CO_2$ emissions (Bayer and Aklin, 2020) and related fuel consumption data in the institutional/viable sector from the IEA Statistics (IEA Statistics, 2011), the $CO_2$ emission factors for on-site fuel combustion from 2000 to 2014 were estimated. From

2014 to 2016, it was conservatively projected that $CO_2$ emission factors would remain constant. Total emissions from 2000 to 2010 for each nation were retrieved from the Global Atmospheric Research Emissions Database v4.3.1 (Ellerman et al., 2016) and v4.2 FT2012 (Shindell, 2015) for air pollutants ($N_2O$, $CH_4$, $NO_x$, $SO_2$, and $PM_{2.5}$) and other important GHGs. The 1A1a figures were scaled to 1A1a1 (public power generation sector), and the 1A4 values were adjusted to 1A4a (commercial/institutional sector) using the emission data ratio of carbon dioxide for the two sectors from the IEA Statistics. This was made to align $CO_2$ emission components with the sectors investigated. This is based on the assumption that emissions of $N_2O$, $CH_4$, $NO_x$, $SO_2$, and $PM_{2.5}$ are directly related to $CO_2$ $CO_2$emissions, which depend on the energy sources used to power different sectors (as explained by Eqs. 5.2 and 5.3) (Shindell, 2015; Sicurella et al., 2012).

$$SO_2\left(CH_4,\text{etc.}\right)\text{emissions}_{sec\,tor\,1A1a1} = \frac{CO_2\ \text{emissions}_{sec\,tor\,1A1a1}}{CO_2\ \text{emissions}_{sec\,tor\,1A1a}} \times SO_2\left(CH_4,\text{etc.}\right)\text{emissions}_{sec\,tor\,1A1a} \tag{5.2}$$

$$SO_2\left(CH_4,\text{etc.}\right)\text{emissions}_{sec\,tor\,1A4a} = \frac{CO_2\ \text{emissions}_{sec\,tor\,1A4a}}{CO_2\ \text{emissions}_{sec\,tor\,1A4}} \times SO_2\left(CH_4,\text{etc.}\right)\text{emissions}_{sec\,tor\,1A4} \tag{5.3}$$

## 5.15 CO-BENEFITS FOR CLIMATE AND HEALTH

The average health impact per ton values for $CH_4$, $NO_x$, $SO_2$, and $PM_{2.5}$ were adjusted from 2007 to 2016 USD. Using Social Cost of Atmospheric Release (SCAR) measures, five emissions were estimated with average discount rate of 3% (Shi et al., 2013). According to the SCAR approach, the health impact estimates have an 80% degree of uncertainty, which is due to the variability in epidemiologic concentration–response functions and the predicted concentrations with changing emission levels (Yuan et al., 2013). The climate impacts of GHG emissions were calculated using the similar SCC standards as those in the U.S. study. The total energy and financial savings from projects with LEED certification since 2000 are shown in Table 5.1 per country.

## 5.16 DIFFERENT ELEMENTS TO ACHIEVE GREEN BUILDINGS

Technical, behavioral, and managerial components constitute the three main categories of important success elements for green buildings.

*Table 5.1* Total savings of energy from LEED-certified construction projects by country during the period of 2000–2016 (in millions of U.S. dollars) (MacNaughton et al., 2018)

| *Country* | *Fuel oil*[a] | *Natural gas* | *Electricity* | *Other on-site combustion*[b] | *Total (country)* |
|---|---|---|---|---|---|
| USA | 164.00 | 790.00 | 5380.00 | 380.00 | 6710.00 |
| China | --- | 8.29 | 52.30 | 8.62 | 69.30 |
| Brazil | --- | 0.78 | 250.00 | 0.00 | 251.00 |
| India | 2.27 | 1.55 | 57.20 | 11.20 | 72.30 |
| Germany | 24.00 | 33.60 | 181.00 | 0.05 | 239.00 |
| Turkey | 0.00 | 18.90 | 134.00 | 9.93 | 163.10 |
| Total | **190.00** | **853.00** | **6050.00** | **410.00** | **7510.00** |

[a] Brazil and China's specific household fuel oil cost was unknown.
[b] Due to the variety of fuel types, the IEA database does not provide a precise commercial heat price.

## 5.16.1 Technological elements

Advancements in renewable energy technology are essential for achieving green construction goals and accreditation (Li et al., 2013). This occurs as a result of the environmental problems caused by the depletion of conventional energy sources, such as coal. Consequently, many governments now emphasize developing renewable energy sources and utilizing them in other industries, a focus reflected in relevant public policies. The building industry is no different. Buildings that use renewable energy consume less energy and produce fewer emissions. One major challenge is the absence of infrastructure needed to link generated electricity to the power grid. Renewable energy sources frequently used in buildings include solar PV, solar water heating, geothermal heat pumps, and small wind turbines (Praene et al., 2012; Marszal et al., 2011; Berggren et al., 2013). In fact, using renewable energy is essential to achieving (net) zero energy buildings (Baniyounes et al., 2012; Dagdougui et al., 2012). Utilizing a solar desiccant cooling system can reduce energy costs by as much as 60% (Ghaffarian Hoseini et al., 2013).

Dagdougui et al. (2012) developed a dynamic model to optimize hybrid renewable energy systems. This prototype was used in a case study that had solar PV, wind, and biomass energy sources. This study found substantial environmental advantages even without an energy storage system. However, the implementation of these technological advancements in green buildings still faces substantial obstacles, including cost, maintenance, and complexity of renewable energy systems (Kibert, 2016; Webster and Dunn, 2011). Managing C&D waste is essential for achieving green building standards (Poon et al., 2013). Reducing resource consumption and increasing resource usage effectiveness are important aspects of sustainable building design (Newsham et al., 2009; Tingley and Davison, 2012). Common strategies include reducing, recycling, and reusing building and demolition waste (Hughes et al., 2011; Lu and Tam, 2013). Government agencies frequently promote the use of green building technology and materials to reduce waste generated during C&D (Robichaud and Anantatmula, 2011; Ofori-Boadu et al., 2012).

### 5.16.2 Managerial elements

Green buildings involve numerous managerial considerations, including corporate, project, and market levels. Managing green building projects requires specialized project management skills. Key changes include hiring green building experts as consultants, using green building rating systems like LEED, giving staff members opportunity for ongoing training and education, and involving external stakeholders (Liu et al., 2012). Ofori-Boadu et al. surveyed top green contractors in the United States and concluded on their project management methods for LEED projects (Beheiry et al., 2006). Common project management practices include investing in LEED, being a member of the Council of Green Building, promoting green building developments, establishing a dedicated green buildings division, collaborating with freelancers who hold LEED certification, and being familiar with the LEED system and related documentation.

At the corporate level, implementing an environmental management system can reduce energy use by 90%, C&D waste by 63%, water use by 70%, accident rates by 20%, and quality complaints by 80%. Top management commitment is necessary for the planning green building developments. The project team benefits from green specifications, such as databases on green items and relevant technical standards, which enhance the ability to secure the resources required for sustainable construction (Lam et al., 2011).

### 5.16.3 Behavioral and cultural elements

The development of green buildings must also take into account cultural and behavioral elements. Therefore, it is vital to increase the degree of consideration given to principles of green building and sustainable development among all stakeholders (such as clients, designers, contractors, and end users). Hoffman and Henn (2008) claim that numerous social and psychological issues hinder the use of green buildings, including egocentrism, overestimating the future, presumptive associations, and favorable illusions. According to McCunn and Gifford's (2012) study, there is no statistically substantial connection between sustainable workplace elements and employees' attitudes and conduct toward the environment. This finding is in agreement with the study by Rashid et al. (2012), who also noted that green construction does not necessarily enhance a company's image directly (Cole and Brown, 2009).

Chau et al. (2010) used discrete choice experiments to evaluate Hong Kong residents' views on green residential property features. The study showed that residents prefer to allocate more funds toward energy-saving measures rather than other environmentally friendly architectural characteristics such as water efficiency, indoor air quality, and noise reduction. The development of green buildings relies heavily on the attitudes and actions of end users. Utilizing education or government initiatives could be an effective strategy to enhance public understanding of sustainability concerns and their willingness to invest in eco-friendly building components (Cole et al., 2010).

## 5.17 CONCLUSIONS

This study conducted a comprehensive evaluation of available global research on green buildings. Based on an extensive literature review, most research on green buildings focuses on technical solutions and environmental sustainability aspects, such as energy use, water utilization, and GHG emissions. Despite there is a significant body of literature highlighting the importance of social and economic sustainability, few studies address these topics. More extensive research should be conducted on the social impact of green construction. The life cycle assessment approach, commonly employed in the technical and environmental aspects of green design, can also serve as a valuable tool for promoting social sustainability. It is a positive development that leading green building grading tools like BREEAM, LEED, and GBCA by Green Star are starting to incorporate these capabilities into their latest version. However, more research in these areas is necessary to support the creation of the new rating tools and to aid clients' or end users' decision-making processes. LEED-certified buildings have already prevented between 172 and 405 premature deaths in the United States and have avoided the release of 33MT of $CO_2$ into the atmosphere in India, Germany, Turkey, China, and Brazil, with nearly one-fifth of these advantages occurring in 2016 alone.

The review also revealed a shift in focus beyond the building itself to the interactions between the building and its users. Several studies have documented the impacts of thermal comfort and Indoor Environmental Quality (IEQ) on occupant satisfaction, performance, and health. However, there are concerns that studies on green buildings often overlook the tenants. Providing residents with knowledge and training will help regulate their behavior when utilizing buildings, which could significantly impact the building's overall performance. The costs and benefits of green construction are also being discussed. To enable customers and project teams to make informed, data-driven decisions, more thorough studies are required.

## REFERENCES


Abowitz, D. A., and Toole, T. M.: Mixed method research: Fundamental issues of design, validity, and reliability in construction research. *Journal of Construction Engineering and Management*, 136(1), 108–116 (2010).

Aden, N., Fridley, D., and Zheng, N.: China's coal: Demand, constraints, and externalities (No. LBNL-2334E). Lawrence Berkeley National Lab (LBNL), Berkeley, CA (United States) (2009).

Akadiri, P. O., and Olomolaiye, P. O.: Development of sustainable assessment criteria for building materials selection. *Engineering, Construction and Architectural Management*, 19(6), 666–687 (2012).

Ali, H. H., and Al Nsairat, S. F.: Developing a green building assessment tool for developing countries: Case of Jordan. *Building and Environment*, 44(5), 1053–1064 (2009).

Allen, J. G., MacNaughton, P., Satish, U., Santanam, S., Vallarino, J., and Spengler, J. D.: Associations of cognitive function scores with carbon dioxide, ventilation, and volatile organic compound exposures in office workers: A controlled exposure study of green and conventional office environments. *Environmental Health Perspectives*, 124(6), 805–812 (2016).

Alwaer, H., and Clements-Croome, D. J.: Key performance indicators (KPIs) and priority setting in using the multi-attribute approach for assessing sustainable intelligent buildings. *Building and Environment*, 45(4), 799–807 (2010).

Aminuddin, A. M. R., Rao, S. P., and Thing, H. W.: Thermal comfort field studies in two certified energy efficient office buildings in a tropical climate. *International Journal of Sustainable Building Technology and Urban Development*, 3(2), 129–136 (2012).

Baniyounes, A. M., Liu, G., Rasul, M. G., and Khan, M. M. K.: Analysis of solar desiccant cooling system for an institutional building in subtropical Queensland, Australia. *Renewable and Sustainable Energy Reviews*, 16(8), 6423–6431 (2012).

Bayer, P., and Aklin, M.: The European Union emissions trading system reduced $CO_2$ emissions despite low prices. *Proceedings of the National Academy of Sciences*, 117(16), 8804–8812 (2020).

Beheiry, S. M., Chong, W. K., and Haas, C. T.: Examining the business impact of owner commitment to sustainability. *Journal of Construction Engineering and Management*, 132(4), 384–392 (2006).

Berardi, U.: Clarifying the new interpretations of the concept of sustainable building. *Sustainable Cities and Society*, 8, 72–78 (2013).

Berggren, B., Hall, M., and Wall, M.: LCE analysis of buildings: Taking the step towards net zero energy buildings. *Energy and Buildings*, 62, 381–391 (2013).

Bharti, G., Shukla, B. K., Srinivasan, V., and Bansal, V.: Experimental study on eccentrically loaded rectangular footing on sand for embedded condition to enhance serviceability of structures. *Materials Today: Proceedings*, 61, 517–522 (2022a).

Bharti, G., Singh, A., Shukla, B. K., Varadharajan, S., Kirthanashri, S. V., Bishetti, P., and Tripathi, A.: Effect of self-healing concrete on modular structure. In *Proceedings of Indian Geotechnical and Geoenvironmental Engineering Conference (IGGEC) 2021*, National Institute of Technology, Jalandhar, India, vol. 2, pp. 323–331. Singapore: Springer Nature Singapore (2022b).

Bianchini, F., and Hewage, K.: How "green" are the green roofs? Lifecycle analysis of green roof materials. *Building and Environment*, 48, 57–65 (2012).
Bisoniya, T. S., Kumar, A., and Baredar, P.: Experimental and analytical studies of earth–air heat exchanger (EAHE) systems in India: A review. *Renewable and Sustainable Energy Reviews*, 19, 238–246 (2013).
Buonocore, J. J., Luckow, P., Norris, G., Spengler, J. D., Biewald, B., Fisher, J., and Levy, J. I.: Health and climate benefits of different energy-efficiency and renewable energy choices. *Nature Climate Change*, 6(1), 100–105 (2016).
Chau, C. K., Tse, M. S., and Chung, K. Y.: A choice experiment to estimate the effect of green experience on preferences and willingness-to-pay for green building attributes. *Building and Environment*, 45(11), 2553–2561 (2010).
Coelho, A., and De Brito, J.: Influence of construction and demolition waste management on the environmental impact of buildings. *Waste Management*, 32(3), 532–541 (2012).
Cole, R. J., and Brown, Z.: Reconciling human and automated intelligence in the provision of occupant comfort. *Intelligent Buildings International*, 1(1), 39–55 (2009).
Cole, R. J., Brown, Z., and McKay, S.: Building human agency: A timely manifesto. *Building Research & Information*, 38(3), 339–350 (2010).
Colton, M. D., Laurent, J. G. C., MacNaughton, P., Kane, J., Bennett-Fripp, M., Spengler, J., and Adamkiewicz, G.: Health benefits of green public housing: Associations with asthma morbidity and building-related symptoms. *American Journal of Public Health*, 105(12), 2482–2489 (2015).
Dagdougui, H., Minciardi, R., Ouammi, A., Robba, M., and Sacile, R.: Modeling and optimization of a hybrid system for the energy supply of a "Green" building. *Energy Conversion and Management*, 64, 351–363 (2012).
Danatzko, J. M., Sezen, H., and Chen, Q.: Sustainable design and energy consumption analysis for structural components. *Journal of Green Building*, 8(1), 120–135 (2013).
Deuble, M. P., and de Dear, R. J.: Green occupants for green buildings: the missing link? *Building and Environment*, 56, 21–27 (2012).
Dewlaney, K. S., and Hallowell, M.: Prevention through design and construction safety management strategies for high performance sustainable building construction. *Construction Management and Economics*, 30(2), 165–177 (2012).
Djongyang, N., Tchinda, R., and Njomo, D.: Thermal comfort: A review paper. *Renewable and Sustainable Energy Reviews*, 14(9), 2626–2640 (2010).
Drochytka, R., Zach, J., Korjenic, A., and Hroudová, J.: Improving the energy efficiency in buildings while reducing the waste using autoclaved aerated concrete made from power industry waste. *Energy and Buildings*, 58, 319–323 (2013).
Ellerman, A. D., Marcantonini, C., and Zaklan, A.: The European Union emissions trading system: Ten years and counting. *Review of Environmental Economics and Policy*, 10(1), 89–107 (2016).
Fowler, K. M., Rauch, E. M., Henderson, J. W., and Kora, A. R.: Re-assessing green building performance: A post occupancy evaluation of 22 GSA buildings (No. PNNL-19369). Pacific Northwest National Lab (PNNL), Richland, WA (United States) (2010).

Ghaffarian Hoseini, A., Dahlan, N. D., Berardi, U., GhaffarianHoseini, A., and Makaremi, N.: The essence of future smart houses: From embedding ICT to adapting to sustainability principles. *Renewable and Sustainable Energy Reviews*, 24, 593–607 (2013).

Golić, K., Kosorić, V., and Furundžić, A. K.: General model of solar water heating system integration in residential building refurbishment: Potential energy savings and environmental impact. *Renewable and Sustainable Energy Reviews*, 15(3), 1533–1544 (2011).

Gou, Z., Lau, S. S. Y., and Chen, F.: Subjective and objective evaluation of the thermal environment in a three-star green office building in China. *Indoor and Built Environment*, 21(3), 412–422 (2012).

Henry, A., and Frascaria-Lacoste, N.: Comparing green structures using life cycle assessment: A potential risk for urban biodiversity homogenization? *The International Journal of Life Cycle Assessment*, 17, 949–950 (2012).

Hoffman, A. J., and Henn, R.: Overcoming the social and psychological barriers to green building. *Organization & Environment*, 21(4), 390–419 (2008).

Hughes, B. R., Chaudhry, H. N., and Ghani, S. A.: A review of sustainable cooling technologies in buildings. *Renewable and Sustainable Energy Reviews*, 15(6), 3112–3120 (2011).

Hurukadli, P., Bharti, G., and Shukla, B. K.: Behavior of fiber reinforced polymer laminates strengthening prestressed concrete beams. *Materials Today: Proceedings*, https://doi.org/10.1016/j.matpr.2022.10.036 (2022).

IEA Statistics: CO2 emissions from fuel combustion-highlights. IEA, Paris, Cited July (2011).

International Gas Union. Wholesale gas price survey: 2015 Edition (2013).

Jaillon, L., Poon, C. S., and Chiang, Y. H.: Quantifying the waste reduction potential of using prefabrication in building construction in Hong Kong. *Waste Management*, 29(1), 309–320 (2009).

Jo, J. H., Golden, J. S., and Shin, S. W.: Incorporating built environment factors into climate change mitigation strategies for Seoul, South Korea: A sustainable urban systems framework. *Habitat International*, 33(3), 267–275 (2009).

Kibert, C. J.: *Sustainable Construction: Green Building Design and Delivery*. Hoboken, NJ: John Wiley & Sons (2016).

Korkmaz, S., Riley, D., and Horman, M.: Piloting evaluation metrics for sustainable high-performance building project delivery. *Journal of Construction Engineering and Management*, 136(8), 877–885 (2010).

Lam, P. T., Chan, E. H., Chau, C. K., Poon, C. S., and Chun, K. P.: Environmental management system vs green specifications: How do they complement each other in the construction industry? *Journal of Environmental Management*, 92(3), 788–795 (2011).

Li, D. H., Yang, L., and Lam, J. C.: Zero energy buildings and sustainable development implications: A review. *Energy*, 54, 1–10 (2013).

Liu, A. M., Lau, W. S., and Fellows, R.: The contributions of environmental management systems towards project outcome: Case studies in Hong Kong. *Architectural Engineering and Design Management*, 8(3), 160–169 (2012).

Lu, W., and Tam, V. W.: Construction waste management policies and their effectiveness in Hong Kong: A longitudinal review. *Renewable and Sustainable Energy Reviews*, 23, 214–223 (2013).

MacNaughton, P., Cao, X., Buonocore, J., Cedeno-Laurent, J., Spengler, J., Bernstein, A., and Allen, J.: Energy savings, emission reductions, and health co-benefits of the green building movement. *Journal of Exposure Science and Environmental Epidemiology*, 28(4), 307–318 (2018).

MacNaughton, P., Satish, U., Laurent, J. G. C., Flanigan, S., Vallarino, J., Coull, B., Spengler, J. D. and Allen, J. G.: The impact of working in a green certified building on cognitive function and health. *Building and Environment*, 114, 178–186 (2017).

Marszal, A. J., Heiselberg, P., Bourrelle, J. S., Musall, E., Voss, K., Sartori, I., and Napolitano, A.: Zero energy building: A review of definitions and calculation methodologies. *Energy and Buildings*, 43(4), 971–979 (2011).

Mateus, R., and Bragança, L.: Sustainability assessment and rating of buildings: Developing the methodology SBToolPT-H. *Building and Environment*, 46(10), 1962–1971 (2011).

McCunn, L. J., and Gifford, R.: Do green offices affect employee engagement and environmental attitudes? *Architectural Science Review*, 55(2), 128–134 (2012).

Mekhilef, S., Safari, A., Mustaffa, W. E. S., Saidur, R., Omar, R., and Younis, M. A. A.: Solar energy in Malaysia: Current state and prospects. *Renewable and Sustainable Energy Reviews*, 16(1), 386–396 (2012).

Nelson, J., and Grayson, D.: World business council for sustainable development (WBCSD). In Grayson, D., and Nelson, J. (eds.), *Corporate Responsibility Coalitions*, pp. 300–317. London: Routledge (2017).

Newsham, G. R., Mancini, S., and Birt, B. J.: Do LEED-certified buildings save energy? Yes, but…. *Energy and Buildings*, 41(8), 897–905 (2009).

Ofori-Boadu, A., Owusu-Manu, D. G., Edwards, D., and Holt, G.: Exploration of management practices for LEED projects: Lessons from successful green building contractors. *Structural Survey*, 30(2), 145–162 (2012).

Omer, A. M.: Energy, environment and sustainable development. *Renewable and Sustainable Energy Reviews*, 12(9), 2265–2300 (2008).

Outlook, A. E.: Energy information administration. *Department of Energy*, 92010(9), 1–15 (2010).

Topping, C.: Average electricity usage in the UK: how many kWh does your home use. OVO Energy. (2021).

Owen, A., Brockway, P., Brand-Correa, L., Bunse, L., Sakai, M., and Barrett, J.: Energy consumption-based accounts: A comparison of results using different energy extension vectors. *Applied Energy*, 190, 464–473 (2017).

Pacheco, R., Ordóñez, J., and Martínez, G.: Energy efficient design of building: A review. *Renewable and Sustainable Energy Reviews*, 16(6), 3559–3573 (2012).

Pan, Y., Yin, R., and Huang, Z.: Energy modeling of two office buildings with data center for green building design. *Energy and Buildings*, 40(7), 1145–1152 (2008).

Petrovic-Lazarevic, S.: The development of corporate social responsibility in the Australian construction industry. *Construction Management and Economics*, 26(2), 93–101 (2008).

Poon, C. S., Yu, A. T., Wong, A., and Yip, R.: Quantifying the impact of construction waste charging scheme on construction waste management in Hong Kong. *Journal of Construction Engineering and Management*, 139(5), 466–479 (2013).

Praene, J. P., David, M., Sinama, F., Morau, D., and Marc, O.: Renewable energy: Progressing towards a net zero energy island, the case of Reunion Island. *Renewable and Sustainable Energy Reviews*, 16(1), 426–442 (2012).

Rahman, S. M., and Khondaker, A. N.: Mitigation measures to reduce greenhouse gas emissions and enhance carbon capture and storage in Saudi Arabia. *Renewable and Sustainable Energy Reviews*, 16(5), 2446–2460 (2012).

Rashid, M., Spreckelmeyer, K., and Angrisano, N. J.: Green buildings, environmental awareness, and organizational image. *Journal of Corporate Real Estate*, 14(1), 21–49 (2012).

Ries, R., Bilec, M. M., Gokhan, N. M., and Needy, K. L.: The economic benefits of green buildings: A comprehensive case study. *The Engineering Economist*, 51(3), 259–295 (2006).

Robichaud, L. B., and Anantatmula, V. S.: Greening project management practices for sustainable construction. *Journal of Management in Engineering*, 27(1), 48–57 (2011).

Ross, B., López-Alcalá, M., and Small III, A. A.: Modeling the private financial returns from green building investments. *Journal of Green Building*, 2(1), 97–105 (2007).

Sadineni, S. B., Madala, S., and Boehm, R. F.: Passive building energy savings: A review of building envelope components. *Renewable and Sustainable Energy Reviews*, 15(8), 3617–3631 (2011).

Sargent, J. F.: Global research and development expenditures: Fact sheet, pp. 7–5700. Congressional Research Service (2017).

Sarkis, J., Meade, L. M., and Presley, A. R.: Incorporating sustainability into contractor evaluation and team formation in the built environment. *Journal of Cleaner Production*, 31, 40–53 (2012).

Shi, Q., Zuo, J., Huang, R., Huang, J., and Pullen, S.: Identifying the critical factors for green construction: An empirical study in China. *Habitat International*, 40, 1–8 (2013).

Shindell, D. T.: The social cost of atmospheric release. *Climatic Change*, 130, 313–326 (2015).

Shukla, B. K., and Gupta, A.: Mix design and factors affecting strength of pervious concrete. In *Advances in Structural Engineering and Rehabilitation: Select Proceedings of TRACE 2018*, Amity University, Noida, pp. 125–139. Springer Singapore (2020).

Shukla, B. K., Rashid, H. O., Bhowmik, A. R., and Sharma, P. K.: A review on energy-efficient and sustainable urban buildings. *Advances in Renewable Energy and Sustainable Environment: Select Proceedings of NCRESE 2019*, National Institute of Technology, Kurukshetra, pp. 245–251 (2020).

Sicurella, F., Evola, G., and Wurtz, E.: A statistical approach for the evaluation of thermal and visual comfort in free-running buildings. *Energy and Buildings*, 47, 402–410 (2012).

Singh, M. K., Mahapatra, S., and Atreya, S. K.: Adaptive thermal comfort model for different climatic zones of North-East India. *Applied Energy*, 88(7), 2420–2428 (2011).

TASKGROUP, B. E. E.: Building Energy Performance Metrics (2015).

Ter M ors, S., Hensen, J. L., Loomans, M. G., and Boerstra, A. C.: Adaptive thermal comfort in primary school classrooms: Creating and validating PMV-based comfort charts. *Building and Environment*, 46(12), 2454–2461 (2011).

Kats, G.: *Green Building Costs and Financial Benefits* (pp. 2–8). Massachusetts Technology Collaborative, Boston, MA (2003).

Thiel, C. L., Needy, K. L., Ries, R., Hupp, D., and Bilec, M. M.: Building design and performance: A comparative longitudinal assessment of a Children's hospital. *Building and Environment*, 78, 130–136 (2014).

Tingley, D. D., and Davison, B.: Developing an LCA methodology to account for the environmental benefits of design for deconstruction. *Building and Environment*, 57, 387–395 (2012).

Turner, C., Frankel, M., and Council, U. G. B.: Energy performance of LEED for new construction buildings. *New Buildings Institute*, 4(4), 1–42 (2008).

US EIA: How much energy is consumed in residential and commercial buildings in the United States. US Energy Information Administration (EIA), Washington, DC (2014).

Varadharajan, S., Kirthanashri, S. V., Tiwari, A., Jaiswal, A., and Shukla, B. K.: Use of waste marble dust and waste glass powder and bacterial solution for manufacture of ecofriendly concrete. In *Proceedings of Indian Geotechnical and Geoenvironmental Engineering Conference (IGGEC) 2021*, National Institute of Technology, Jalandhar, vol. 2, pp. 195–207. Singapore: Springer Nature Singapore (2022).

Wang, Y., and Wu, C.: Energy prices and exchange rates of the US dollar: Further evidence from linear and nonlinear causality analysis. *Economic Modelling*, 29(6), 2289–2297 (2012).

Wargocki, P., Wyon, D. P., Sundell, J. A. N., Clausen, G., and Fanger, P. O.: The effects of outdoor air supply rate in an office on perceived air quality, sick building syndrome (SBS) symptoms and productivity. *Indoor Air*, 10(4), 222–236 (2000).

Webster, C. B., and Dunn, B. C.: Creating a model of sustainability through the design, construction, and operations of a new high school. *College Publishing*, 6(3), 1–20 (2011).

Wong, K. D., and Fan, Q.: Building information modelling (BIM) for sustainable building design. *Facilities*, 31(3/4), 138–157 (2013).

Yeheyis, M., Hewage, K., Alam, M. S., Eskicioglu, C., and Sadiq, R.: An overview of construction and demolition waste management in Canada: A lifecycle analysis approach to sustainability. *Clean Technologies and Environmental Policy*, 15, 81–91 (2013).

Yuan, X., Wang, X., and Zuo, J.: Renewable energy in buildings in China: A review. *Renewable and Sustainable Energy Reviews*, 24, 1–8 (2013).

Zabaneh, G. A.: Zero net house: Preliminary assessment of suitability for Alberta. *Renewable and Sustainable Energy Reviews*, 15(6), 3237–3242 (2011).

Zhang, D., and Paltsev, S.: The future of natural gas in China: Effects of pricing reform and climate policy. *Climate Change Economics*, 7(04), 1650012 (2016).

Zhang, Y., and Altan, H.: A comparison of the occupant comfort in a conventional high-rise office block and a contemporary environmentally-concerned building. *Building and Environment*, 46(2), 535–545 (2011).

Zhang, Y., Wang, J., Hu, F., and Wang, Y.: Comparison of evaluation standards for green building in China, Britain, United States. *Renewable and Sustainable Energy Reviews*, 68, 262–271 (2017).

Zuo, J., Jin, X. H., and Flynn, L.: Social sustainability in construction: An explorative study. *International Journal of Construction Management*, 12(2), 51–63 (2012a).

Zuo, J., Zillante, G., Wilson, L., Davidson, K., and Pullen, S.: Sustainability policy of construction contractors: A review. *Renewable and Sustainable Energy Reviews*, 16(6), 3910–3916 (2012b).

Chapter 6

# Ventilation in green buildings through solar chimney

*Rajeev Shrivastava, Sachin Kumar Nikam, and Arun Kumar Wamankar*

## 6.1 INTRODUCTION

Because of the tremendous pollution generated by conventional energy sources and rising oil prices, researchers and specialists are seeking non-contaminating and renewable sources. Amid global energy crisis, solar energy emerged as the foremost solution explored by scientists [1, 2].

Solar energy does not require advanced technology to meet human needs and is also affordable. However, its inaccessibility and need to store solar energy so that it may be used when required are two of its significant disadvantages [3, 4]. Solar energy (which has been transformed into heat) is predominantly used for providing hot water and space heating. Additionally, it is applied in a variety of other areas, including solar air conditioners, solar ponds, solar cookers, water purification and distillations, and solar power plants. Solar energy is harnessed by numerous technologies, including chimneys, solar lakes, solar cells, to generate electricity [5]. Solar chimneys (SCs) are a primary method used for generating electricity from solar energy [6]. An SC consists of three main components: a wind turbine, a solar collector, and a chimney. The concept of SCs are invented for heating the space between the chimney floor and the glass cover; as an output of the greenhouse effect, the hot air's density alleviates. At the chimney's bottom, the flowing air circulates the wind turbine installed while it reaches the top of the SC [7]. The system operates based on fluid motion driven by pressure variations and upward buoyancy, which depend on density differences. The collectors' semi-permeable nature allows it to transmit the solar radiation when exposed to the sun [8]. The chimney of the system functions as a motor, enabling the air to be evacuated as it gains kinetic energy and heats up in the collector. As the air accelerates, it further heats up the collector. Further, due to the opaque structure, there is an elevation in the ground temperature the time when from the collector (semi-permeable feature), the solar radiation reaches the ground. The system generates energy by converting the kinetic energy (which strikes the blades of the turbine) into electrical energy of the air. A turbine is mounted at a specific height above ground

 DOI: 10.1201/9781003521341-6

within the chimney. As the solar radiation enters the system, the collector, acting as a cover, heats up the air. Because of the reduced density of heated air, the air beneath the collector moves upward. The system starts functioning as soon as the heated air is transmitted toward the ground.

Under the collector, the air of the system is exposed to an upward buoyancy force because of the ground's temperature variance. This pressure causes the air beneath the collector to be pushed toward the collector's center, the only location it can move. The pressure variance between the entrance and exit of the chimney, which is mounted vertically at the center of the collector, forces the air upward. Due to these effect as a results, the temperature and speed of the air increase with the rise in air's kinetic energy within the system. To a greater extent no mechanism is required. The height difference between the entrance and exit of the tower creates a vacuum effect at the center of the collector, drawing the air through the chimney [9]. The heating and cooling modes of the SC, make it superior to other natural ventilation systems thanks to the integration of damping and openings, which increases the sustainability of the structure. Additionally, its structural features offer convenience when used in conjunction with other ventilation systems or integrated into buildings [10].

Globally, almost half (47%) of all deaths occur in aged care centers or similar facilities. Due to the improper utilization or design of ventilation systems, elders are at relativity high health risk [11, 12]. When current ventilation systems are not aptly used, they can contribute to the spread of COVID-19, according to several studies [13, 14]. For instance, virus particles that accumulate in the ventilation system filters may spread throughout the entire building. The World Health Organization (WHO), recommends that the isolation rooms of existing buildings have natural ventilation with at least six air changes per hour (ACH), though the necessary value for new buildings is considerably greater [15].

Because natural ventilation systems are passive, SCs are a practical and feasible option for buildings due to their high performance [16, 17]. However, no studies in the literature have found the utilization of SCs in aged care centers or similar other institutions, not their potential for preventing the COVID-19 cross-infection. SCs enhance natural ventilation through thermal buoyancy backed by solar energy. The three main components of SCs are (i) air inlets/outlets for air flow, (ii) a solar heat absorption wall, and (iii) a glazing wall that allows solar penetration [18, 19]. The performance of SCs is determines by the rate of airflow through its chimney cavity. Previously developed analytical models for multi-chamber systems overlooked the complex it is of frameworks with several horizontal air inlet vents. Most of them have ignored the influence of the connected chambers (i.e., rooms), leading to overestimated outputs [20].

One of the key principles of building ventilation systems is to avoid reverse or back flow, as it can compromise ventilation performance and cause the spread of used air from one room to another. In a high-rise office buildings with SCs, reverse airflow or inclined absorption wall have been observed [21]. Moreover, in an multi-zone buildings with SCs, reverse flow at the other windows or vents could accelerate the spread of COVID-19 or similar infections. It remains unclear SCs can provide sufficient natural ventilation under typical weather conditions to meet the WHO's recommendation of six ACH for airborne infection isolation rooms. Under typical weather conditions in Mexico, SCs was found to provide natural ventilation of 7.9–11.4 ACH for a $27m^3$ room, according to a study by Zavala-Guillen et al. [22]. Hashemi et al. [23] optimized the design of SCs for energy savings in a house, considering factors such as environmental conditions (e.g., solar radiation), attic and chimney, and room configuration. The situation becomes complicated in aged care center with long corridors and multiple rooms. As depicted in a typical SC shown in Figure 6.1, the design includes heat-insulating materials, an absorption wall, tuyeres/vents, and a glazing wall.

Airflow is influenced by the variation in air density between the internal and external environments and by the external wind. SCs components can drive airflow in a space using either direct or indirect solar energy [24]. The stale air escapes through specially designed openings due to the thermo-siphoning effect. A roof-top SC can increase the ventilation rate by 1.16%–45.0% for cooling a single room, according to Jiménez-Xamán et al. [25], people tend to spend more time outdoors in the past.

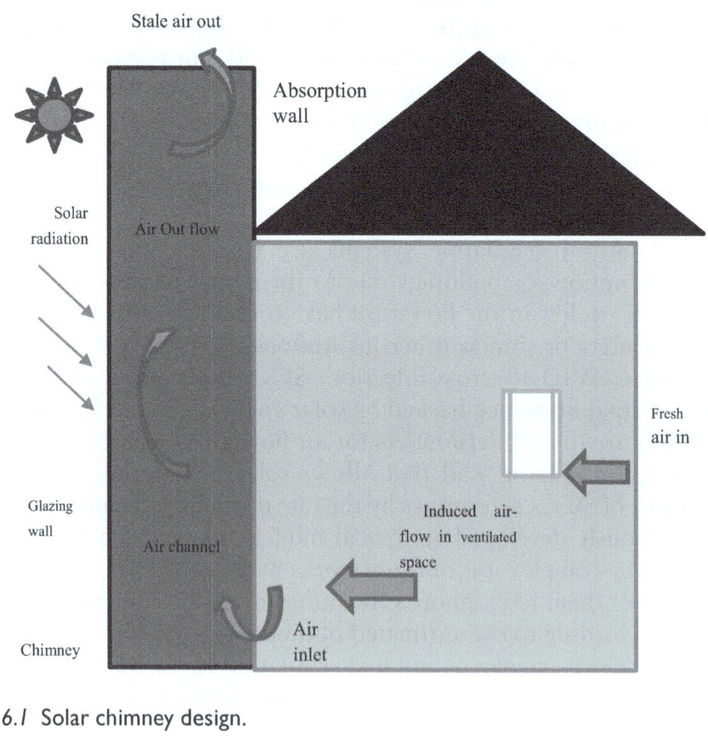


*Figure 6.1* Solar chimney design.

But this trend has shifted significantly with cultural, social, and technological advancements. For example, it is customary for European and U.S. population (approximately 90%) to spend a significant amount of their time indoors, with around 5% of the population spend time inside vehicles. Globally, buildings account approximately for ~40% of primary energy consumption and about one-third of $CO_2$ emissions. Hence, improving energy efficiency in buildings, particularly by reducing the energy consumption HVAC systems and lighting, is crucial. Integrating high-efficiency HVAC systems and passive cooling technologies is essential for meeting low-energy requirements. The ventilation concept involves introducing exterior air into the indoor room environment through inlet gaps.

This concept is associated with indoor air quality and thermal comfort conditions experienced by occupants. Compared to mechanical and electro-mechanical systems, passive ventilation systems have proven to be the best alternative due to their advantages such as, low air-contaminant emission level, electricity savings, and reduced maintenance costs. Natural ventilation systems results rely on two driving sources (i.e., buoyancy or wind), which cause a temperature variance between indoor and outdoor air (known as the chimney or stack effect). SCs function similarly to natural ventilation systems by absorbing solar energy and releasing it into the airflow within a vertical flat plate channel. This causes the airflow to move upward, increasing the air temperature. The heated air is then expelled, diluting contaminant concentration and introducing fresh air into the rooms. The implementation and function of standalone chimneys for of cooling buildings may be apt, and they can also help reduce electricity grid stress during peak demand periods [25]. To assess the solar radiation absorption on a surface inclined relative to a horizontal surface, beam radiation ratio values, Rb, are needed. These values are calculated for the collector's vertical and horizontal surfaces. Equation (6.1) will be used to calculate the $R_b$ values at the surface azimuth angles of –120°, 0°, and 120° for wings 1, 2, and 3, respectively:

$$R_b \frac{\cos\theta}{\cos\theta_z} \tag{6.1}$$

Here, $\theta$ is radiation's incident angle given as follows:

$$\cos\theta = -\sin\delta\cos\varphi\cos\gamma + \cos\delta\sin\varphi\cos\gamma\cos\omega + \cos\delta\sin\gamma\sin\omega \tag{6.2}$$

This is expressed in terms of $\omega$ for each of the wings. The zenith angle values are $\theta$ for each wing at the halfway point of an hourly time period from 7.00am to 5.00pm solar time.

The zenith angle is demonstrated as follows:

$$\theta_z = \cos^{-0}\left[\cos\delta\cos\varphi\cos\omega + \sin\delta\sin\varphi\right] \tag{6.3}$$

## 6.2 FUNDAMENTALS OF VENTILATION AND PASSIVE COOLING

In the early 1970s, the "passive" phase has been used introduced by the U.S. researchers. This concept refers to systems that function naturally without the use of mechanical devices powered by electricity. To achieve cooling, heat energy must be transferred from indoors to outdoors, along with maintaining moisture levels lower than those found in the natural environment. The term "passive cooling" refers to various architectural techniques utilized to cool a building without mechanical systems. When applied, these systems may enhance efficiency, often using a fan with low-energy consumption or fans. The upper atmosphere or sky fans serves as a natural heat sink (sky). Some passive systems require the interaction of the occupant, such as opening and closing windows. Proper building layout and construction team should consider microclimate effects [26]. Strategies for preventing heat include shading or using dull colors or reflective surfaces for the building's exterior, managing internal thermal gains, and using insulated wrappers. Building cooling strategies should be conceived on three levels: reducing heat gains, protecting from solar radiation and heat, discharging internal heat through ventilation to outdoor heat sinks via radiative, earth, or evaporative cooling [26]. High temperature during summer needs more energy consumption. For instance, the mean temperature in Iraq during summer (April to September) ranges from 40°C to >50°C. Both heat transfer and heat sink mechanisms are crucial for cooling.

Heat rejection transfers indoor heat to outdoor sinks, thus maintaining lower indoor temperatures than outdoors. Heat modification is linked to materials that have the potential to absorb and store heat. In building constructions, thermal mass materials like brick and concrete function as cold storage at night and heat storage during the day. Although they are linked, the primary focus from an ideological standpoint is on the difference between passive cooling and ventilation, as well as the principles of bioclimatic design. The use of these materials is highly effective for buildings with constant occupancy, such as houses. Many third-world countries require passive cooling due to the high ambient temperatures, making the energy consumption associated with active cooling techniques unaffordable. For example, Iraq lies in the subtropical region, where summers are hot and dry, and the major portion of the area is desert [27].

## 6.3 ARCHITECTURE OF SOLAR CHIMNEYS

Tall and wide constructed SCs that face the sun are typically constructed with dark-colored, matte surfaces to maximize solar irradiance absorption. As the chimney heats up, the air inside becomes warmer, rises up into the attic and escapes through the top. This convection procedure draws additional air in at the bottom of the chimney, facilitating passive ventilated in structures, and developers prefer not to utilize energy-intensive mechanical ventilation [28]. SCs are especially useful in hot and humid topical settings. They are most efficient when they become tall and wide but not too deep, as this proportions maximizes both the volume of air that can collect energy from the sun and the surface area in contact with the airflow inside the chimneys. To enhance the contact area, multiple chambers can be used and minerals with strong thermal conduction qualities, such as metal, can be utilized to maximize the temperatures achieved within chimneys. Similarly to Trombe wall designs, low-emissivity coatings and glass can be employed to limit thermal losses to the exterior. The chimney must be insulated from building to prevent heat gains from being transmitted into controlled areas. In colder weather, the chimneys can be closed at the top to direct accumulated heat into the structure [29].

House windows, walls, and flooring can be constructed to collect, preserve, and transmit hear energy from the sun during winter, while they can be engineered to reject solar radiation during summer. This approach is also called instilling solar panels or climatological design. One of the most effective solutions for the home heating is solar thermal technology, which can significantly improve a structure's energy efficiency and, in some situations, provide 100% of a home's heating needs. Houses that are heated nearly entirely by sunlight, such as those with south-facing windows that offer only a percentage of the heat exchanger, are all examples of passive energy homes. SCs are an efficient way to incorporate a ventilation system in buildings with solar energy access. They are most commonly used in settings where a small amount of indoor fluctuations is acceptable, as well as in moderate climates. This technology does not require a large wall and can provide better air quality than a non-load-bearing wall by ensuring adequate ventilation for heaters. SCs have been studied by a number of academicians for a variety of purposes, including natural building convection and electricity production [29].

## 6.4 BRIEF INTRODUCTION OF GREEN BUILDING DESIGN

Achieving true green building requires a high degree of social resource integration. Architects can exercise control over the most basic forms and spaces to practice the green concept [30]. Glare evaluation takes into account the luminance, the relative positions of glare sources within the visual field, size, and the overall backdrop luminance or illuminance according to contemporary thresholds. This ensures that the appropriate illumination levels are provided for the intended indoor tasks, while still meeting the requirements from selected regulating standards.

Recently, good lighting conditions have become a key consideration in the design of high-performance buildings. Although the level of discomfort or perception varies among individuals, the aforementioned factors help quantify it. On-site field measurements can provide data related to these parameters, but an alternative approach involves computer-based illumination simulations. Visual comfort is critical and plays a substantial role in enhancing well-being, health, and productivity [31–33]. Metrics based on contrast (luminance ratio in the field of view) or saturation (comprehensive illuminance on the eye) are commonly used for the evaluating the lighting intensity of indoor environments. Glare, being highly subjective, varies significantly from person to person, and it is an important consideration in the life-cycle concept of green building design.

Direct luminaries within the field of view can lead to visual discomfort. Since the 1950s, several efforts have been made to quantify this situation. The significant parameters while evaluating glare in the indoor environment include (i) the relative location of the glare source from the viewer, (ii) glare size (e.g., window size), (iii) the luminance of the glare source ($L_g$), and (iv) the background luminance of the view ($L_b$). High-illuminance lighting or excessive daylight from exterior windows can cause glare in indoor environments when light is reflected from windows or directed toward areas where occupants often sit or work. When daylight is taken into consideration during the conceptual design phrase of a building, simulation is often used to assess lighting quality and guide indoor lighting design according to thresholds prescribed in relevant country standards [31].

## 6.5 GREEN BUILDING

Green building is a complex system that considers multi-objective attribute optimization, where energy consumption should be reduced through passive design while maintaining adequate indoor comfort. The concept of "green building" refers to conservation of resources, reduction of pollution, and protection of the environment throughout a building's entire life cycle. This approach aims to create comfortable, healthy, and effective spaces for occupants while fostering a harmonic relationship with nature [34]. On the contrary, when passive strategies alone do not fulfill the needs for normal interior comfort, a few suitable active measures can be utilized to supplement the system minimizing building's energy consumption as much as possible.

## 6.6 DESIGN CONCEPT OF GREEN BUILDING

Elevating the overall quality of a building by integrating "green" principles into formal space operation and design can be challenging, even if green building is a renowned concept [35, 36].As an organic whole, the core of building design is constituted by the treatment of natural environment and the buildings that are man-made. This design approach fully utilizes natural resources to achieve the building's usage cultural and aesthetic functions while mitigating the environmental impact caused by construction of building. The significance of eco-environmental protection, highlighted in Rachel Carson's "Silent Spring," (1962) marked the beginning of green building revolution [37]. Wins James, in further advanced this movement with the publication of "Green Architecture: Under Construction" in 1990. "The role of architecture in the environment" in article [38–40]. In 1990, the evolution of the British BREEAM (Building Research Establishment Environmental Assessment Method) introduced an environmental evaluation strategy focusing on three aspect: construction and design, operation management, and performance of building [41, 42]. This effort led to the creation of BREEAM, the world's first green building evaluation system [43, 44].

## 6.7 RELATED WORK

Various parameters related to the optimization of SCs design, passive ventilation, and geometrical parameters that depend on the latitude and longitude of the structures have been examined in previous studies. According to Ref. [45], the SCs inclination is at 45° angle and the absorptivity of the absorber plate is 0.95. The influence of radiation on heat transfer's mass flow rate to air via convection is alleviated by setting the convective heat transfer coefficients of the inner wall to zero. For RH values up to 0.5, an inclination of 45° is optimal for Bhubaneshwar's Jatni region (20.148°N latitude). The mass flow rate of the chimney can be increased by enlarging the gap between the absorber plate and the glass. Analysis shows that depending on the amount of sunlight and humidity, warm wet air can increase the mass flow rate by 10%–15%. The absorber plate has an emissivity of 0.95, the glass plate has an emissivity of 0.05, and the glass plate's transmissivity is 0.84. An 18.75% increase in air change per hour (ACPH) is observed. At a solar heat flux (SHF) of 500 W/m$^2$ at 40% RH, there the mass flow rate increases by10%. Under the same solar flux, a 25.74% increase in ACPH is observed at 80% RH, with a corresponding 13.5%rise in the mass flow rate.

According to Ref. [46], an SCs at any inclination angle with a horizontal bottom suction opening develops optimal thermal performance in Maysan. The ventilation rate is significantly influenced by the solar radiation angle $\alpha = 145°$, with an additional effect from window shading applied to the room model $\beta = 60°$. At a solar radiation level of 700 W/m$^2$, the ACH at the horizontal bottom inlet reaches a remarkable value of 165.

According to Ref. [47], an SC was designed on southern side of the house to maximize solar radiation absorption. The building sketch was created using SketchUp with a height of 3 m and an area of 220 m$^2$.

In Bandar-Abbas, the coldest and hottest days have outside temperatures of "2°C" and "52°C," respectively. This city experiences a temperate climate from November to April. A structure was planned with a 220-$m^2$ area and a 3-m height was planned. The internal heat gain from lighting and equipment is 4.5 W/$m^2$. Thermal comfort indices ranged from 0.16 to 0.31 in winter and from 0.34 to 0.78 in summer. For Bandar-Abbas' hot, humid environment, the average thermal comfort indices improved from (i) 0.75 to 0.67 in summer and from (ii) 0.07–0.04 in winter. In contrast, in Tehran's cold, semi-arid environment, the average thermal comfort indices reduced through Phase-Change Material (PCM) from 1.24 to 1.16 in the summer but increased from 1.1 to 0.85 in the winter.

A three-dimensional quasi-steady Computational Fluid Dynamics (CFD) model has been developed by Ref. [48] and validated against experimental results. An optimization method using the Exploration module of ANSYS17.2 FLUENT Design was used to enhance indoor occupant comfort by increasing air velocity due to buoyancy effects. The optimized SC, with dimensions of height=1.85m, breadth=2.65 m, air gap=0.28 m, and an inclination=75°, successfully improves thermal comfort under solar intensities >500 W/$m^2$. The generated air motion in the activity zone reaches 0.28 m/s at a solar radiation of 500 W/$m^2$, reducing the operative temperature by ~0.9°C. At higher solar radiation levels, the interior air velocity falls between two solar radiation levels: (i) 850 W/$m^2$=0.52 m/s and (ii) 700 W/$m^2$=0.47 m/s. At a low solar radiation of 250 W/$m^2$, the integrated buoyant air motion reaches 0.14 m/s in the activity zone, although the operating temperature is least impacted by it.

According to Ref. [49], the SHF for solar radiation ranges from 200 to 1,000W/$m^2$ under conditions of no wind and wind speeds of 1.0–8.0 m/s. From the SC with an SHF, the power output attained was 400 W/$m^2$. The chimney heights used were 5 and 9 m, respectively, with the room dimension being 2-D 4×4 $m^2$ and a gap ranging from 0.5–1 m. It was observed that the greatest exit mass flow rate was achieved by mounting the exit cavity at the ceiling level. The power output results were as follows: (i) SC with SHF of 400 W/$m^2$ in and chimney height of 5 m produced 32.8 W/$m^2$. (ii) SC with SHF of 400 W/$m^2$ and a chimney height of 9 m produced 85.2 W/$m^2$.

According to Gao et al. [50], the mass rate trend changes significantly increasing wind speed and channel width ratio $w_2/w_1$. The ventilation rate increased to its maximum as the channel width ratio $w_2/w_1$ changes. Under various outdoor wind speeds, an elevation in the inclination angle $\theta$ leads to an increase in the ventilation rate. The SCs has a height of 1,025 mm, length of 925 mm, heat flux of 600 W/$m^2$, and a width ranging from 20 to 110 mm. The height of the room is KJ=3,000 mm. The room length is CK=3,000 mm. The maximum mass flow rate, 0.378 kg/s, is achieved under the following conditions: (i) channel width ratio=$w_2/w_1$=0.2; (ii) outdoor wind speed=$U$=1.0 m/s. Ventilation rate's increase by 166% for $U$=2.0 m/s, 127% for $U$=3.0 m/s, and 212% for $U$=1.0 m/s. When the (i) channel width ratio of

$w_2/w_1$is 0.267 and (ii) outdoor wind speed $U$ is 3.0 m/s, the maximum mass flow rate of 0.591 kg/s is reached. Inclination angles increase from 30° to 90°, together with the solar radiation of $I$=600 W/m$^2$. Finally, it is decided that AB is a valid solution because it forms the reasonable calculation domain.

According to Hong et al. [51] the assessed house model has heating and cooling demands of 26.8 and 48.6 kWh/m$^2$, respectively with two floors and the total floor area of 220 m$^2$. To increase natural ventilation, SCs is mounted on the west wall with a height of 8 m, breadth of 1.6, and depth of 1 m. The chimney's top outlet has an effective discharging area of 1.6 m$^2$. Inside the room, the SCs inlet is connected to a wide rectangular duct measuring 2.77 m×0.13 m ×1 m. The SCs can reduce the Variable Refrigerant Flow (VRF) energy consumption by 2.3%, or 129.3 kWh, in case of heating and cooling, and ventilation energy by 77.8%, or 481.6 kWh. During natural ventilation operations, the VRF energy consumption can be reduced by about 15.9%, or 20.8 kWh. For the studied house, SCs total energy-saving capacity is 9% of the overall annual HVAC energy consumption. According to Ref. [52], air velocity was measured at nine testing points on the five grates at intake grate level, situated in the center of the nine elementary fields and were split into a square shape by the grate. Air velocity was also measured at the SCs outlet level, where the grate was split into ten testing points in the middle of ten elementary fields. Air humidity and air temperature were assessed at specific points located at a height of 1.2 m. The chimneys were made of brick, with a length of 0.2 m, width of 1 m, and height of 2 m. SCs have a 5-cm insulation where the volume of the superstructure is 0.47 m long, 1.86 m wide, and 1.45 m high. Higher values were observed when sunlight directly hit glazing. The values indicate variations from 0.05 to 0.12m$^3$/s. The air volume stream values decreased from 4 p.m. to 4 a.m. The values are higher after building a SCs, which ranges from 0.004 to 0.008 m$^3$/s. According to Ref. [53], the average volume flow rate at the total air outlet for model 2was 0.278 m$^3$/s. The average volume flow rate of the total air outlet for model 3was 0.253 m$^3$/s. The height of the vertical part of the chimney was 1.5 m and chimney's width (vertical part) was 1.0 m. The ambient air inlet to the horizontal part was 1.5 m×0.14 m. This flow rate corresponds to 2–5.6 ACH for a typical room of 27 m$^3$. For model 2, when the air flow rate is 0.230, then there is a 7% reduction in air circulation. For model 3, when the air flow rate is 0.249, then there is a 7.6% elevation in air circulation.

## 6.8 CONCLUSION

The study emphasizes the importance of considering air as a medium for radiative heat transfer when utilizing SCs. It is significant to note that the density impacts the air's mass flow rate, and the chimney performance is substantially impacted by the radiation mode of heat transfer, especially under sunlight and humid air. Furthermore, while modeling the SCs, it must be taken into account that density changes with humidity and temperature, especially in humid climate zones. The study suggests that moist air can increase the mass flow rate depending on sunlight and humidity. Therefore, SCs is helpful in reducing energy consumption in semi-cold and dry areas. Thus, the SCs are suitable for passive cooling and ventilation in buildings under humid conditions.

## REFERENCES


1. Abed A.A., Ahmed O.K., Weis M.M., Performance analysis of a bi-fluid photovoltaic/trombe wall under Iraqi climate. *AIP Conf. Proc.*, vol. 2213, no. 1, p. 02015 (2020).
2. Ravi Kumar K., Krishna Chaitanya N.V.V., Sendhil Kumar N., Solar thermal energy technologies and its applications for process heating and power generation: A review. *J. Clean. Prod.*, vol. 282, p. 125296 (2021).
3. Lin W., Ma Z., Wang S., Sohel M.I., Lo Cascio E., Experimental investigation and two-level model-based optimisation of a solar photovoltaic thermal collector coupled with phase change material thermal energy storage. *Appl. Therm. Eng.*, vol. 182, p. 116098 (2021).
4. Ahmed O.K., Hamada K.I., Salih A.M., Enhancement of the performance of photovoltaic/trombe wall system using the porous medium: Experimental and theoretical study. *Energy*, vol. 171, pp. 14–26 (2019).
5. Ahmed O.K., Bawa S.M., The combined effect of nanofluid and reflective mirrors on the performance of PV/thermal solar collector. *Therm. Sci.*, vol. 23, no. 2A, pp. 573–587 (2019).
6. Guo P., Wang Y., Li J., Wang Y., Thermodynamic analysis of a solar chimney power plant system with soil heat storage. *Appl. Therm. Eng.*, vol. 100, pp. 1076–1084 (2016).
7. Guo P., Li T., Xu B., Xu X., Li J., Questions and current understanding about solar chimney power plant: A review energy convers. *Manag*, vol. 182, pp. 21–33 (2019).
8. Cuce E., Cuce P.M., Carlucci S., Sen H., Sudhakar K., Hasanuzzaman M., Daneshazarian, R., Solar chimney power plants: A review of the concepts, designs and performances. *Sustainability*, vol. 14, pp. 1450 (2022). https://doi.org/10.3390/su14031450
9. Zhang H., Tao Y., Shi L., Solar chimney applications in buildings. *Encyclopedia*, vol. 1, pp. 409–422 (2021). https://doi.org/10.3390/encyclopedia1020034
10. Aitken G.E., Holmes A.L., Lbrahim J.E., COVID-19 and residential aged care: Priorities for optimising preparation and management of outbreaks. *Med. J.*, vol. 214, pp. 6–8 (2021).
11. Guo Y., Zhang N., Hu T.R., Wang Z.Y., Zhang Y.P., Optimization of energy efficiency and COVID-19 pandemic control in different indoor environments. *Energy Build*, vol. 261, p. 111954 (2022).

12. Wang J., Huang J., Feng Z., Cao S.-J., Haghighat F., Occupant-density-detection based energy efficient ventilation system: Prevention of infection transmission. *Energy Build.*, vol. 240, p. 110883 (2021).
13. Correia G., Rodrigues L., da Silva M.G., Goncalves T., Airborne route and bad use of ventilation systems as non-negligible factors in SARS-CoV-2 transmission. *Med. Hypotheses*, vol. 141, p. 109781 (2020).
14. Berry G., Parsons A., Morgan M., Rickert J., Cho H., A review of methods to reduce the probability of the airborne spread of COVID-19 in ventilation systems and enclosed spaces. *Environ. Res.*, vol. 203, p. 111765 (2022).
15. Wang Q, Zhang G, Wu Q, Shi L., Ventilating aged-care center based on solar chimney: Design and theoretical analysis. *Energy Build.*, vol. 1, no. 266, pp. 112–145 (2022). doi: 10.1016/j.enbuild.2022.112145.
16. Taghdisi A., Ghanbari Y., Eskandari M., Energy-conservation considerations through a novel integration of sunspace and solar chimney in the terraced rural dwellings. *Int. J. Energy Econ. Policy.*, vol. 10, pp. 1–13 (2020).
17. Khosravi M., Fazelpour F., Rosen M.A., Improved application of a solar chimney concept in a two-story building: An enhanced geometry through a numerical approach. *Renewable Energy*, vol. 143, pp. 569–585 (2022).
18. Jimenez-Xaman C., Xaman J., Moraga N.O., Hernandez-Perez I., Zavala-Guillen I., Arce J., Jimenez M.J., Solar chimneys with a phase change material for buildings: An overview using CFD and global energy balance. *Energy Build.*, vol. 186, pp. 384–404 (2019).
19. Shi L., Ziem A., Zhang G.M., Li J., Setunge S., Solar chimney for a real building considering both energy-saving and fire safety: A case study. *Energy Build.*, vol. 221, p. 110016 (2020).
20. Shi L., Zhang G.M., An empirical model to predict the performance of typical solar chimneys considering both room and cavity configurations. *Build. Environ.*, vol. 103, pp. 250–261 (2016).
21. Raji B., Tenpierik M.J., Bokel R., van den Dobbelsteen A., Natural summer ventilation strategies for energy-saving in high-rise buildings: A case study in the Netherlands. *Int. J. Vent.*, vol. 19, no. 1, pp. 25–48 (2020).
22. Zavala-Guillen I., Xaman J., Hernandez-Perez I., Hernandez-Lopez I., Jimenez-Xaman C., Moreno-Bernal P., Sauceda D., Ventilation potential of an absorber-partitioned air channel solar chimney for diurnal use under Mexican climate conditions. *Appl. Therm. Eng.*, vol. 149, pp. 807–821 (2020).
23. Hashemi S.H., Dinmohammad M., Hashemi S.A., Optimization of Trombe wall system parameters using TWsim software. *Model. Earth Syst. Environ.*, vol. 7, no. 2, pp. 917–924 (2021).
24. Shi L., Zhang G.M., Cheng X.D., Guo Y., Wang J.H., Chew M.Y.L., Developing an empirical model for roof solar chimney based on experimental data from various test rig. *Build. Environ.*, vol. 110, pp. 115–128 (2016).
25. Jiménez-Xamán C., Xamán J., Gijón-Rivera M., Zavala-Guillén I., Noh-Pat F., Simá E., Assessing the thermal performance of a rooftop solar chimney attached to a single room. *J. Build. Eng.*, vol. 31, pp. 101–380 (2021).
26. Al-Shamkhee D., Al-Aasam A.B., Al-Waeli A.H.A., Abusaibaa G.Y., Moria H., Passive cooling techniques for ventilation: An updated review. *Renew. Energy Environ. Sustain.*, vol. 7, no. 23, p. 16 (2022).

27. Abdullah H.K., Aboud S.M., Reduction of heat transfer through walls in buildings by using insulation. *Int. J. Comput. Appl.*, vol. 141, pp. 975–8887 (2016).
28. Waheba J.H., Enhance the thermal performance of Iraqi house using new passive systems in summer season. *Int. J. Sci.: Basic Appl. Res.* (2016). https://gssrr.org/index.php?journal=JournalOfBasicAndApplied.
29. Kumar G., Diwakar V., Review on solar chimney in natural ventilation. *Res. J. Eng. Technol. Med. Sci.*, vol. 5, no 02 (2022). Available at www.rjetm.in/
30. Xiao F., Lei L., Lin S., Basic green building design: A formal space design strategy in response to climate. *Archit J.*, vol. 1, pp. 100–104 (2019).
31. Wang T.-H., Huang Y., Park, J., Development of daylight glare analysis method using an integrated parametric modelling approach: A comparative study of glare evaluation standards. *Buildings*, vol. 12, p. 1810 (2022). https://doi.org/10.3390/buildings12111810
32. Park J., Loftness V., Aziz A., Wang T.H., Critical factors and thresholds for user satisfaction on air quality in office environments. *Build. Environ.*, vol. 164, pp. 106–310 (2019).
33. Park J., Loftness V., Wang T.H., Examining in situ acoustic conditions for enhanced occupant satisfaction in contemporary offices. *Buildings*, vol. 12, p. 1305 (2022).
34. Lin G., Green building design based on the concept of energy saving and environmental protection: Comment on "key technology of green building design and operation". *Concr. Cem. Prod.*, vol. 8, p. 97 (2020).
35. Fanbin M., Qiqun Z., Zhanying Z., Bo W., Hengrui M., Abdullah Heba M., et al., An intelligent hybrid wavelet-adversarial deep model for accurate prediction of solar power generation. *Energy Rep.*, vol. 7, pp. 2155–2164 (2019).
36. Yu W., Jijiang H., Wenying C., Distributed solar photovoltaic development potential and a roadmap at the city level in China. *Renew Sustain. Energy Rev.*, vol. 141, p. 110772 (2021).
37. Zhao G., Li M., Jian L., He Z., Shuang J., Yuping S., et al., Analysis of fire risk associated with photovoltaic power generation system. *Adv. Civ. Eng.* Vol. 2018, pp. 1–7 (2018).
38. Kang M.-S., Park Y.-K., Kim K.-T., Economic feasibility through the optimal capacity calculation model of an energy storage system connected to solar power generator. *Energy Environ.*, vol. 31, no. 5, pp. 860–869 (2020).
39. Chengkang G., Sulong Z., Nan A., Hongming N., Huan Y., Chengbo G., Comprehensive comparison of multiple renewable power generation methods: A combination analysis of life cycle assessment and ecological footprint. *Renew Sustain. Energy Rev.*, vol. 147, p. 111255 (2021).
40. Boukelia T.E., Arslan O., Bouraoui A., Thermodynamic performance assessment of a new solar tower-geothermal combined power plant compared to the conventional solar tower power plant. *Energy*, vol. 232, p. 121109 (2021).
41. Palacios A., Barreneche C., Navarro M.E., Ding Y., Thermal energy storage technologies for concentrated solar power: A review from a materials perspective. *Renew Energy*, vol. 156, pp. 1244–1265 (2020).
42. Cavalcanti E.J.C., Lima M.S.R., de Souza G.F., Comparison of carbon capture system and concentrated solar power in natural gas combined cycle: Exergetic and exergoenvironmental analyses. *Renew Energy*. vol. 156, pp. 1336–1347 (2020).

43. Xian Z., Jia-Xing W., Zhe C., Shuo S., Shuo M., Jing-Li F., What is driving the remarkable decline of wind and solar power curtailment in China? Evidence from China and four typical provinces. *Renew Energy*, vol. 174, pp. 31–41 (2021).
44. Ramadoni S., Indah S., Renewable energy systems based on micro-hydro and solar photovoltaic for rural areas: A case study in Yogyakarta, Indonesia. *Energy Rep.*, vol. 7, pp. 472–490 (2021).
45. Dahire H., Kannan S.R., Saw S.K., Effect of humidity on the performance of rooftop solar chimney. *Therm. Sci. Eng. Prog.*, vol. 27, pp. 101–026 (2022). https://doi.org/10.1016/j.tsep.2021.101026
46. Mohamed A. Q., et al., Numerical study for the ventilation with solar chimney under effect of different location and the shape of the section opening window. *IOP Conf. Ser. Mater. Sci. Eng.*, vol. 881, pp. 12–164 (2022).
47. Nateghi S.K., Jahangir M.H., Performance evaluation of solar chimneys in providing the thermal comfort range of the building using phase change materials. *Cleaner Mater.*, vol. 5, pp. 100–120 (2022). https://doi.org/10.1016/j.clema.2022.100120
48. Abdeen A., Serageldin A.A., Ibrahim M.G.E., El-Zafarany A., Ookawara S., Murata R., Solar chimney optimization for enhancing thermal comfort in Egypt: An experimental and numerical study. *Solar Energy*, vol. 180, pp. 524–536 (2019).
49. Layeni A.T., Waheed M.A., Adewumi B.A., Bolaji B.O., Nwaokocha C.N., Giwa S.O., Computational modelling and simulation of the feasibility of a novel dual purpose solar chimney for power generation and passive ventilation in buildings. *Sci. Afr.*, vol. 8, no. 2, p. e00298 (2020).
50. Gao N., Yan Y., Sun R., Lei Y., Natural ventilation enhancement of a roof solar chimney with wind-induced channel. *Energies*, vol. 15, p. 6492 (2020). https://doi.org/10.3390/en15176492
51. Hong S., He G., Ge W., Wu Q., Lv D., L, Z., Annual energy performance simulation of solar chimney in a cold winter and hot summer climate. *Build. Simul.*, vol. 12, no. 5, pp. 847–856 (2019). doi: 10.1007/s12273-019-0572-y
52. Nakielska M., Pawłowski, K., Increasing natural ventilation using solar chimney. *E3S Web Conf.*, vol. 7, pp. 85–796 (2017). doi: 10.1051/e3sconf/20171401051
53. Al-Kayiem H.H., Heng Y.M., Experimental investigation of rooftop solar chimney for natural ventilation. *ARPN J. Eng. Appl. Sci.*, vol. 10, no. 21, pp. 10249–10254 (2015).

Chapter 7

# Development of blue–green infrastructure to mitigate high building energy demand due to the urban heat island

## A case of Varanasi city

*Aishwarya Dwivedi and Rajat Soni*

### 7.1 INTRODUCTION

The urban heat island (UHI) effect refers to warmer urban areas as compared to their rural surroundings (Oke, 1982), turning urban centers into 'islands' of higher temperature amid the comparatively cooler rural landscape. Urbanization induces rapid land-use and land-cover changes over an area along with increased anthropogenic activities, which increase the thermal mass of the respective region, resulting in greater heat gain and storing (Wesley & Brunsell, 2019; Li et al., 2013; Manoli et al., 2019; Shepherd, 2013). In this way, UHI becomes the most apparent human-induced climate change aspect and an important phenomenon while understanding urban local climate (Jia et al., 2017). Globally, climate change is projected to cause severe damage to human and wildlife ecosystems, increasing exposure and vulnerability to various climatic hazards, including extreme weather events (IPCC, 2023). Climate change has already had diverse adverse impacts on human systems, including water security and food production, health, and well-being, cities, settlements, and infrastructure (IPCC, 2023). UHI and climate change have a compounding effect on urban systems and infrastructure, as warming trends intensify the already elevated temperatures in heat island areas. The severity of UHI becomes even more prominent considering the vulnerability of Indian cities to extreme climatic events, especially extreme heat and drought. According to Mohanty and Wadhawan (2021), nearly 74% of all Indian districts are vulnerable to extreme climate events, representing a composite measure of exposure, sensitivity, and adaptive capacity. As a result, UHI has become one of the most significant climate hazards in Indian cities, affecting the efficient functioning of various urban systems.

One of the most direct and evident impacts of UHI is the increase in building cooling energy demand. Previous studies show that for every 1°C increase in atmospheric temperature, building cooling energy demand can

 DOI: 10.1201/9781003521341-7

rise by up to 11.5% (Ma & Yu, 2020). This increase in energy demand is expected to rise further considering the ongoing climate change and global warming (Hosseini et al., 2022). Therefore, it is important to mitigate the impact of UHI on buildings to create a more sustainable urban form. Urban Blue–Green Infrastructure (UBGI) plays an important role in mitigating the negative effects of UHI. UBGI includes blue areas such as water bodies, streams, and drainage channels, and green areas such as parks, forests, tree covers, and grasslands. The presence of UBGI elements helps moderate the natural climate of an area, acting as a nature-based solution for multiple climatic hazards. UBGI increases the surface albedo, provides shade, and enhances evapotranspiration, thereby preventing ambient temperatures from reaching higher levels. In this way, UBGI creates cool islands within urban areas (Yu & Hien, 2006; Zhu et al., 2023). This study focuses on the cooling effect of UBGI and examines the impact on building energy demand. First, a UBGI network is delineated for Varanasi city by incorporating natural drainage channels of the urban watershed. Second, the reduction in cooling energy demand is estimated using quantified cooling factors for regions influenced by the UBGI network. The study therefore presents a methodological framework for delineating UBGI and analyzing its role in reducing building cooling energy demand.

## 7.2 STUDY AREA

Varanasi city derives its physical, cultural, and social identity from the water bodies surrounding it. While the River Ganga provides religious and cultural significance to the city, the rivers Varuna and Assi define the historical boundaries of the urban settlement and give the city its name, 'Varanasi'. This signifies the deep and multifaceted relationship between the city and its natural environment, where urban dimensions and these rivers are characteristically interwoven. Moreover, the city has been an important center of economy and politics since the ancient period, with its history dating back to 800 BCE (Singh, 2018). As a part of an important national transport corridor (Delhi–Kolkata), the city has experienced rapid development and shows immense potential for future growth. Water, therefore, becomes an important economic asset for the city, making well-being and conservation of water resources crucial for sustainable urban development. However, past trends of urban development have not favored the preservation of natural land covers. Increasing pressure on the city infrastructure and scarcity of land resources have led to challenges such as overexploitation, encroachments, and pollution of water bodies and green spaces One of the major indicators of the depleting urban environment in Varanasi city is the consistently increasing frequency of extreme temperatures. As a result of rising ambient temperatures, the cooling demand in urban areas has also

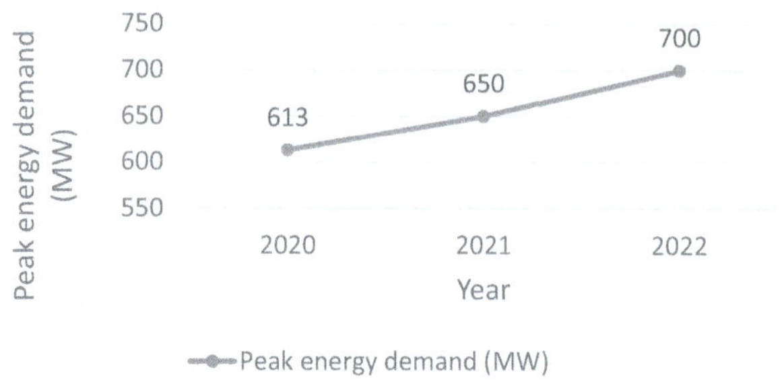


*Figure 7.1* Energy demand trend for Varanasi city.

increased significantly in recent years. A major share of urban electricity demand for the city (60%) comes from residential and commercial building electricity consumption. These areas are highly sensitivity to the UHI effect and are major drivers of the surge in electricity demand during periods of higher ambient temperatures (Hirano & Fujita, 2012; Roxon et al., 2020). The city experiences excessive energy demand during the summer season, and the trend in recent years shows a consistent increase in peak energy demand, as shown in Figure 7.1 (Central Electricity Authority, 2020). Increasing peak energy demand not only leads to higher energy expenditure but also increases the rate of deterioration of electricity infrastructure (Zhang et al., 2020).

## 7.3 METHODOLOGY

The methodology for this study consists of two stages. The first step involves generating a UBGI network for the urban area. Natural drainage lines serve as the site for intervention to create the UBGI network. This process begins with identifying primary drainage channels within the urban watershed area and dividing the urban area into smaller sub-watersheds. The primary drainage lines for the designated area are identified through spatial and hydrological analyses, using Shuttle Radar Topography Mission Digital Elevation Model (SRTM DEM) and QGIS software. The demarcation of primary drainage lines helps in strategically locating green infrastructure as these areas are naturally suitable for creating wetlands and green surfaces. The drainage lines are further classified into three categories depending upon the flow conditions (as detailed in Table 7.2):

1. Continuous flow
2. Limited flow
3. Obstructed flow

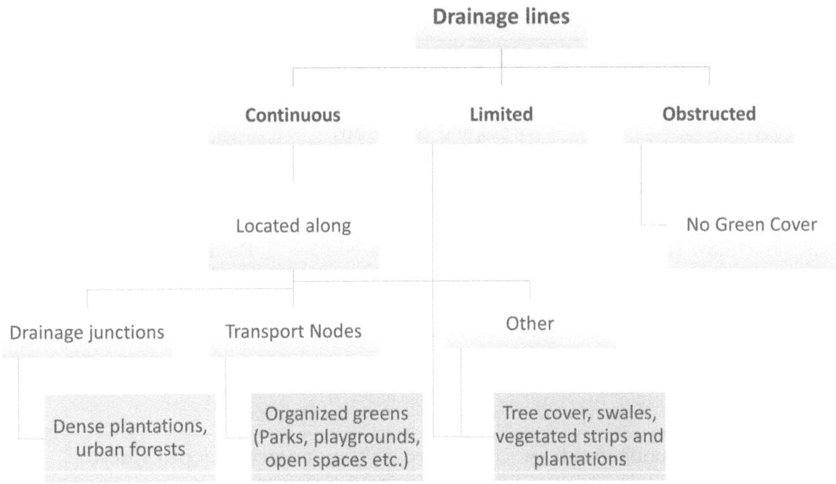


*Figure 7.2* Delineation of blue–green infrastructure.

Green infrastructure is proposed along these drainage corridors, considering the availability of open areas and built patterns. This is achieved through a land-use and land-cover analysis within a buffer area of 100 m along identified drainage lines. Furthermore, the green elements are divided into different types and located according to the flowchart shown in Figure 7.2.

The green elements obtained, combined with the drainage lines, constitute the UBGI network for Varanasi city. The UBGI elements are categorized as UBGI-1, UBGI-2, and UBGI-3, representing dense urban forests, organized greens, and sparse vegetation, respectively. In the second step, the impact of the UBGI network on cooling energy demand is identified. This is accomplished by using quantified factors derived from previous studies to estimate reductions in cooling energy demand. These factors are categorized into three categories, depending upon the magnitude of the cooling effect of different types of green areas (Table 7.1).

These factors are used to calculate the overall impact of the UBGI network on cooling energy demand within each watershed, assuming the effective influence zone of the UBGI network extends 500 m (Eq. 7.1). The calculations assume a homogenous cooling energy demand across the urban area, with results values expressed as potential percentage reduction in energy demand.

$$\text{Cooling effect} = \frac{\text{Cooling factor} \times \text{area affected (ha)}}{\text{Total area (ha)}} \tag{7.1}$$

Furthermore, the potential cooling effects are evaluated for two case scenarios: a realistic and an optimistic scenario, both of which are explained in detail in the following chapter.

*Table 7.1* Quantified factors for the cooling effect of UBGI

| *S. No.* | *Cooling factor* | *Green area type* | *References* |
|---|---|---|---|
| 1 | High (50%) | Large shading trees, dense grass, and coniferous forest (UBGI-1) | Sawka et al. (2013) |
| 2 | Moderate (20%) | Pasture and average grass (UBGI-2) | Akbari et al. (1997) and Akbari and Taha (1992) |
| 3 | Low (10%) | Poor grass, cultivated row crops (UBGI-3) | Pandit and Laband (2010) and Yu and Hien (2006) |

*Table 7.2* Categorization for drainage line continuity

| *S. No.* | *Flow type* | *Definition* |
|---|---|---|
| 1 | Continuous flow | Major roads/transit corridors/open spaces overlapping drainage lines |
| 2 | Limited flow | Roads/transit corridors/open spaces partially overlapping or parallel to drainage lines |
| 3 | Obstructed flow | No overlapping/parallel roads/open spaces |

## 7.4 IDENTIFYING PRIMARY DRAINAGE CHANNELS AND UBGI

To identify and delineate possible locations of green infrastructure elements, primary drainage lines are marked by analyzing the natural hydrological patterns and urban built form. The natural drainage lines for the city are identified through the Channel Network and Drainage Basins tool in QGIS using SRTM DEM (NASA JPL, 2013). The drainage pattern is then translated to the urban fabric by overlapping urban physical form (e.g., transport infrastructure) over primary drainage lines. The lines are ranked according to their continuity of flow and analyzed based on the level of freedom of movement. The movement index is categorized into three levels, as shown in Table 7.2.

Natural drainage lines of Strahler order greater than three are taken into consideration for this analysis. This approach helps translate the natural drainage lines into the urban built environment and distinguishes the most hydrologically sensitive areas of the city. It also provides an overview of the drainage scenario of the city by categorizing the primary drainage lines according to their freedom of movement and gives a basis for future interventions to improve urban drainage. The watershed basin analysis also divides the total urban watershed area into sub-watersheds. Delineating these sub-watersheds is an important step for calculating the runoff volume directed to a particular drainage line, which informs the requirements for drainage infrastructure. A total of 11 sub-watersheds are delineated in this way, as shown in Figure 7.3.

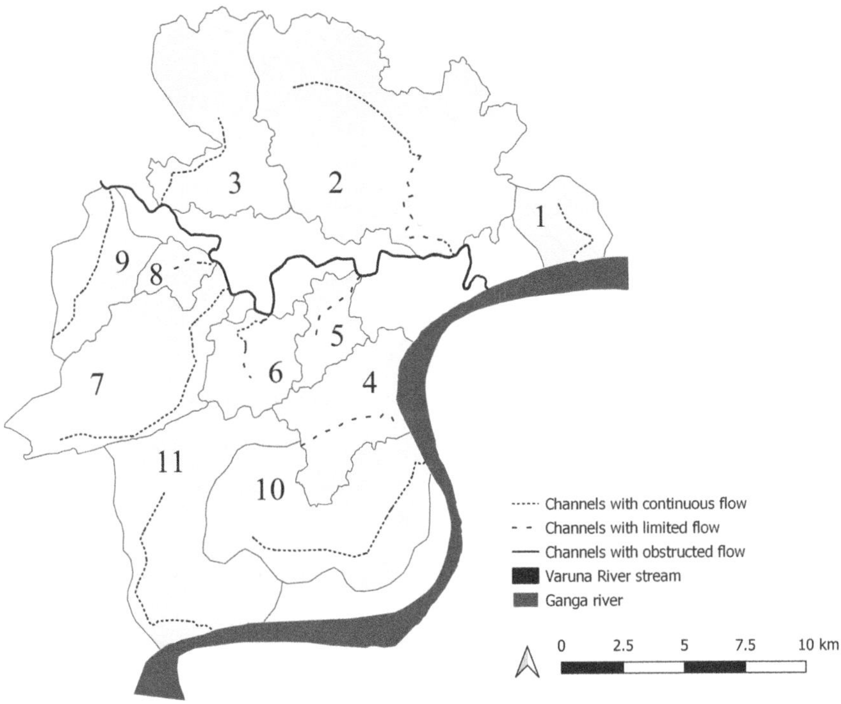


*Figure 7.3* Primary drainage lines and sub-watersheds for Varanasi city.

Figure 7.3 gives the longitudinal lines of UBGI intervention. To define the green areas, it is required to determine the lateral dimension of the drainage network, thus creating a green corridor. The total available open area along these lines is identified within a buffer of 100 m, which defines the effective area of intervention. Green elements are delineated within this intervention zone according to the type of flow movement, as shown in Figure 7.1. Hence, for drainage junctions and transport nodes, specific green infrastructure types are delineated, considering the optimal area for detention basins (2 ha) and organized greens (5 ha: equivalent to a community park). The remaining length of the lines is delineated with small green elements such as tree cover, vegetated strips, swales, plantations, and open bare land. This results in a network of permeable and semi-permeable surfaces, in place of completely impermeable surfaces of roads and pavements. The locations of these UBGI elements are shown in Figure 7.4.

The UBGI network created is detailed in Table 7.3, which shows the lengths of each type of green infrastructure with different values of cooling factors. UBGI-3 forms 87.5% of the total green areas delineated, followed by UBGI-2 (7.4%) and UBGI-1 (5%). This represents the scarcity of open spaces in dense urban settings such as Varanasi city. The current scenario

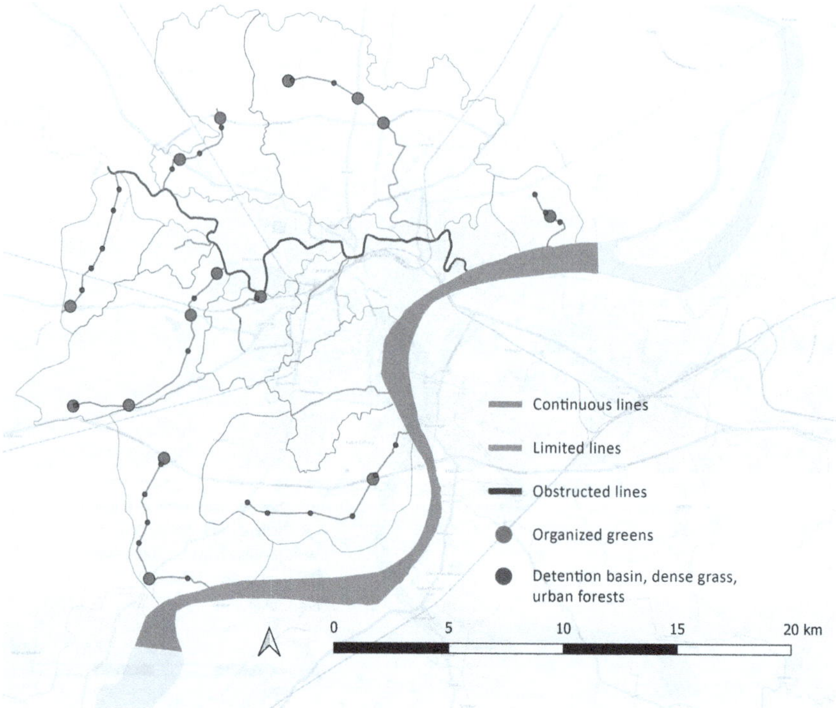


*Figure 7.4* Locations of designated UBGI elements.

*Table 7.3* Sub-watersheds and UBGI properties

| *S. No.* | *Sub-watershed* | *Total area (ha)* | *UBGI-equipped drainage length (m)* | | |
|---|---|---|---|---|---|
| | | | *CF=50%* | *CF=20%* | *CF=10%* |
| 1 | SW1 | 1131.5 | 445.47 | 355.9 | 2463.34 |
| 2 | SW2 | 7059.4 | 316.98 | 865.67 | 11503.47 |
| 3 | SW3 | 822.5 | 0 | 0 | 3105.84 |
| 4 | SW4 | 1364.5 | 151.74 | 265.32 | 3044.38 |
| 5 | SW5 | 3478.9 | 454.37 | 1510.13 | 8842 |
| 6 | SW6 | 520.3 | 0 | 0 | 1913.31 |
| 7 | SW7 | 3078.8 | 480.68 | 885 | 3723.74 |
| 8 | SW8 | 1648.6 | 804.85 | 358.57 | 5549.5 |
| 9 | SW9 | 2074.9 | 0 | 0 | 4684.41 |
| 10 | SW10 | 4036.2 | 830.14 | 363.41 | 8274.37 |
| 11 | SW11 | 4292.5 | 0 | 562.03 | 7925.09 |

represents the maximum possible cooling effect that can be achieved within the present urban fabric. It is also important to note that none of the proposed interventions are located in the core urban center of Varanasi city because of its highly dense morphology. As a result, the cooling effect varies across different urban localities, corresponding to its respective sub-watersheds.

## 7.5 IMPACT OF UBGI ON COOLING ENERGY DEMAND

The green infrastructure delineated above represents the change in green cover density of an urban area and corresponds to the cooling effect in its influencing zone. However, different green covers have a different scale of impact on the cooling demand. To understand the corresponding effect on building energy demand, two case scenarios are considered as follows:

1. **Scenario 1**: A series of green elements with varying degrees of cooling factors (10%–50%) depicting a realistic scenario, adapted considering the available area and urban form.
2. **Scenario 2**: A completely dedicated green corridor with highly dense and shading trees, resulting in a maximum cooling factor (50%), depicting an optimistic scenario, adapted to prioritize UHI mitigation.

The optimistic scenario depicts the complete adoption of the UBGI approach into the urban spatial pattern, retrofitting and developing green infrastructure in existing and newly built forms, respectively. The calculated cooling effect for each watershed, according to Eq. (7.1), is given in Table 7.4.

The above analysis presents a methodological approach for mitigating the increasing UHI magnitude and its potential impact on building cooling energy demand. Table 7.4 shows the degree of change in cooling demand across different sub-watersheds of the urban area. Although this change varies largely across different categories, the delineation of UBGI showcases a visible decrease in the cooling demand of buildings. This decrease ranges from 2.1% to 4.8% for a realistic case scenario and from 8.3% to 20.3% for an optimistic case scenario. The results can further be understood through the graph shown in Figure 7.5.

The optimistic scenario naturally showcases a much higher degree of reduction in cooling demand than the realistic scenario. Although the exact reduction in cooling demand would depend on many other factors such as land-use pattern, urban morphology, and climatic parameters, the present study validates the importance of UBGI as a tool for mitigating the increasing cooling demand in buildings due to rising ambient temperatures. This also shows that developing green land covers within a very small surface area relative to the complete watershed, but integrating them with specific hydrologically sensitive areas could significantly mitigate the challenges of increased surface temperature. This cooling effect of UBGI can further be enhanced when combined with other passive green building strategies, such as green roofs, increased albedo, and shading devices, to create a much more temperature-resilient building framework. Hence, future studies

*Table 7.4* Impact of UBGI on cooling demand for different sub-watersheds

| *S. No.* | *Sub-watershed* | *Impact on cooling demand* | |
|---|---|---|---|
| | | *Realistic scenario (%)* | *Optimistic scenario (%)* |
| 1 | SW1 | 4.77 | 14.43 |
| 2 | SW2 | 2.10 | 8.99 |
| 3 | SW3 | 3.78 | 18.88 |
| 4 | SW4 | 3.18 | 12.68 |
| 5 | SW5 | 4.06 | 15.53 |
| 6 | SW6 | 3.68 | 18.39 |
| 7 | SW7 | 2.57 | 8.27 |
| 8 | SW8 | 6.24 | 20.36 |
| 9 | SW9 | 2.26 | 11.29 |
| 10 | SW10 | 3.26 | 11.73 |
| 11 | SW11 | 2.11 | 9.89 |

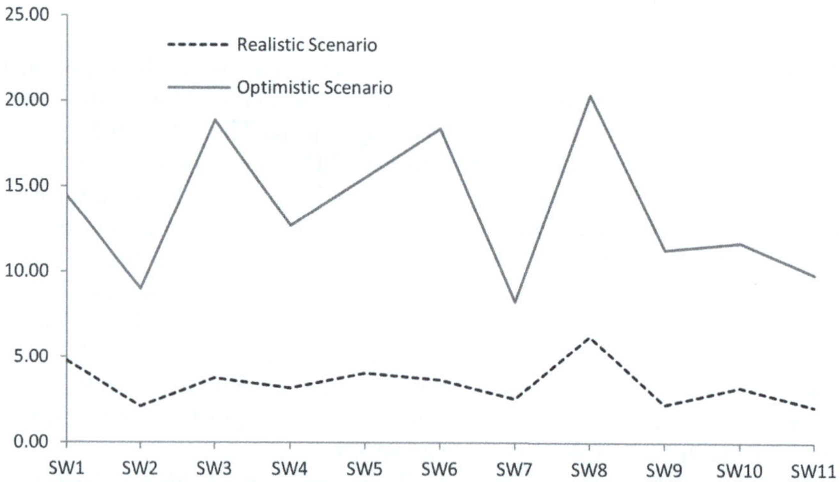


*Figure 7.5* Percentage decrease in cooling demand across urban sub-watersheds.

should incorporate this composite application of multiple green building and nature-based solutions to derive an optimal urban climate scenario.

## 7.6 CONCLUSIONS AND FUTURE SCOPE

Urban areas are considered the engines of growth for society, serving as centers of economy, culture, and knowledge. However, urbanization depletes the natural environment of the region and tends to accumulate heat, leading to the phenomena of the UHI effect, which results in higher ambient temperatures in urban areas. One of the major impacts of UHI on the built environment is an increase in energy demand due to higher cooling load in buildings. A consistent threat of UHI not only consumes large quantities of energy but also decreases the overall efficiency and life cycle of built infrastructure and equipment. Thus, reducing this impact of UHI on building energy demand becomes an important aspect of sustainable urban development. UBGI serves as an important mitigation tool in this context. UBGI maintain the natural climatic qualities of urban areas and provides a cooling effect to the surrounding environment. UBGI elements have shown a significant improvement in building energy performance and overall environmental quality. However, the application of UBGI in dense and developing urban settings is a challenging task. The present study offers a methodological approach to the development of UBGI in urban areas, delineating green areas along the primary drainage channels of the urban watershed as a way of strategically locating the UBGI network. In this way, the blue and green areas sustain each other, presenting an optimum land-suitability scenario for urban areas. The UBGI developed in this way reduce building energy demand by limiting the magnitude of UHI and high temperatures. The study quantifies the cooling factor of different green areas to estimate the potential reduction in cooling demand in the influencing areas of the UBGI network. The results show a 2.1%–4.8% cooling load reduction in the realistic scenario and an 8.3%–20.3% reduction in an optimistic scenario. Thus, the UBGI represents a viable way to mitigate the impacts of UHI on building energy demand and forms an important part of the climate-resilient and sustainable urban future. The methodological framework presented in this study also helps in assessing urban areas using the sub-watershed method to identify the most vulnerable and resilient areas. Future research can explore appropriate UHI mitigation methods for each of these sub-watersheds and assess a composite effect of different adaptations on cooling energy demand. Furthermore, the theoretical framework presented in this study needs to be validated through real-world data, which can be undertaken in future studies.

## REFERENCES

Akbari, H., & Taha, H. (1992). The impact of trees and white surfaces on residential heating and cooling energy use in four Canadian cities. *Energy*, *17*(2), 141–149. https://doi.org/10.1016/0360-5442(92)90063-6

Akbari, H., Kurn, D. M., Bretz, S. E., & Hanford, J. W. (1997). Peak power and cooling energy savings of shade trees. *Energy and Buildings*, *25*(2), 139–148. https://doi.org/10.1016/S0378-7788(96)01003-1

Central Electricity Authority (2020). *Report on Nineteenth Electric Power Survey of India (Volume-III) Part - II (Mega Cities)*. https://cea.nic.in/old/reports/others/planning/pslf/19th_EPS_Mega%20City_Part%20II.pdf

Hirano, Y., & Fujita, T. (2012). Evaluation of the impact of the Urban Heat Island on residential and commercial energy consumption in Tokyo. *Energy*, *37*(1), 371–383. https://doi.org/10.1016/j.energy.2011.11.018

Hosseini, M., Javanroodi, K., & Nik, V. M. (2022). High-resolution impact assessment of climate change on building energy performance considering extreme weather events and microclimate: Investigating variations in indoor thermal comfort and degree-days. *Sustainable Cities and Society*, *78*. https://doi.org/10.1016/j.scs.2021.103634

IPCC. (2023). Summary for policymakers. In Pörtner, H.-O., & Belling, D. (eds.), *Climate Change 2022: Impacts, Adaptation and Vulnerability* (pp. 3–34). Cambridge University Press. https://doi.org/10.1017/9781009325844.001

Jia, P., Zhuang, D., & Wang, Y. (2017). Impacts of temperature and precipitation on the spatiotemporal distribution of water resources in Chinese mega cities: The case of Beijing. *Journal of Water and Climate Change*, *8*(4), 593–612. https://doi.org/10.2166/wcc.2017.038

Li, Y., Zhu, L., Zhao, X., Li, S., & Yan, Y. (2013). Urbanization impact on temperature change in China with emphasis on land cover change and human activity. *Journal of Climate*, *26*(22), 8765–8780. https://doi.org/10.1175/JCLI-D-12-00698.1

Ma, Y. X., & Yu, C. (2020). Impact of meteorological factors on high-rise office building energy consumption in Hong Kong: From a spatiotemporal perspective. *Energy and Buildings*, *228*. https://doi.org/10.1016/j.enbuild.2020.110468

Manoli, G., Fatichi, S., Schläpfer, M., Yu, K., Crowther, T. W., Meili, N., Burlando, P., Katul, G. G., & Bou-Zeid, E. (2019). Magnitude of urban heat islands largely explained by climate and population. *Nature*, *573*(7772), 55–60. https://doi.org/10.1038/s41586-019-1512-9

Mohanty, A., & Wadhawan, S. (2021). *Mapping India's Climate Vulnerability – A District Level Assessment*. Council on Energy, Environment and Water. https://www.ceew.in/sites/default/files/ceew-study-on-climate-change-vulnerability-index-and-district-level-risk-assessment.pdf

NASA JPL. (2013). NASA Shuttle Radar Topography Mission Global 1 arc second. NASA EOSDIS Land Processes DAAC. Accessed March 27, 2023. https://doi.org/10.5067/MEaSUREs/SRTM/SRTMGL1.003

Oke, T. R. (1982). The energetic basis of the urban heat island. *Quarterly Journal of the Royal Meteorological Society*, *108*(455), 1–24. https://doi.org/10.1002/qj.49710845502

Pandit, R., & Laband, D. N. (2010). Energy savings from tree shade. *Ecological Economics*, *69*(6), 1324–1329. https://doi.org/10.1016/j.ecolecon.2010.01.009

Roxon, J., Ulm, F. J., & Pellenq, R. J. M. (2020). Urban heat island impact on state residential energy cost and CO2 emissions in the United States. *Urban Climate*, *31*. https://doi.org/10.1016/j.uclim.2019.100546

Sawka, M., Millward, A. A., Mckay, J., & Sarkovich, M. (2013). Growing summer energy conservation through residential tree planting. *Landscape and Urban Planning*, *113*, 1–9. https://doi.org/10.1016/j.landurbplan.2013.01.006

Shepherd, J. M. (2013). Impacts of urbanization on precipitation and storms: Physical insights and vulnerabilities. In Pielke, R. A. (ed.), *Climate Vulnerability* (pp. 109–125). Elsevier. https://doi.org/10.1016/B978-0-12-384703-4.00503-7

Singh, R. P. B. (2018, January 5). Urbanisation in Varanasi and interfacing historic urban landscapes. National Seminar on Urbanization in Indian History.

Wesley, E. J., & Brunsell, N. A. (2019). Greenspace pattern and the surface urban heat island: A biophysically-based approach to investigating the effects of urban landscape configuration. *Remote Sensing*, *11*(19), 2322. https://doi.org/10.3390/rs11192322

Yu, C., & Hien, W. N. (2006). Thermal benefits of city parks. *Energy and Buildings*, *38*(2), 105–120. https://doi.org/10.1016/j.enbuild.2005.04.003

Zhang, Y., Asce, S. M., Ayyub, B. M., & Asce, P. E. M. (2020). Electricity system assessment and adaptation to rising temperatures in a changing climate using Washington metro area as a case study. *Journal of Infrastructure Systems*, *26*(2). https://doi.org/10.1061/(ASCE)IS.1943-555X.0000550

Zhu, S., Causone, F., Gao, N., Ye, Y., Jin, X., Zhou, X., & Shi, X. (2023). Numerical simulation to assess the impact of urban green infrastructure on building energy use: A review. *Building and Environment*, *228*, 109832. https://doi.org/10.1016/j.buildenv.2022.109832

Chapter 8

# Role of urban green spaces in reducing urban heat island effect

## A case of central India's four metropolitan cities

*Shubhangi Choudhary and Rajat Soni*

### 8.1 INTRODUCTION

The world is rapidly urbanizing, and according to projections, 3.47 billion people—or 62.5% of the urban population—will reside in metropolises by 2035. In the following 15 years, the population of cities will reach approximately 1 billion (UN-Habitat, 2020). In India, almost 25.72% of the population resided in urban regions in 1991, which increased to 27.86% in 2001 and around 31.16% in the following decade (Chaudhry et al., 2015). By 2026, this urban component is expected to rise to 40% (TNCPO, 2014). This significant urban expansion has led to the degradation of natural resources due to the increasing urban population, combined with the sprawl in urban and peri-urban areas, the expansion of transportation networks, urban infrastructure, and industrial activities (Yigitcanlar and Dizdaroglu, 2015). The current situation demonstrates that the environment is frequently sacrificed due to urban growth (Bhupatiraju and Banerjee, 2019). Therefore, the idea of urban green spaces (UGSs) is essential for achieving sustainability, as it provides numerous benefits, including enhancing aesthetics, reducing pollution, improving physical and mental health, reducing the urban heat island (UHI) effect, managing storm water, reducing urban floods, and protecting biodiversity (Turaga et al., 2020). Research on how developing nations manage and incorporate green spaces in urban planning has a vast scope (Byomkesh, Nakagoshi and Dewan, 2012). In developing nations like India, which is on the verge of becoming the most populous country in the world (Gaigbe-Togbe et al., 2022), green spaces in cities are critical in determining the livability index of regions (Scaria et al., 2020).

This chapter focuses on the existing guidelines for UGSs and the parameters for reviewing green space standards. Based on the analysis, suggestions are offered for enhancing green spaces in the urban planning processes in four cities: Bhopal, Indore, Gwalior, and Jabalpur. These four metropolitan cities in central India have significant similarities in terms of demography, location, geography, and climatic conditions. This study examines the current state and challenges for green spaces in the Indian context and the


DOI: 10.1201/9781003521341-8

minimum amount of green cover required for healthy living. It also discusses the parameters that can be considered for a rational division of green spaces and building cover, spatial patterns, and temperature variations in the selected region. Finally, a conclusion is drawn through a comparative analysis of these cities.

## 8.2 UGSs IN URBAN PLANNING

UGS, as defined by the World Health Organization (WHO), refers to all land within urban areas that is covered by vegetation of any kind (WHO, 2017). Bhat and Dash (2019) defined UGS as any area that is accessible to the public and open to the sky and has a permeable surface (Bhat and Dash, 2019). India does not have an accurate definition of UGS, unlike most modern nations (Sangwan et al., 2022). Green spaces are still categorized under recreational or open spaces (TNCPO, 2014). In this chapter, UGS is considered as areas covered with man-made or natural vegetation within city limits, with sparse and dense vegetation being considered major parameters for measuring UGS.

UGS has a significant effect on the urban climate, which is often distinguished from the rural climate by higher temperatures, weaker winds, and change in solar radiation depending on the level of pollution. Urban regions frequently have air pollution levels 5–25 times greater than neighboring rural areas (Heidt and Neef, 2008). Today, almost every city in the world is typically 1°C–4°C warmer than the nearby rural regions, demonstrating as the "heat island" effect (Mohajerani, Bakaric and Jeffrey-bailey, 2017). Increasing greenhouse gases in metropolitan cities have exacerbated the UHI effect (Aram et al., 2019). The evapotranspiration of trees helps reduce temperature in cities, making trees and vegetation crucial for altering climatic conditions and reducing UHI by lowering air temperature (Qiu et al., 2017). For example, a small park with both trees and bushes per year can capture up to 68 metric tons of dust per hectare per year. A study in Ankara, Turkey, recorded temperature variations in several areas in the summer season. The results of the study showed that areas with fewer green spaces had the highest average temperature, while areas with more green spaces had the lowest average temperature (Çay, 2016). This underscores the need to stabilize the UHI effect, which can be achieved through green space allocation in cities. According to estimates, 1 hectare of forest may sequester up to 6 tons of carbon dioxide per year, or around 6 kg per tree (Secon, 2007). Given that human respiratory carbon emissions per person range from 192 to 328 kg per year, urban areas need a minimum of eight trees per inhabitant to sequester respiratory $CO_2$. Urban planners must retain a minimum of 33% of the area as green space to see overall improvements in human well-being and community health (Ramachandra et al., 2017a).

Various standards have been implemented to establish uniformity in UGS planning (Byrne and Sipe, 2010). The standards strategy has evolved over the years and addresses how much green space is appropriate per individual (Maryanti et al., 2016). The United Kingdom, the first nation to adopt the standards method since the late 1800s, pioneered this concept (Theobald, 1984). The demand for UGS is directly proportional to the population density of an area (Jafrin and Beza, 2018). In a study by Maryanti et al. (2016), different target sizes for green space per 1,000 inhabitants were used for different cities, yielding varied results in terms of green area required per person. For example, Greater London has 4 ha—40 $m^2$/person—whereas Bristol has 1 ha –10 $m^2$/person. The WHO recommends that every city should have at least 9.5 $m^2$ of green space per person. Research shows that the world's most livable cities have the greenest space for their population (Huang and Yang, 2017; Brilhante and Klaas, 2018). In 2022, The Economist Intelligence Unit voted Vienna as the most livable city (Economist Intelligence Unit, 2022). Vienna has 120 $m^2$ of UGS per person, whereas Singapore, the third most densely populated city, has 66 $m^2$ of UGS per person. Singapore should serve as a spatial planning model because of its success in preserving UGS in a congested and small metropolitan region (Siemens Economist Intelligence Unit, 2011).

The standards approach, developed by urban planners, helps understand the type and amount of green space needed in a city (Maryanti et al., 2016). Five different standards approaches have been widely employed globally: population-to-area ratio, area percentage, catchment area, facility requirements, and local standards (Veal, 2013). The population-to-area ratio technique is the most accepted approach by urban planners in most nations. Standards are frequently created by national or international organizations and accepted by professionals (Maryanti et al., 2016). Figure 8.1 shows the comparison of a few standards in both the Indian and international contexts (TNCPO, 2014; Town and Country Planning, 2014, 2015).

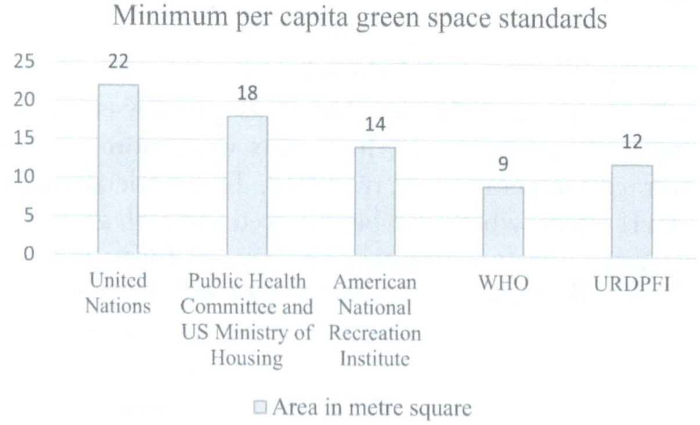


*Figure 8.1* Comparison of minimum per capita green space standards.

Cities pay little attention to managing and conserving trees in contrast to air and water. The consistent reduction in urban tree cover makes this clear. A recent study in Bangalore revealed an 88% decline in urban vegetation from 1973 to 2017. Another study found that there was only one tree available for every seven persons, which is insufficient to sequester respiratory carbon (Tv and Aithal, 2016). A 2020 land-use survey observed that 93% of the landscape was covered with paved surfaces, reducing the green or open spaces. This lack of green space deprives city dwellers of clean air, water, and environment, increasing greenhouse gas emissions and water scarcity and making the city uninhabitable (Ramachandra et al., 2017b). Hence, the minimum standards recommended by the WHO must be applied in urban planning.

### 8.2.1 Current state of UGS in India

The WHO recommends planning regions with the aim that every single inhabitant has access to a nearby open space within a 15-minute walk. Indian UGS typology is given in Table 8.1 (TNCPO, 2014; Town and Country Planning, 2014). However, many cities fall short of the minimum WHO standard due to the high population densities in metropolitan areas. Table 8.2 presents population data for the 13 largest cities as of the 2018 Estimated Census, sourced from the India Environment and Population Portal. The table also includes details on each city's geographic area, green cover, and the amount of green space per person (Ramaiah and Avtar, 2019). Among these cities, Kolkata, Bangalore, Chennai, Surat, and Jaipur do not meet the standards.

The Indian government has launched numerous programs to manage forests and protect ecosystems, including the "National Mission of Green India," which is designed to address environmental and sustainability-related issues (Rattani, 2018). In urban areas, the Urban Regional Development Planning Formulation and Implementation (URDPFI) guidelines (Town and Country Planning, 2014) and the Urban Greening Guidelines (TNCPO, 2014) acknowledge the advantages of UGS for urban sustainability and offer important recommendations on inclusive planning in cities.

The URDPFI guidelines prescribe a land-use share of 12%–14% for recreational uses (including parks, playgrounds, gardens, open spaces, and other natural features) in small and medium-sized cities and 14%–16% in large cities. Recent programs introduced by the Indian government such as the Atal Mission for Rejuvenation and Urban Transformation (AMRUT) have highlighted vegetated areas in towns and cities to enhance green/open spaces in cities. The Smart City Mission, however, aims to protect and improve open spaces including parks, playgrounds, and recreational areas (Sen and Mukherjee, no date; Krishnaveni, 2020).

*Table 8.1* Indian UGS typology

| *Documents* | *Context* | *Category* | *Classification* | *Hierarchy* |
|---|---|---|---|---|
| URDPFI (1996) | Open spaces | Recreational facilities | Parks and open spaces<br>Sports centers and playgrounds<br>Botanical and zoological parks<br>Water bodies/other natural features<br>Places of tourist interest | Housing cluster<br>Sector<br>Community<br>District<br>Sub-city center |
| URDPFI (2014) | Open spaces | Recreational space<br>Organized green<br>Other common spaces (such as vacant lands/open spaces including flood plains and forest cover, in plain areas) | Recreational | Recreational<br>P-1 playgrounds/stadium/sports complex<br>P-2 parks and gardens—public open spaces<br>P-3 multi-open space (Maidan)<br>Organized green<br>Housing cluster<br>Neighborhood<br>Community<br>District/zone<br>Sub-city center |
| Urban Green Guideline (2014) | Green spaces | Urban green spaces | Reserved forest<br>Protected forest<br>District park<br>Neighborhood park | Tot-lots<br>Playgrounds<br>Neighborhood park<br>Community park green belts (buffer)<br>Green strip |

*Table 8.2* Major cities of India with per capita green space

| *City* | *Populations in millions* | *Geographical area ($km^2$)* | *Green cover percentage ($km^2$; 2017)* | *Per capita green space ($m^2$; 2018)* |
|---|---|---|---|---|
| Delhi | 28.50 | 1484.00 | 20.00 (296.80) | 10.41 |
| Mumbai | 23.50 | 603.00 | 36.48 (220.00) | 9.36 |
| Kolkata | 15.20 | 1380.00 | 7.30 (100.74) | 6.61 |
| Bangalore | 13.90 | 2196.00 | 2.09 (46.03) | 3.31 |
| Hyderabad | 11.57 | 650.00 | 1.66 (10.79) | 0.93 |
| Chennai | 9.88 | 1189.00 | 15.00 (178.35) | 18.05 |
| Ahmedabad | 8.41 | 464.00 | 17.00 (78.88) | 9.38 |
| Surat | 6.55 | 326.50 | 11.84 (38.66) | 5.90 |
| Gandhinagar | 6.33 | 649.00 | 54.00 (188.46) | 29.77 |
| Jaipur | 3.71 | 467.00 | 5.43 (24.75) | 6.67 |
| Nagpur | 2.94 | 285.90 | 18.00 (51.42) | 17.49 |
| Mysore | 1.70 | 128.40 | 20.19 (25.92) | 15.25 |
| Chandigarh | 1.05 | 114.00 | 35.00 (39.90) | 38.00 |

*Table 8.3* Categories of UGS in Indian cities as per URDPFI guidelines

| *S. no.* | *Category* | *Population served per unit* | *Area requirement (ha)* |
|---|---|---|---|
| 1 | Housing area park | 5,000 | 0.50 |
| 2 | Neighborhood park | 15,000 | 1.00 |
| 3 | Community park | 1,00,000 | 5.00 |
| 4 | District park | 5,00,000 | 25.00 |
| 5 | Sub-city park | 10,00,000 | 100.00 |

### 8.2.2 UGS categorization

UGSs are classified as housing areas, neighborhoods, communities, districts, and sub-city parks based on their spatial hierarchy by the URDPFI, which recognizes them as a component of organized green spaces within the social infrastructure of the city (Table 8.3) (Town and Country Planning, 2014).

The population ratio standard is commonly accepted due to its simplicity, but it is not regarded as successful because it does not take distribution factors or user accessibility into account. Therefore, accessibility in terms of proximity radius is also reviewed (Lahoti, Lahoti and Saito, 2019). The categorization of green areas in selected four regions according to the latest development plans of cities is presented in Table 8.4. These spaces are categorized under recreational space, without any specific provision for green spaces. The classification of categorized spaces is only done in the case of Bhopal. Hence, a better understanding at the local level is required (Directorate of Town and Country Planning, 2005, 2021a–f; Indore Development Authority, 2021).

*Table 8.4* Categorization of green spaces according to the typology

| *City* | *Categorization of space* | *Classification of categorized space* | *Hierarchy of categorized space* |
|---|---|---|---|
| Bhopal | Recreational space | G-1: Parks and gardens<br>G-2: Playfield and stadium<br>G-3: Zoological/botanical garden<br>G-4: Lakefront development<br>G-5: City forest/afforestation | Community parks<br>City parks<br>Regional parks |
| Indore | Recreational space | Regional park<br>City/other parks<br>Stadium<br>Green buffer | No further classification |
| Jabalpur | Recreational space | City park<br>Playground/stadium<br>afforestation/city forest | No further classification |
| Gwalior | Recreational Space and special purpose | Parks/afforestation/playground/ stadium/exhibition ground<br>and<br>Reserved forest/cantonment area/hillock/zoo/fort | No further classification |

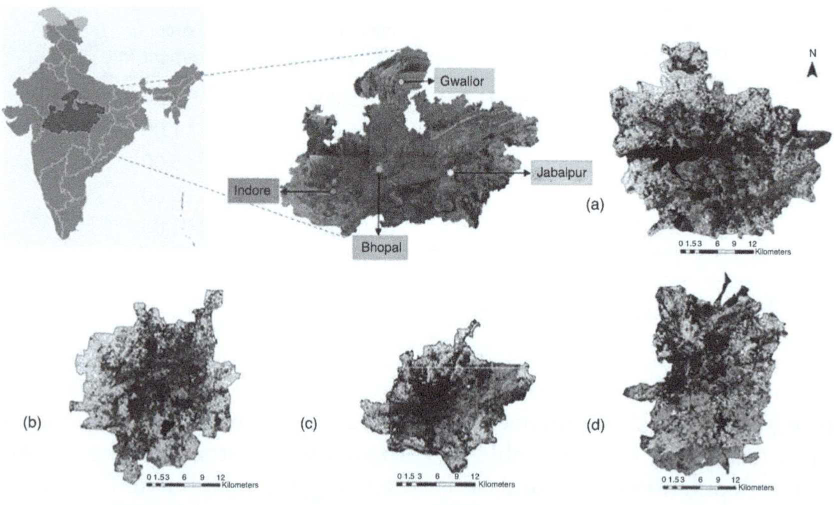


*Figure 8.2* Location of four cities in Madhya Pradesh (a) Gwalior, (b) Indore, (c) Bhopal, and (d) Jabalpur as seen in Landsat 8 images.

## 8.3 APPROACH AND METHODOLOGY

### 8.3.1 Study area

The study focuses on Central Indian cities, namely, Bhopal, Indore, Jabalpur, and Gwalior (Figure 8.2). These cities collectively account for 33% of the urban population in the state and are all metropolitan areas (Wilcox, 2015).

*Table 8.5* General geographic characteristics of study areas

| *City name* | *Bhopal* | | *Indore* | *Jabalpur* | *Gwalior* |
|---|---|---|---|---|---|
| *Year* | *2021* | *2031* | *2021* | *2021* | *2021* |
| Demography | 26 lakhs | 36 lakhs | 28 lakhs | 16 lakhs | 16 lakhs |
| Geographical location | 23.2599°N, 77.4126°E | 23.2599°N, 77.4126°E | 22.7196°N, 75.8577°E | 23.1815°N, 79.9864°E | 26.2124N, 78.1772°E |
| Total geographical area (km$^2$) | 813.92 | 1016.09 | 505.25 | 245.17 | 426.52 |
| Total proposed recreational area (km$^2$) | 22.12 | 101.29 | 48.17 | 7.02 | 11.40 |

They act as centers for administrative (Bhopal), academic (Indore), economic and industrial (Jabalpur), and cultural (Gwalior) activities in the central part of India. A summary of these cities are given in Table 8.5.

Bhopal, also known as "City of Lakes," is the state capital of Madhya Pradesh and one of the 32 mission cities under AMRUT. However, routine silting and irregular encroachment into the city's major lakes violate the National Green Tribunal's (NGT) regulations. The NGT established a 300-m buffer zone for the upper lake in 2014, which was reduced to 50m in 2018.

The buffer zone for the Kaliasot reservoir, originally set at 200m, has been reduced to 33m (Saba and Yadav, 2021). The TNCP Authority has delineated a recreational zone for Bhopal, which involves five major zones (Table 8.5). The purpose of this zone is to increase green cover in the Bhopal Planning Area. Under BDP 2031, the development of a 10% recreational zone is mandated in each mixed-use zone (Directorate of Town and Country Planning, 2021a).

Indore is considered the commercial capital of the state. Its economy is rapidly growing, and it simultaneously encourages urbanization. It also functions as a center for commerce, medical care, education, and industrial development(Rajput, 2020). Although the Indore Development Authority has proposed various schemes for open spaces, implementation has been hindered due to lack of public participation (Indore Development Authority, 2021).

Jabalpur has the third largest population in Madhya Pradesh as per the 2011 census. The city sprawls across the plains of the Narmada River's tributaries and is situated on its banks (Garg and Sharma, 2018). Once known for its abundance of lakes and aquatic bodies, most of these water sources are now extinct, and the remaining ones are severely polluted. Urban ecology has been impacted by urban expansion over the past 25years (Tiwari, 2017).

Gwalior urban agglomeration comprises the Gwalior Municipal Corporation and Gwalior Military Cantonment. Gwalior is a tourism hub and is surrounded by many historical monuments. The Gwalior Development Authority has delineated a 15-m green buffer along Murar and Swarna Rekha rivers. Additionally, the special purpose zones in the city include areas classified as reserved forests (Directorate of Town and Country Planning, 2021d).

As per the sector-wise allocation of funds from AMRUT Mission, Bhopal, Indore, Jabalpur, and Gwalior have been allocated 25, 115.21, 75, and 24 crore rupees, respectively, for green spaces, accounts for only 1.3%, 2.7%, 2.4%, and 2% of the total funds allocated to each city, respectively (Wilcox, 2015). The total proposed recreational space in these four cities is mentioned in Table 8.5.

### 8.3.2 Methodology

Research employed both secondary and satellite data. The United States Geological Survey Earth Explorer was used to obtain multispectral satellite data for December 2013 and January 2023. Landsat 8 (OLI/TIRS) imagery, with a resolution of 30 m, was used to analyze changes in green areas. The boundaries of the maps were prepared by geo-referencing the 2021 Development Plans of the cities on the Open Street Map in QGIS. City Development Plan reports from Tirunelveli City Planning Commission (TNCP) were utilized to map land-use/land cover data (Directorate of Town and Country Planning, 2021a,d,f; Indore Development Authority, 2021). Arc GIS software was used for satellite image processing, while QGIS was used to compute, classify, and analyze the data. The average temperature data for the cities were sourced from the Indian Meteorological Department (https://mausam.imd.gov.in/).

### 8.3.3 UGS extraction

Due to data availability, Landsat images were employed in large-scale mapping and monitoring of UGSs (Shahtahmassebi et al., 2021; Wikantiyoso, Sulaksono and Suhartono, 2021). This research used the normalized differentiated vegetation index (NDVI) method to delineate the vegetated areas and analyze their distribution.

The NDVI technique has been extensively used for UGS imaging. The NDVI was calculated using the following equation:

$$\text{NDVI} = \text{NIR} - R/\text{NIR} + R \tag{8.1}$$

Here, NIR is the reflectance in the near-infrared band and $R$ is the reflectance in the red band (Akbar et al., 2019).

In Landsat 8,

$$\text{NDVI} = B5 - B4/B5 + B4 \tag{8.2}$$

where B5 is the reflectance in band 5 and B4 is the reflectance in band 4. The NDVI value ranges from +1 to −1, with values close to +1 typically

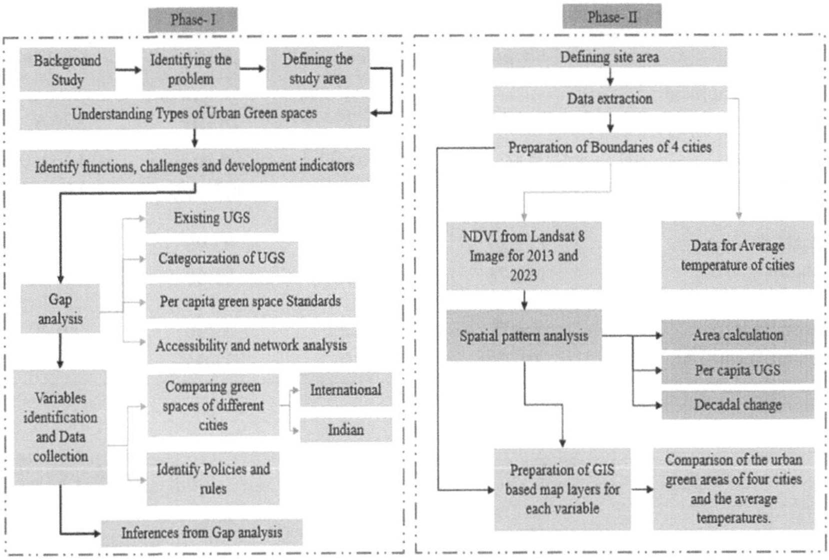


*Figure 8.3* Methodology.

indicating dense vegetation, while lower and negative values indicate built-up areas, waterbodies, and barren land. UGSs are classified into six categories, namely, water body, built-up, fallow land, cropland, sparse vegetation, and dense vegetation (Akbar et al., 2019).

### 8.3.4 Methodology

The detailed methodology is illustrated in Figure 8.3.

## 8.4 FINDINGS AND ANALYSIS

### 8.4.1 Spatial pattern of UGS

Table 8.6 illustrates the spatial coverage of UGSs, showing a decadal change in green cover across the cities. A negative trend is observed in all four cases, with vegetated surface reductions of 6.83, 1.97, 14, and 6.98, respectively. In most cities, agricultural land is being converted for other uses, increasing unorganized sprawl and decreasing vegetation.

A report by the Global Earth Society for Environment, Energy, and Development revealed that the forest area of Bhopal decreased from 35% to 9% between 2009 and 2019, with around 4 lakh trees being cut during this period. This deforestation has also led to climatic changes, with the city

*Table 8.6* The current status and decadal change of UGS in the study area

| *City name* | | *Bhopal* | *Indore* | *Jabalpur* | *Gwalior* |
|---|---|---|---|---|---|
| Total geographical area (km²) | | 813.92 | 505.25 | 245.17 | 426.52 |
| NDVI value (minimum and maximum) | 2013 | −0.06 to 0.52 | −0.08 to 0.53 | −0.04 to 0.51 | −0.12 to 0.52 |
| | 2023 | −0.21 to 0.60 | −0.11 to 0.52 | −0.007 to 0.49 | −0.18 to 0.53 |
| Total UGS in planning area (km²) | 2013 | 212.29 | 157.02 | 85.95 | 162.96 |
| | 2023 | 156.76 | 147.22 | 51.65 | 133.73 |
| Percentage of green space (%) | 2013 | 26.08 | 31.07 | 35.06 | 38.20 |
| | 2023 | 19.25 | 29.10 | 21.06 | 31.22 |

*Figure 8.4* Place: Habibganj Railway Station, right side, area: 10 acre.

*Figure 8.5* Place: Jagran Lakecity University, Area: 20 acre.

experiencing its hottest temperature in 20 years on June 9, 2019, at 46° C. The green cover reduction can be seen in two major examples mentioned in Figure 8.4 (reduced green cover for infrastructural development in Public–Private Partnership with the Bansal Group) and Figure 8.5 (reduced green cover due to formation of an institute). The Bhopal Development Authority has proposed various strategies for inclusive planning of green areas in the city (Directorate of Town and Country Planning, 2021b). Densely vegetated

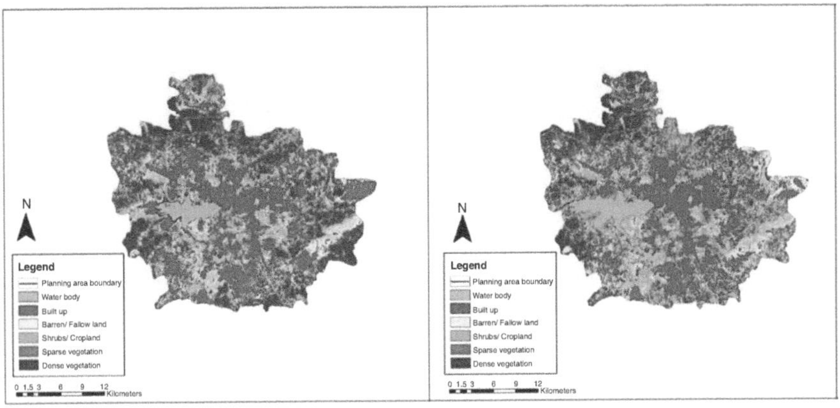


*Figure 8.6* UGS of Bhopal derived from NDVI values on ArcGIS.

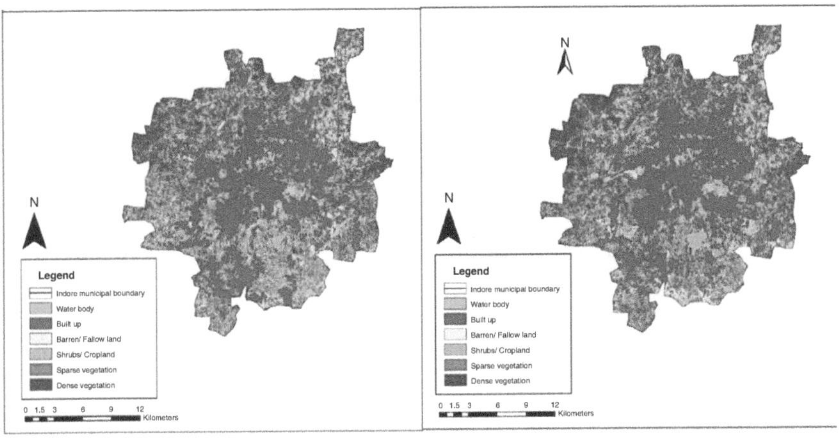


*Figure 8.7* UGS of Indore derived from NDVI values on ArcGIS.

areas in the northern and eastern regions of the city are now being used as cropland, based on Google Earth Images (Figure 8.6).

In Indore, the city core is becoming denser, with very minimal green areas remaining (Figure 8.7). Further projects, like constructing a 5-km road in the Bhavarkuan region, will involve cutting 900 trees. Similarly, for the Indore Metro proposal, authorities have reported a compensation of 1.91 crores to the forest department for preparing a stretch of forest in Dhar due to the deforestation caused by constructing the metro.

In Jabalpur, there has seen a drastic reduction in UGS, specifically in the northern and north-western areas, where the densely vegetated areas have been converted to croplands. The cantonment area in the city retains most shrubs and remains unchanged, whereas the city core has the least green areas (Figure 8.8).

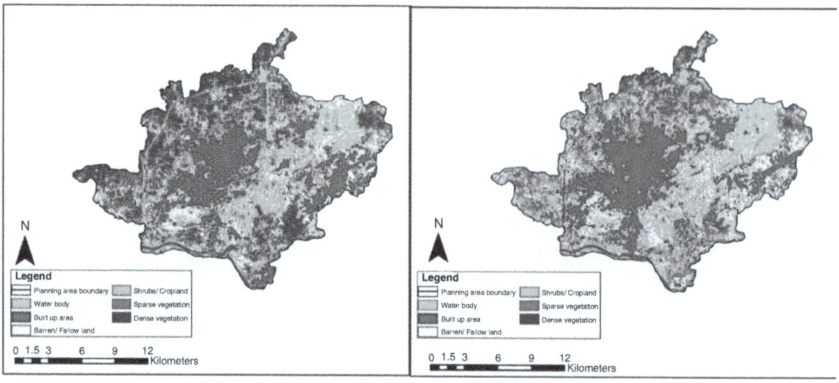


*Figure 8.8* UGS of Jabalpur derived from NDVI values on ArcGIS.

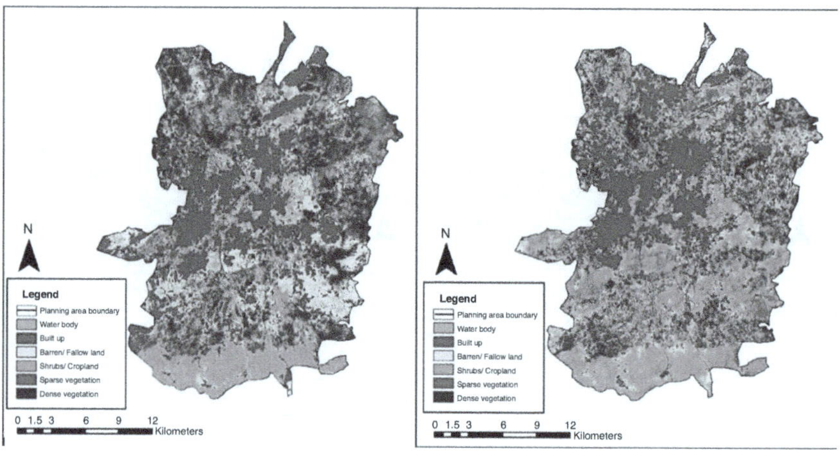


*Figure 8.9* UGS of Gwalior derived from NDVI values on ArcGIS.

In Gwalior, the increase in the urban sprawl to address the housing shortage is the reason for the decrease in UGS coverage. The population in this region has grown significantly from 10 lakhs in 2005 to 16 lakhs in 2021 (Directorate of Town and Country Planning, 2021d). Consequently, over the past two decades, housing developments particularly in the northern and north-eastern regions of the city have increased (Figure 8.9).

Comparing the four cities reveals a considerable gap in the UGS provisions. Bhopal has the maximum green areas, whereas Jabalpur has the least (Table 8.6). This difference is partly due to the municipal boundaries of. Bhopal has an area of 813.92 $km^2$, whereas Jabalpur only has 245.17 $km^2$.

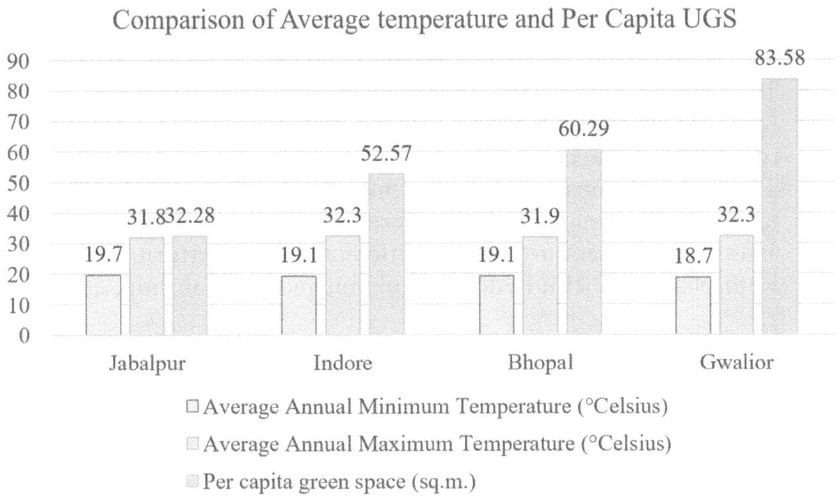


*Figure 8.10* Comparing average temperature and per capita UGS of the study area.

### 8.4.2 Average temperature comparison of cities

The average temperature of cities is compared with green space per capita, showing that cities with more green areas tend to have lower temperatures across seasons (Figure 8.10) (Indian Meteorological Department, 2023a–d). Jabalpur, with the least green space, has the highest average minimum temperature (19.7°C), while Gwalior, with the most green space, has the lowest average minimum temperature (18.7°C). Indore and Gwalior have high average maximum temperatures (32.3°C), which can be due to the dense built-up areas in the two cities. Gwalior's rocky terrain also contributes to high temperatures. In contrast, Bhopal has distributed green areas, due to which the temperature is lower (31.9°C). Jabalpur's huge cantonment area contributes to lower temperature (31.8°C) than Indore and Gwalior.

## 8.5 LIMITATIONS

This study provides an analytical approach to UGS distribution and spatial planning but has a few limitations:

- A ward analysis for distribution could be done for a micro-level analysis. However, time constraints and limited data availability prevented this. Further research could focus on calculating the proportional vegetation index of cities on a ward level (Anguluri and Narayanan, 2017).
- The accessibility of the green areas was not considered in this study. Network analysis, by mapping out parks, afforested land, green buffer, and reserved forests, could provide insights into accessibility.

Accessibility does not guarantee availability for use. More research is needed on their usage and accessibility of UGSs to better understand how communities interact (Kaur et al., 2022).

- The 30 m resolution of Landsat images used here provides only a rough idea of UGSs. Satellite images with higher resolution could capture UGS more accurately.
- This study compared average temperatures. Future studies could focus on UHI mapping by analyzing land surface temperatures to assess the temperature rise in different wards (Setturu, Rajan and Ramachandra, 2013; Reddy, Manikiam and Jeevalakshmi, 2017).

## 8.6 CONCLUSION

This study used a geospatial approach to analyze UGS distribution in Central Indian cities. The findings show that UGS decreased in these cities between 2013 and 2023, consistent with earlier research. Various studies have shown that cities with effective urban planning approaches have seen increased UGS (Littke, 2015). To maintain urban ecology, these cities must implement green space management policies. Systematically collecting quantitative and qualitative data on the vegetation and greenery in urban regions could help achieve sustainable urban development.

Despite following WHO and URDPFI recommendations, Bhopal still faces challenges with the usability and accessibility of green spaces (Saba and Yadav, 2021). In Indore, dense settlements in the city center are expanding toward municipal boundaries, with rapid residential and commercial growth encroaching on productive agricultural and fallow land (Gupta, Swain and Kumari, 2014; Rajput, 2020). Jabalpur also faces a decrease in agricultural land (Garg and Sharma, 2018). The Smart City Mission in Gwalior aims to increase open spaces in the city core from 1% to 13% but connectivity remains unaddressed (Turaga et al., 2020).

An interdependence between air temperature and green cover has been observed in the subtropical cities of Central India, which is severely impacting their sustainable growth. A reduction in the number of trees is directly proportional to increasingtemperatures. The cooling effect of vegetation depends on various factors like evapotranspiration and heat absorption by built-up and fallow land (Qiu et al., 2017). However, in a city like Bhopal, water bodies significantly increase the cooling effect. Urban planners must therefore differentiate between cooling techniques based on urban planning and choose the most practical and cost-effective approach.

This study suggests organized sustainable urban development which can minimize the urban sprawl, pay attention to existing UGS, bring policies of retrofitting in core city areas, reduce deforestation, and increase public participation in preserving trees. Hence, with suitable UGS planning, the temperature in the environment can be lowered, making vegetation an economic and efficient way to lessen the UHI effect in urban areas.

## REFERENCES


Akbar, T. A. et al. (2019) 'Investigative spatial distribution and modelling of existing and future urban land changes and its impact on urbanization and economy', *Remote Sensing*, 11(2). doi: 10.3390/rs11020105.

Anguluri, R. and Narayanan, P. (2017) 'Role of green space in urban planning: Outlook towards smart cities', *Urban Forestry and Urban Greening*, 25, pp. 58–65. doi: 10.1016/j.ufug.2017.04.007.

Aram, F. et al. (2019) 'Urban green space cooling effect in cities', *Heliyon*, 5(4), p. e01339. doi: 10.1016/j.heliyon.2019.e01339.

Bhat, A. S. S. and Dash, M. (2019) 'The need to evaluate urban green spaces in the city core in Bengaluru', *International Journal of Recent Technology and Engineering*, 8(2 Special issue 3), pp. 908–913. doi: 10.35940/ijrte.B1171.0782S319.

Bhupatiraju, S. and Banerjee, U. K. (2019) 'Assessment and strategies for urban green spaces: Case of Hyderabad', *Proceedings of the Annual International Conference on Architecture and Civil Engineering*, 11(3), pp. 374–382. doi: 10.5176/2301-394X_ACE19.574.

Brilhante, O. and Klaas, J. (2018) 'Green city concept and a method to measure green city performance over time applied to fifty cities globally: Influence of GDP, population size and energy efficiency', *Sustainability (Switzerland)*, 10(6). doi: 10.3390/su10062031.

Byomkesh, T., Nakagoshi, N. and Dewan, A. M. (2012) 'Urbanization and green space dynamics in Greater Dhaka', *Landscape and Ecological Engineering*, 8, pp. 45–58. doi: 10.1007/s11355-010-0147-7.

Byrne, J. A. and Sipe, N. (2010) 'Green and open space planning for urban consolidation: A review of the literature and best practice', Urban Research Program.

Çay, R. D. F. A. (2016) 'Benefits of urban green spaces for citizens : Ankara Case study', Conference: 4th International Conference "Ecology of Urban Areas 2014", Zrenjanin/Serbia Authors.

Chaudhry, B. P. et al. (2015) 'Valuation of urban environmental amenities in developing countries : A case study from Chandigarh, India', *Global Journal of Science Frontier Research Environment & Earth Science*, 13(2), pp. 1–13.

Directorate of Town and Country Planning, M. P. (2005) 'Jabalpur development plan'.

Directorate of Town and Country Planning, M. P. (2021a) 'Bhopal development plan-2031, (draft) volume – II (planning proposals and recommendations)'.

Directorate of Town and Country Planning, M. P. (2021b) 'Bhopal development plan-2031, (draft) volume – III (Development Promotion Regulations)'.

Directorate of Town and Country Planning, M. P. (2021c) 'Bhopal development plan -2031, (draft) volume – I'.

Directorate of Town and Country Planning, M. P. (2021d) 'Gwalior development plan (draft 2035)', p. 78.

Directorate of Town and Country Planning, M. P. (2021e) 'Indore development plan 2021'.

Directorate of Town and Country Planning, M. P. (2021f) 'Jabalpur development plan'.

Economist Intelligence (2022) 'The global liveability index 2022'.

Gaigbe-Togbe, V. et al. (2022) World population prospects 2022. Available at: https://www.un.org/development/desa/pd/sites/www.un.org.development.desa.pd/files/wpp2022_summary_of_results.pdf.

Garg, V. and Sharma, A. (2018) 'Urban sprawl analysis using GIS applications for Jabalpur city', *International Research Journal of Engineering and Technology*, 5(5), pp. 4171–4176.

Gupta, A., Swain, S. and Kumari, M. (2014) 'Urban growth trend analysis of indore city (2005–2014) through index based models', *15th Esri India User Conference 2014 Urban*, Delhi, pp. 1–7.

Heidt, V. and Neef, M. (2008) 'Benefits of urban green space for improving urban climate', pp. 84–85.

Huang, C. and Yang, J. (2017) 'Green spaces as an indicator of urban health : Evaluating its changes in 28 mega-cities', *Remote Sensing*, 9(12), p. 1266. doi: 10.3390/rs9121266.

Indian Meteorological Department (2023a) 'Climate of smart city: Bhopal', p. 31.

Indian Meteorological Department (2023b) 'Climate of smart city: Indore', p. 55.

Indian Meteorological Department (2023c) 'Climate of smart city: Gwalior', p. 31.

Indian Meteorological Department (2023d) 'Climate of smart city: Jabalpur', p. 31.

Indore Development Authority (2021) 'Indore master development plan 2021'.

Jafrin, M. and Beza, B. B. (2018) 'Developing an open space standard in a densely populated city : A case study of Chittagong City', *Infrastructures*, 3(3), pp. 1–25. doi: 10.3390/infrastructures3030040.

Kaur, N. et al. (2022) 'Geospatial analysis of the distribution of urban green spaces: A study of four Indian cities', *Cities & Health*, 6(3), pp. 443–459. doi: 10.1080/23748834.2021.1941722.

Krishnaveni, K. S. (2020) *Smart Living for Smart Cities: Case Studies*, Advances in 21st Century Human Settlements. Springer Nature, New York.

Lahoti, S., Lahoti, A. and Saito, O. (2019) 'Urban Forestry & Urban Greening Benchmark assessment of recreational public Urban Green space provisions : A case of typical urbanizing Indian City, Nagpur', *Urban Forestry & Urban Greening*, 44(August), p. 126424. doi: 10.1016/j.ufug.2019.126424.

Littke, H. (2015) 'Planning the green walkable city: Conceptualizing values and conflicts for urban green space strategies in Stockholm', *Sustainability (Switzerland)*, 7(8), pp. 11306–11320. doi: 10.3390/su70811306.

Maryanti, M. R. et al. (2016) 'The urban green space provision using the standards approach: Issues and challenges of its implementation in Malaysia', *Sustainable Development and Planning VIII*, 1, pp. 369–379. doi: 10.2495/sdp160311.

Mohajerani, A., Bakaric, J. and Jeffrey-bailey, T. (2017) 'The urban heat island effect, its causes, and mitigation, with reference to the thermal properties of asphalt concrete', *Journal of Environmental Management*, 197, pp. 522–538. doi: 10.1016/j.jenvman.2017.03.095.

Qiu, G. Y. et al. (2017) 'Experimental studies on the effects of green space and evapotranspiration on urban heat island in a subtropical megacity in China', *Habitat International*, 68, pp. 30–42. doi: 10.1016/j.habitatint.2017.07.009.

Rajput, A. S. (2020) 'Peri-urban areas using spatial metrics : Case study of indore', *Nagarlok*, LIII(Part I), pp. 1–15.

Ramachandra, T. V. et al. (2017a) 'Green spaces in Bengaluru: Quantification through geospatial techniques', *Indian Forester*, 143(4), pp. 307–320. Available at: https://www.researchgate.net/publication/315670439_green_spaces_in_bengaluru_quantification_through_geospatial_techniques.

Ramachandra, T. V., Shivamurthy, V., and Aithal, B. H. (2017) 'Frequent floods in Bangalore : Causes and remedial measures frequent floods in Bangalore : Causes and remedial measures. Report number: ETR123. doi: 10.13140/RG.2.2.17517.90088.

Ramaiah, M. and Avtar, R. (2019) 'Urban green spaces and their need in cities of rapidly urbanizing India : A review', *Urban Science*, 3(3), p. 94.

Rattani, V. (2018) 'Coping with climate change: An analysis of India's national action plan on climate change', *Centre for Science and Environment*, 1(May), pp. 213–224. Available at: https://cdn.cseindia.org/attachments/0.55359500_1519109483_coping-climate-change-NAPCC.pdf.

Reddy, S. N., Manikiam, B. and Jeevalakshmi, D. (2017) 'Land surface temperature retrieval from LANDSAT data using emissivity estimation', *International Journal of Applied Engineering Research*, 12(20), pp. 9679–9687. Available at: https://www.ripublication.com/ijaer17/ijaerv12n20_57.pdf.

Saba, B. and Yadav, J. (2021) 'The politics of green and blue space availability: Case of Bhopal', *IOP Conference Series: Earth and Environmental Science*, 796(1). doi: 10.1088/1755-1315/796/1/012035.

Sangwan, A. et al. (2022) 'Urban green spaces prospects and retrospect's', *Urban Green Spaces (Working Title)*, (March). doi: 10.5772/intechopen.102857.

Scaria, T. S. et al. (2020) 'Dimensions of liveability and identification of key parameters applicable for Kerala', *International Journal of Science, Engineering and Management*, 5(4), pp. 50–55.

Secon (2007) 'Urban trees in Bangalore City : Literature review and pilot study on the role of trees in mitigating air pollution and the heat island effect'.

Sen, A. and Mukherjee, P. (no date) 'Smartgreen urbanism : Ahmedabad city's perspective', pp. 42–47.

Setturu, B., Rajan, K. S., and Ramachandra, T. V. (2013) 'Geostatistics : An overview land surface temperature responses to land use land cover dynamics', *Geoinformatics & Geostatistics*, pp. 1–10. doi: 10.4172/2327-CITATIONS.

Shahtahmassebi, A. R. et al. (2021) 'Remote sensing of urban green spaces: A review', *Urban Forestry and Urban Greening*, 57, p. 126946. doi: 10.1016/j.ufug.2020.126946.

Siemens Economist Intelligence Unit (2011) 'Asian Green City Index'.

Theobald, W. (1984) 'A history of recreation resource planning: The origins of space standards', *Leisure Studies*, 3(2), pp. 189–200. doi: 10.1080/02614368400390161.

Tiwari, S. (2017) 'Identification of ecological open spaces of urban areas for smart city development: A case of Jabalpur city in Madhya Pradesh in India', *International Journal of Landscape Planning and Architecture*, 3(2), pp. 23–32.

TNCPO, G. (2014) 'Urban green guidelines, 2014', Ministry of Urban Development.

Town and Country Planning (2014) 'URDPFI guidelines', Ministry of Urban Development, 2A and 2B, pp. 9–10.

Town and Country Planning (2015) 'Urban and regional development plans formulation and implementation (URDPFI) guidelines volume I', Government of India Ministry of Urban Developmen, vol. 1, pp. 65–69. doi: 10.1201/9781482293500-22.

Turaga, R. M. R. et al. (2020) 'Exploring the role of urban green spaces in "smartening" cities in India', *Impact Assessment and Project Appraisal*, 38(6), pp. 479–490. doi: 10.1080/14615517.2019.1690864.

Tv, R. and Aithal, B. H. (2016) 'Decaying lakes of Bengaluru and today's irrational decision makers. Source : https://wgbis.ces.iisc.ac.in/energy/water/paper/Decaying_lakes_of_Bengaluru/#:~:text=The%20surroundings%20of%20these%20lakes,at%20one%20of%20these%20sites.

UN-Habitat (2020) 'Global state of metropolis 2020: Population data booklet'. Available at: https://www.ptonline.com/articles/how-to-get-better-mfi-results.

Veal, A. J. (2013) 'Open space planning standards in Australia in search of origins'.

WHO (2017) 'Urban green spaces: A brief for action', WHO, p. 24. Available at: https://iris.who.int/bitstream/handle/10665/344116/9789289052498-eng.pdf?sequence=1.

Wikantiyoso, R., Sulaksono, A. G. and Suhartono, T. (2021) 'Detection of potential green open space area using landsat 8 satellite imagery', *ARTEKS : Jurnal Teknik Arsitektur*, 6(1), pp. 149–154. doi: 10.30822/arteks.v6i1.730.

Wilcox, W. (2015) 'Madhya Pradesh', *State Politics in India*, pp. 127–174. doi: 10.4324/9781003159971-10.

Yigitcanlar, T. and Dizdaroglu, D. (2015) 'Ecological approaches in planning for sustainable cities a review of the literature', *Global Journal of Environmental Science and Management*, 1(2), pp. 159–188. doi: 10.7508/gjesm.2015.02.008.

Chapter 9

# Reactive power planning with load uncertainty using FACTS

*Sandeep Yadav and Sanju Saini*

## 9.1 INTRODUCTION

It is challenging to optimise power systems with reactive power planning. All sources of reactive electricity must be managed to ensure effective generation. Reactive power is supplied by generators, tap-changing transformers, static capacitors, etc. Active power loss is decreased by reactive power optimisation. Active power efficiency is decreased by reactive power. When combined with reactive power sources, static VAR compensator (SVC) and TCSC FACTS devices lower transmission losses and enhance the voltage profile of the power network. The availability of reactive power from all sources must be known in order to optimise a reactive power problem.

Hingorani [1] introduced FACTS in 1988. Static phase shifters and FACTS controllers are used to increase the transmission capacity of gearbox line power [2]. The installation of a FACTS device is described in Ref. [3]. Power flow management is improved by FACTS [4]. Reference [5] discusses FACTS device-based power flow regulation that takes transfer capacity into consideration. A genetic algorithm (GA) is used to deploy various FACTS devices in power grids [6], demonstrating the advantages of multi-type FACTS devices for system loadability [7]. Reference [8] outlines a hybrid genetic algorithm-FACTS device power flow optimisation approach, and several FACTS devices are used to address the power flow problem. In deregulated power markets, a number of FACTS devices are useful [9,10].

The demand for distributed generation (DG) has increased due to economic and environmental factors. System reliability, greenhouse gas emissions, energy losses, and investment flexibility are affected [11]. However, existing distribution networks, not designed for connecting power-producing units, pose challenges for distribution network operators as DG increases. Because of its fluctuation, renewable energy sources make DG passive network integration dynamic. Understanding the behaviour of DG-integrated systems is crucial for planning power distribution networks [12–14].

DOI: 10.1201/9781003521341-9

The distribution network is impacted by demand. Industrial distribution systems, when heavily burdened, are prone to collapse [15]. Distribution system connections for generators have an impact on voltage profiles and load-dependent power flow [16]. Studies on load characteristics using power system modelling are advised by the IEEE Task Force in Ref. [17].

FACTS-enhanced transfer devices are described in Ref. [17]. Effective reactive power delivery and adaptable series and shunt FACTS devices are covered in Ref. [18]. Sensitivity analysis and linear programming are used to locate and size an SVC in a connected power system [19]. The optimal TCSC for loadability and gearbox loss is placed by a GA in Ref. [20]. Loss-sensitive VAR source location is described in Ref. [21]. The load capacity of FACTS devices is increased via computational methods [22]. Reorganising the network decreases active power loss [23]. The capacitor of a radial distribution system is described in Ref. [24].

Intelligent search-based optimisation algorithms like seeker optimisation algorithm (SOA), harmony search algorithm (HSA), DEBM, and gravitational search algorithm (GSA) address the single-objective optimal reactive power dispatch (ORPD) problem. Real power loss and voltage stability enhancement index decrease. Reference [23] discusses UPFCs distributing reactive power sources to stabilise load bus voltage. The ORPD problem [24] maximises VSM by managing and scheduling VAR generation. Particle swarm optimiser (PSO) and FSSPSO improved wind farm ORPD. Reference [25] used EP, PSO, DE, and HDE evolutionary computation algorithms to solve the optimal active-reactive power dispatch problem. Optimised load flow shares reactive power Jeremy [26,27] discusses the tangent vector approximation for loss control in low-voltage regions. The local voltage stability indicator DSY, which is positively associated with VSM, underlies one ORPD objective function [28]. ORPD mixes heuristics. SFLA, NM-SFLA, MTLA, and DDE hybrids are utilised to solve the ORPD [29]. Speedy hybrids perform well. Penalty functions facilitate continuous and differentiable ORPD models [30]. Chance-limited ON-OFF power shedding reduced active power losses [31]. Nodal power injections and unpredictable branch outages are significant concerns in this study.

Static or moving loads are typical in power systems. Static loads include impedance, current, and power. Load modelling is complex since stability research load buses include fluorescent and incandescent lighting, freezers, heaters, compressors, motors, furnaces, and more. External factors like location, temperature, time of day, and season might make load composition prediction difficult. Power flow studies often use power supply load characteristics [32]. Most load flow analyses use the constant power load model, which assumes active and reactive powers are voltage independent. Distribution hub voltages fluctuate because most buses are unregulated. Load characteristics are necessary for distribution system analysis.

## 9.2 MATHEMATICAL MODELLING

The amount of wind-generated electricity depends on the farm's location and wind speed. The uncertainty in wind speed must be taken into consideration when generating electricity from wind turbines. The Weibull distribution [25,26] is a well-liked models for wind speed.

$$\mathrm{PDF}(U)=\left(\frac{k}{C}\right)\left(\frac{U}{C}\right)^{k-1}e^{-\left(\frac{U}{C}\right)^{K}} \tag{9.1}$$

$$f(U)=1-e^{-\left(\frac{U}{C}\right)^{K}} \tag{9.2}$$

$$\left\{\begin{array}{l} Q_{Gi}^{\min}\le Q_{gi}\le Q_{Gi}^{\max} \\ P_{Gi}^{\min}\le P_{gi}\le P_{Gi}^{\max} \end{array},\ i\in N_b\right\} \tag{9.3}$$

$$T_k^{\min}\le T_k\le T_k^{\max},\ k\in N_T \tag{9.4}$$

### 9.2.1 FACTS devices

New transmission circuits increase transmission line power transfer. However, planning constraints may prevent new transmission lines. FACTS devices enhance power transfer without requiring system modifications. This study uses an SVC and thyristor-controlled series compensator.

## 9.3 RESULTS AND DISCUSSION

The load bus L-index is established first, followed by the assignment VAR compensation techniques. Load buses with high L-indices may be VAR compensation equipment. Buses 24, 25, 26, 29, and 30 have greater L-index values than other load buses when using the proposed strategy, indicating these vehicles as candidates for VAR compensators. Following the assignment of VAR adjustment devices, the VAR compensator capacities should be zero. Figure 9.1 shows the case-A Pareto optimal front of the two objective functions. Modern methods are compared to active power losses. Solution #1 is illustrated in Figures 9.1 and 9.2. As shown, the solution outperforms the SOA, comprehensive learning PSO, local differential evolution, HSA, simple GA (SGA), conventional PSO (CePSO), and GSA. Since the references only provided active power loss mitigation results, Solution#1 is compared to them. Figure 9.2 shows the voltage standard deviation (VSD) of

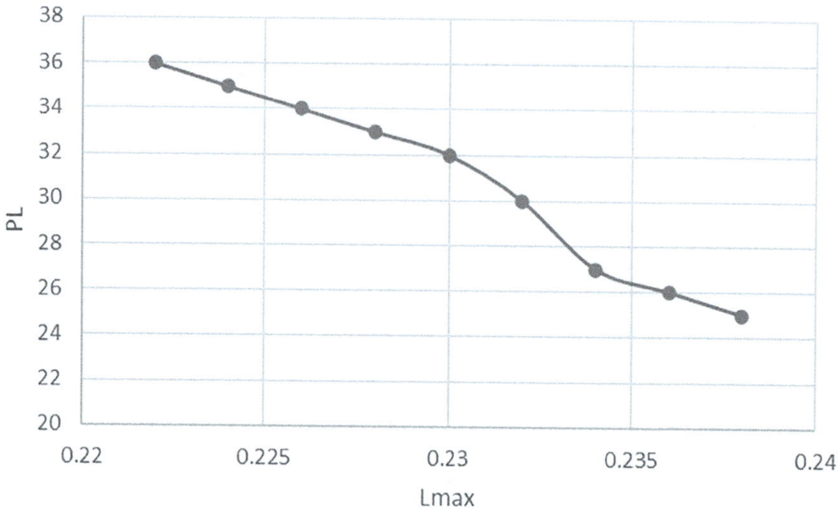


*Figure 9.1* PL and $L_{max}$.

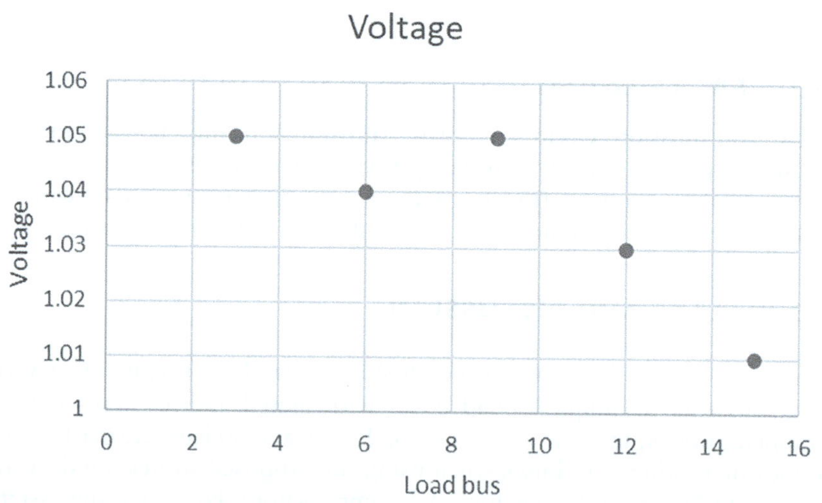


*Figure 9.2* Load bus voltage profiles.

buses 23, 34, and 44 for different load levels. It can be observed that VSD is the highest for the first site that has the higher correlation with other sites. Similarly, the VSD for the second site is more than the third site. Additionally, the values of VSD for the buses near the WTs are influenced by these units. For example, buses 21–27 have a similar VSD to bus 23.

Since operating load bus voltages are close to load bus voltage limits, OPF issues often arise. We need 0.95–1.05 p.u. load bus voltage for our study. Figure 9.2 shows optimal load bus voltage patterns, with similar voltage profiles in other optimisations. Legal bus voltages are unquestionable.

## 9.4 CONCLUSIONS AND FUTURE SCOPE

To decrease active power losses and improve voltage stability, this chapter suggests a multi-objective RPP architecture for electrical grids that are predominately powered by wind. It makes adjustments for load and wind uncertainty. The constraint method addresses the MO-RPP problem. The L-index is used to identify the VAR compensation buses, and planning studies for VAR are then compared.

## REFERENCES

1. Hingorani N. Flexible AC transmission. *IEEE Spect* 1993;30(4): 40–5.
2. Nelson R, Bian J, Williams S. Transmission series power flow control. *IEEE Trans Power Deli* 1995;10(1):504–10.
3. Lie TT, Deng W. Optimal flexible ac transmission systems (FACTS) devices allocation. *Int J Electr Power Energy Syst* 1997;19(2):125–34.
4. Gotham DJ, Heydt GT. Power flow control and power flow studies for system with FACTS devices. *IEEE Trans Power Syst* 1998;13(1):60–5.
5. Xiao Y, Song YH, Sun YZ. Power flow control approach to power systems with em-bedded FACTS devices. *IEEE Trans Power Syst* 2000;17(4):943–50.
6. Gerbex S, Cherkaoui R, Germond AJ. Optimal location of multitype FACTS devices in a power system by means of genetic algorithms. *IEEE Trans Power Syst* 2001;16(3):537–44.
7. Chung TS, Li YZ. A hybrid GA approach for OPF with consideration of FACTS devices. *IEEE Power Eng Rev* 2000;20(8):54–7. doi: 10.1109/39.857456
8. Prasad PN, Abdel Moamen MA. Power flow control and solutions with multiple and multi-type FACTS devices. *Electr Power Syst Res* 2005;74:341–51.
9. Cai LJ. Optimal choice and allocation of FACTS devices in deregulated electricity market using genetic algorithms. *IEEE PES Power Systems Conference and Exposition*, New York, 2004.
10. Bhattacharyy B, Kumar S. Reactive power planning with FACTS devices using gravitational search algorithm. *Ain Shams Eng J* 2015;6:865–871.
11. Ruiz-Romero S, Colmenar-Santos A, Gil-Ortego R, Molina-Bonilla A. Distributed gen-eration: The definitive boost for renewable energy in Spain. *Renew Energy* 2013;53:354–64.
12. Kersting WH. *Distribution System Modeling and Analysis.* New York: CRC Press, 2006.
13. Ding M, Zhang Y, Mao M. Key technologies for microgrids: A review. In: *International Conference on Sustainable Power Generation and Supply*, Nanjing, 2009, pp. 1–5.

14. Roy NK, Pota HR, Hossain MJ. Reactive power management of distribution net-works with wind generation for improving voltage stability. *Renew Energy* 2013;58:85–94.
15. Chakravorty M, Das D. Voltage stability analysis of radial distribution networks. *Electr Power Energy Syst* 2001;23:129–35.
16. Roy NK, Hossain MJ, Pota HR. Effects of load modeling in power distribution system with distributed wind generation. In: *Australasian Universities Power Engineering Conference*, Brisbane, 2011, pp. 1–6.
17. IEEE Task Force. Load representation for dynamic performance analysis. *IEEE Trans Power Syst* 1993;8(2):472–82.
18. Preedavichit P, Srivastava SC. Optimal reactive power dispatch considering FACTS de-vices. *Electr Power Syst Res* 1998;46(3): 251–7. https://dx.doi.org/10.1016/S0378-7796(98)00075-3.
19. Ibrahim E, Abusorrah A. Optimal allocations of SVCs for improvement of power sys-tem performance. *Electr Power Comp Syst* 2003;31(1):27–46. https://dx.doi.org/10.1080/15325000390112044.
20. Abdelaziz AY, El-Sharkawy MA, Attia MA. Optimal location of thyristor-controlled series compensators in power systems for increasing loadability by genetic algorithm. *Electr Power Comp Syst* 2011;39(13):1373–87. https://doi.org/10.1080/15325008.2011.584108
21. Bhattacharyya B, Goswami SK, Bansal RC. Loss sensitivity approach in evolutionary algorithms for reactive power planning. *Electr Power Compon Syst* 2009;37(3):287–99. https://dx.doi.org/10.1080/15325000802454468.
22. Nagalakshmi S, Kamaraj N. Comparison of computational intelligence algorithms for loadability enhancement of restructured power system with FACTS devices. *Swarm Evol Comput* 2012;5:17–27. https://dx.doi.org/10.1016/j.swevo.2012.02.002.
23. Vitorino RM, Jorge HM, Neves LP. Loss and reliability optimization for power distribution system operation. *Electr Power Syst Res* 2013;96:177–84. https://dx.doi.org/10.1016/j.epsr.2012.11.002.
24. Abul'wafa AR. Optimal capacitor allocation in radial distribution systems for loss reduction: A two stage method. *Electr Power Syst Res* 2013;95:168–74. https://dx.doi.org/10.1016/j.epsr.2012.09.004.
25. Thukaram D, Yesuratnam G. Optimal reactive power dispatch in a large power system with AC-DC and FACTS controllers. *IET Gener Transm Distrib* 2008;2:71–81.
26. Rabiee A, Parniani M. Optimal reactive power dispatch using the concept of dynamic VAR source value. In: *Power & Energy Society General Meeting, 2009 PES'09 IEEE*, IEEE, 2009, pp. 1–5.
27. Martinez-Rojas M, Sumper A, Gomis-Bellmunt O, Sudria-Andreu A. Reactive power dispatch in wind farms using particle swarm optimization technique and feasible solu-tions search. *Appl Energy* 2011;88:4678–4686.
28. Huang CM, Chen SJ, Huang YC, Yang HT. Comparative study of evolutionary computation methods for activeereactive power dispatch. *IET Gener Transm Distrib* 2012;6:636–645.
29. de Souza AZ, Leme R, Lo K, de Souza JS, Almeida A. Reactive power redispatch for loss reduction using a modified power flow Jacobian. *IET Gener Transm Distrib* 2012;6:657–664.

30. Rabiee A, Vanouni M, Parniani M. Optimal reactive power dispatch for improving voltage stability margin using a local voltage stability index. *Energy Convers Manag* 2012;59:66–73.
31. Rajan A, Malakar T. Optimal reactive power dispatch using hybrid NeldereMead simplex based firefly algorithm. *Int J Electr Power Energy Syst* 2015;66:9–24.
32. Khorsandi A, Alimardani A, Vahidi B, Hosseinian S. Hybrid shuffled frog leaping algorithm and NeldereMead simplex search for optimal reactive power dispatch. *IET Gener Transm Distrib* 2011;5:249–256.

Chapter 10

# Carbon sequestration potential of urban parks

## A case study of Chinar Park

*Vibhor Bhardwaj, Mariette Rose George, Shivansh Sharma, and Surabhi Mehrotra*

### LIST OF ABBREVIATIONS AND ACRONYMS

| Abbreviation | Meaning |
|---|---|
| $CO_2$ | Carbon dioxide |
| INR | Indian rupee |
| PPMV | Parts per million by volume |
| AQI | Air Quality Index |
| GHG | Greenhouse gas |
| EC | Eddy covariance |
| e.g.- | Example |
| PM | Particulate matter |
| µm | Microns/micrometer |
| lbs | Pounds |
| kgs | Kilograms |

### 10.1 INTRODUCTION

In a period of rising urbanization and industrialization, cities throughout the world face the combined challenge of economic development and environmental sustainability. Carbon emissions, notably those from industrial activity, transportation, and energy generation, have emerged as a major threat to global warming and climate change. Elevated carbon levels have far-reaching consequences, including poor air quality, negative health impacts, and environmental destruction.

The epicenters of carbon emissions are urban centers, and the burden of pollution is mostly felt in heavily populated regions. The complex interactions among carbon emissions, industrial growth, and urbanization highlight the pressing need for creative ways to lessen the negative environmental effects of cities (Intergovernmental Panel on Climate Change, 2005).

In light of this, urban parks and other green areas take on a greater relevance as potential carbon sink. In addition to providing a welcome


DOI: 10.1201/9781003521341-10

break from the concrete jungle, these verdant havens have the capacity to absorb large volumes of carbon dioxide, potentially balancing the carbon footprints left by cities. Urban parks have the potential to reduce urban pollution, but their effectiveness in sequestering carbon is yet unknown. Therefore, further research is needed to fully understand how urban parks might help reduce pollution in cities (Lal, 2008).

Bhopal, the capital city of Madhya Pradesh, India, is known for its rich cultural heritage, but it is facing environmental pollution challenges that have grave consequences for its residents' health. The city is characterized by elevated levels of industrialization, urbanization, and population growth, which have led to increased emissions of carbon dioxide ($CO_2$) and other greenhouse gases (GHGs) along with the historic Bhopal gas catastrophe in 1984. The environmental and health impacts of high pollution levels in Bhopal are alarming, contributing to the city's poor air quality and posing significant risks to the health of its residents. Additionally, the elevated carbon levels also contribute to global warming and climate change, threatening the well-being of people and the city.

There are several sources of carbon emissions in Bhopal, including factories, power plants, and transportation. Madhya Pradesh Pollution Control reports indicate worst areas in terms of the Air Quality Index are Tatya Tope Nagar and Sector D industrial area. There are several urban parks and green covers that can act as a carbon sink and offset the overall footprint of the city, but scientific research on the potential of such parks is inadequate. This study aims to bridge this gap in the knowledge of the efficiency of such parks and directs the way to analyze the sample size of fully grown trees. The objectives of the paper are (i) estimating carbon sequestration in an urban park and (ii) validating the i-Tree Canopy tool and its efficiency in estimating $CO_2$ sequestration through field validation surveys.

## 10.2 LITERATURE REVIEW

### 10.2.1 Climate change

Climate change poses a multifaceted challenge with far-reaching consequences for the planet, and its impact is particularly pronounced in urban environments. Cities, as hubs of economic activity and population concentration, are not only significant contributors to global GHG emissions but also bear the brunt of climate change impacts. The relentless rise in atmospheric $CO_2$ levels, exemplified by the 2021 global average concentration exceeding 416 ppmv, underscores the urgency of addressing emissions reduction efforts. The pace of $CO_2$ level growth has also increased, from 0.7 ppmv per year in the early 1960s to 2.0 ppmv per year between 2000 and 2010. A study by the Indian Institute of Science, Bangalore, on the GHG footprint of major cities in India calculated GHG emissions for various cities (Ramachandra, Aithal, and Sreejith, 2015). To address climate change, urban design can contribute to reducing GHG emissions through strategies such as promoting sustainable transportation options and designing more energy-efficient buildings. Furthermore, the integration of green spaces and urban forests into urban planning frameworks not only enhances the aesthetic appeal of cities but also serves as carbon sinks, sequestering $CO_2$ emissions and mitigating the urban heat island effect. Strategic urban greening initiatives, coupled with investments in biodiversity conservation and ecosystem restoration, can enhance urban resilience to climate change while fostering social and ecological well-being. Low-income and marginalized populations are often more exposed to the effects of climate change due to a lack of resources such as air conditioning, green areas, and transportation choices. Urban planners and designers must ensure that climate change policies are fair and benefit all people in the community, regardless of income or color (Behera et al., 2022).

### 10.2.2 Carbon sequestration and methods of carbon sequestration

There are primarily two ways of carbon sequestration: (i) abiotic sequestration and (ii) biotic sequestration. Abiotic means include geological injection, where the sealing effectiveness of low-permeable sequences (caprock) underneath prospective storage reservoirs is a significant challenge for all storage solutions. $CO_2$ is preferably supplied in a supercritical condition to maximize pore space usage ($CO_2$). In hydrodynamic trapping, $CO_2$ is stopped from leaking through the caprock while waiting for other trapping methods to take effect. However, the process of mineral entrapment, which depends on lithological structure, mineralogy, and hydrogeology, can result in both positive and negative effects, highlighting the necessity for comprehensive understanding and assessment of storage dynamics (Falkowski, 1997). In biotic means, oceanic sequestration occurs when seaweed reaches the deep ocean, sequestering carbon and preventing it from interacting with the atmosphere for millennia. Algae, with their rapid growth rates and carbon uptake capabilities, hold promise as short-term carbon storage solutions and potential biofuel feedstocks. Reforestation and limiting deforestation can enhance carbon sequestration in four main ways: expanding the size of the current forest, enhancing existing forests' carbon density, increasing the use of forest products to replace fossil fuel emissions, and lowering carbon emissions caused by deforestation and environmental deterioration. In addition to increasing environmental awareness and encouraging action against climate change, urban forestry and parks offer a number of physiological, recreational, and psychological benefits from a social perspective (Zhang and Song, 2014). For example, there are little data on the effectiveness of urban vegetation to reduce GHG emissions or concentrations of airborne pollutants for the measurement of the $CO_2$ exchange. For example, eddy covariance measurements of carbon dioxide flux have been used to study the carbon cycle in natural ecosystems (Baldocchi, 2008). This chapter addresses the lack of research in this area.

### 10.2.3 Carbon management

Carbon management is increasingly recognized as a critical component of efforts to mitigate global warming and its associated impacts, which span from rising sea levels to more frequent and severe extreme weather events, and the loss of biodiversity. As carbon emissions continue to drive climate change, it becomes imperative for businesses and governments to take proactive steps to quantify and reduce their carbon footprint (Velasco et al., 2016). To manage carbon efficiently, businesses and governments must calculate their carbon footprint or the volume of GHGs produced by their operations and discharged into the environment. Carbon sequestration in living plants and soils is likely to have an immediate net positive effect on atmospheric carbon dioxide, along with a positive effect on biodiversity and other ecosystem services. The protection of existing mature forests keeps the living plant carbon out of the atmosphere and preserves the current level of biodiversity (Huston and Marland, 2003). Recent research has assessed carbon sequestration potential for application in urban forestry applications (Behera et al., 2022). Similarly, this study aims to elucidate the efficacy of nature-based solutions in mitigating urban carbon emissions and enhancing environmental quality. Building upon this foundation, the present research aims to assess the carbon sequestration potential of a specific study area in Bhopal, employing a combination of the i-Tree tool and physical surveys.

## 10.3 SITE PROFILE: CHINAR PARK, BHOPAL

Chinar Park is one of the urban parks of Bhopal, located on Link Road No. 1 near Vyapam Square and Shivaji Square, coming under Ward No. 33 (Figure 10.1).

Chinar Park is also known as a "lovers" park. Beautifully planned and spacious manicured lawns and well-carved-out flower beds make this park an ideal spot for picnics which makes it one of the leading visiting destinations for tourists and locals. Plant Nursery, where plants are grown for transplanting, is also located. Chinar Park has a collection of many figurines of animals, birds, and natural varieties of many flowers and other shrubs made from scrap parts of automobiles or machines. Though the park is visited throughout the year, the peak time is between January and March, as during these months people come here to lie down in sunlight

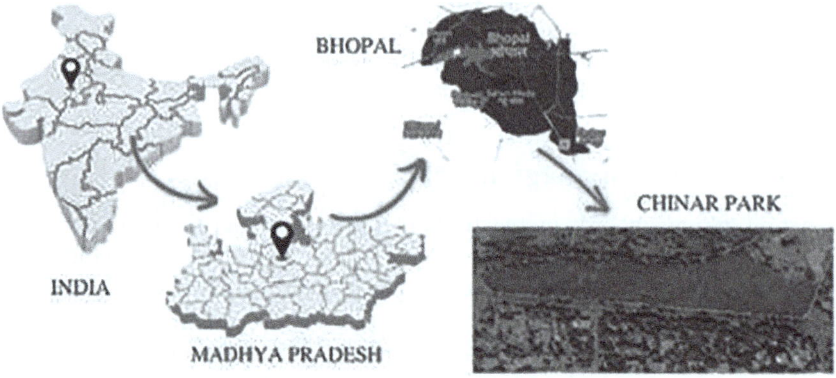


*Figure 10.1* Location of Chinar Park.

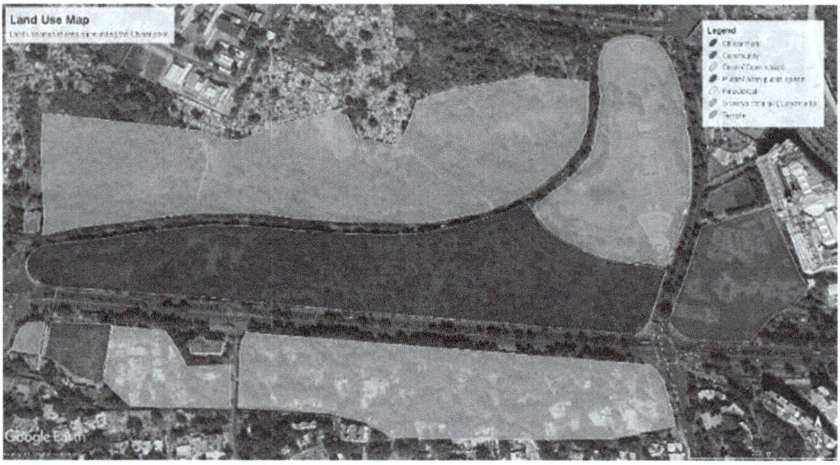


*Figure 10.2* Land use around Chinar Park.

and spend leisure time wandering the flower blooms and migratory birds. It is a tree-filled park offering walking/jogging paths and open grassy areas for picnics and yoga. Nestled within the park's verdant expanse lies a thriving plant nursery, where a diverse array of flora is nurtured for transplanting, adding to the park's botanical richness and ecological diversity. Moreover, the park's unique charm is accentuated by its collection of whimsical figurines crafted from scrap parts of automobiles and machines, showcasing the creative ingenuity of local artisans and the park's commitment to sustainability (Figures 10.2 and 10.3).

Chinar Park is surrounded by residential areas, public/semi-public areas, landmark sites, and open spaces. There are two bus stops nearby.

*Figure 10.3* Surroundings of the park.

*Figure 10.4* Road typology.

It is surrounded by two major roads: a four-lane road connecting Shivaji Square and Vallabh Bhavan Road and six-lane road known as Link Road No. 1. The road has pedestrian walkways with plantations on the median of roads and footpaths (Figures 10.4–10.6).

The park is on a slope. Elevation of the park gradually decreases from 513 to 504 m from one side to another, as observed in the picture (from Vallabh Bhavan side to Link Road No. 1). Vallabh Bhawan is on the upper side of the park, whereas it connected to link road number 1 from its lower side.

Chinar Park is known for its uniqueness, especially when it comes to landscaping. Various elements are used to create the imageability of the park, as follows:

- **Water**: Water enters the park from the hill located above Vallabh Bhawan, and it flows in several streams.
- **Benches**: Benches are provided at every corner of the park.

*Figure 10.5* Traffic density.

*Figure 10.6* Landscape elements.

- **Porous Lanes:** These lanes not only give scenic views to the area but also let water to seep into the soil.
- **Artificial Sheds:** These provided shaded seating to avoid direct sunlight.
- Lighting
- **Figurines:** Made from scrap parts of automobiles and machines, they represent animals, human figures, etc.
- **Flora and Fauna:** The park is rich in vegetation, with various species of trees, shrubs, and herbs. It also hosts stray dogs, a few reptiles, insects, and migratory birds in winter.
- Amenities and Infrastructure

## 10.4 METHODOLOGY

The methodology adopted includes two steps: (i) an i-Tree survey that was conducted first and (ii) a survey conducted by the authors. The following chart shows the methodology followed in the study (Figure 10.7).

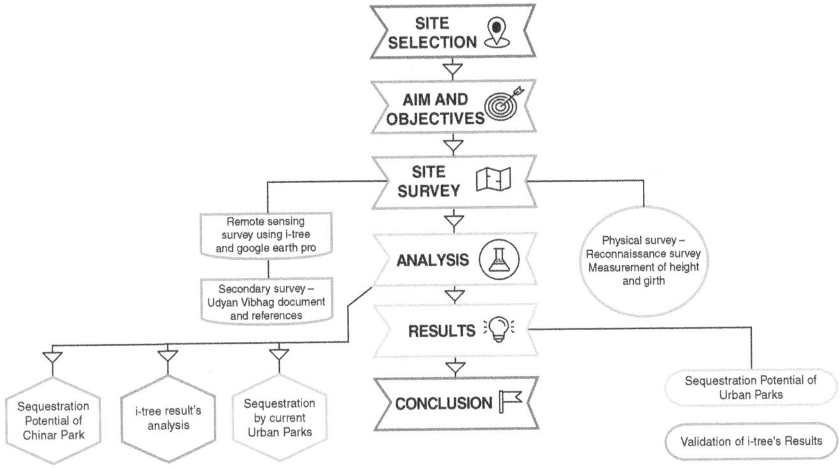


*Figure 10.7* Methodology chart.

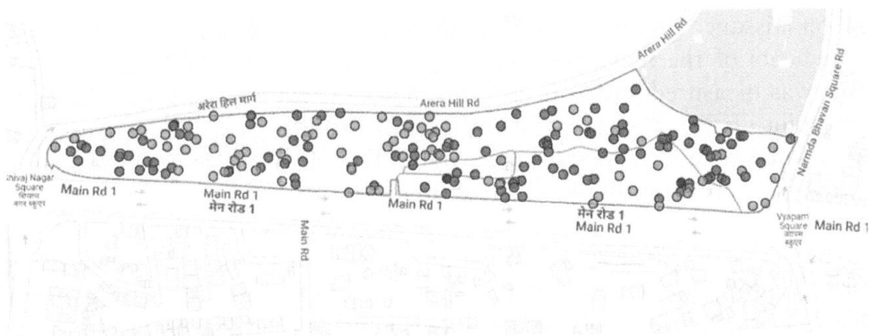


*Figure 10.8* Survey points in i-Tree.

## 10.4.1 Working and survey through i-Tree

The i-Tree Canopy tool is used to estimate tree canopy and $CO_2$ sequestered. By reviewing Google Maps aerial photography at random points, it conducts a cover assessment within a defined project area. It is intended to help users estimate trees and other cover classes (e.g., grass, building, and roads) inside their city or any region they choose. This tool randomly places points (decided by the user) on Google Earth images, and the user then determines which cover class each point belongs to. The user may specify any number of cover classes, and the program will provide an anticipated percentage for each cover class during the interpretation process (Figure 10.8).

The tool gives yearly amounts of air pollutants eliminated by trees through the dry deposition process and related monetary values based on the area defined as the tree cover class. The estimated air pollutants are seven criteria pollutants listed by the U.S. Environmental Protection Agency: carbon monoxide (CO), nitrogen dioxide ($NO_2$), ozone ($O_3$), sulfur dioxide ($SO_2$), and particulate matter (PM), which includes particulate matter less than 2.5 μm (PM2.5) and particulate matter >2.5 and <10 μm (PM10*) . The data we get are based on U.S. estimates, and results in India may vary, but a general estimation of a global perspective can be done using this tool.

### 10.4.2 Physical survey

A sample study was conducted in Chinar Park in Bhopal to determine the total carbon absorbed by a small urban park. The trees in the park were mapped and identified for this investigation. The girth of each tree was measured using a 5-m measuring tape, and its height was measured using a disto-meter. All the trees were mapped and numbered according to species (School of Planning and Architecture, n.d.).

The survey was carried out over three days in Chinar Park. During this reconnaissance survey, traffic flow analysis, species identification, and measurements of the identified species were carried out. The height of each tree was measured with a laser disto-meter, which was aimed at the highest point of the canopy at the eye level of the surveyor. The height of the surveyor was added to the final height. The girth of each tree was measured at 1.3 m from the base. After making appropriate assumptions, the final height and girth were noted. For each species, the mean height and girth were calculated by averaging all the readings (it is comprehensive to note the different heights and girths of distinct species with their quantity). The species were identified using Google Lens and the iNaturalist app, with assistance from the park management for the identification of some species (Figure 10.9).

*Figure 10.9* Conduction of physical survey.

### 10.4.3 Method for estimating current $CO_2$ sequestration

The amount of $CO_2$ sequestered in each tree (to be followed in the chronological order) can be estimated as follows (Eqs. 10.1–10.8) (Nowak, 1994).

Determine the total (green) weight of the tree $W$=Above-ground weight of the tree in pounds

$D$=Diameter of the trunk in inches

$H$=Height of the tree in feet

For trees with $D<11$ inches:

$$W = 0.25(D)^2(H)\text{lbs.} \tag{10.1}$$

For trees with $D \geq 11$ inches:

$$W = 0.15(D)^2(H)\text{lbs.} \tag{10.2}$$

(Note: These two equations could be seen as an "average" of all the species' equations.)

The root system weighs about 20% as much as the above-ground weight of the tree.

$$W_T = 120\% \times W \text{ lbs.} \tag{10.3}$$

Determine the dry weight of the tree

(Note: The average tree is 72.5% dry matter and 27.5% moisture.)

$$W_D = 72.5\% \times W_T \text{ lbs.} \tag{10.4}$$

Determine the weight of carbon in the tree

The average carbon content is 50% of the tree's total volume.

$$W_C = 50\% \times W_D \text{ lbs.} \tag{10.5}$$

Determine the weight of carbon dioxide sequestered in the tree

(The ratio of $CO_2$ to C is 43.999915/12.001115=3.6663.)

$$W_{CO_2} = 3.7 \times W_C \text{ lbs.} \tag{10.7}$$

Convert pounds to kgs

$$1\text{lbs} = 0.45\text{kgs}$$

Determine the weight of $CO_2$ sequestered in the tree per year (in kgs)

$$W_{CO_2}/\text{year} = W_{CO_2}/\text{Age}(\text{kg/year}) \tag{10.8}$$

## 10.5 RESULTS AND ANALYSIS

### 10.5.1 Calculation of total $CO_2$ sequestered

Table 10.1 contains information about our sample of 250 trees, including the identified species, girth, height, and results by applying formulas to calculate the $CO_2$ sequestration of the site.

### 10.5.2 Analysis according to i-Tree

After analyzing 200 sample points for Chinar Park, the values shown in figure were obtained. The total amount of $CO_2$ sequestered by the park, according to i-Tree, is around 1,000 tons of $CO_2$. According to i-Tree estimations, this translates to a monetary advantage of about 45 lakhs, implying that if this much $CO_2$ were to be eliminated artificially, this would be the cost of removal. The analysis also estimates that the park's trees remove 42 tons of carbon dioxide from the air each year, which is equivalent to the emissions of one gasoline-powered car driving for around 20 years. Additionally, the trees in Chinar Park provided an estimated 1.78 lakhs in annual benefits, including improved air quality, reduced stormwater runoff, and energy savings (Table 10.2). Apart from $CO_2$, there are other pollutants removed from the atmosphere by this park, as described in Table 10.3 (Figure 10.10).

The most significant pollutant removed is ozone, with 200 kg removed annually. However, from a monetary perspective, Chinar Park is efficient in removing PM that is <2.5 µm. The removal of PM 2.5 translates to better air quality and fewer smog and visibility issues during the winter months.

*Table 10.1* Calculation of total $CO_2$ sequestered

| S. No. | Common Name | Botanical Name | Girth (inches) | Diameter (inches) | Height (feet) | Green weight | Green weight | Dry weight | Carbon content | Co2 sequestered (lbs) | No. of trees | Total Co2 Sequestered | Total Co2 Sequestered (Kgs) |
|---|---|---|---|---|---|---|---|---|---|---|---|---|---|
| 1 | Amaltas | Golden shower | 5.25 | 1.67 | 36.28 | 25.35 | 30.42 | 22.05 | 11.03 | 40.43 | 5.00 | 202.15 | 91.78 |
| 2 | Amla | Phyllanthus emblica | 83.86 | 26.71 | 33.26 | 3558.38 | 4270.06 | 3095.79 | 1547.90 | 5675.06 | 1.00 | 5675.06 | 2576.48 |
| 3 | Ashoka | Saraca asoca | 38.69 | 12.32 | 22.21 | 505.87 | 607.04 | 440.10 | 220.05 | 806.78 | 38.00 | 30657.49 | 13918.50 |
| 4 | Bamboo | Bambusa vulgaris | 7.87 | 2.51 | 38.81 | 61.01 | 73.21 | 53.08 | 26.54 | 97.30 | 16.00 | 1556.81 | 706.79 |
| 5 | Banyan Tree | Ficusbengalensis | 117.72 | 37.49 | 31.39 | 6618.12 | 7941.75 | 5757.77 | 2878.88 | 10554.85 | 11.00 | 116103.32 | 52710.91 |
| 6 | Bhirra | Chloroxylon | 59.06 | 18.81 | 42.09 | 2233.15 | 2679.78 | 1942.84 | 971.42 | 3561.51 | 1.00 | 3561.51 | 1616.93 |
| 7 | Bottle brush | Callistemon | 35.83 | 11.41 | 26.63 | 520.08 | 624.10 | 452.47 | 226.24 | 829.45 | 1.00 | 829.45 | 376.57 |
| 8 | Copperpod | Peltophorum pterocarpum | 114.17 | 36.36 | 45.37 | 8997.64 | 10797.17 | 7827.95 | 3913.97 | 14349.80 | 10.00 | 143497.98 | 65148.08 |
| 9 | Drumstick | Moringa Oleifera | 45.28 | 14.42 | 45.37 | 1414.91 | 1697.90 | 1230.98 | 615.49 | 2256.56 | 1.00 | 2256.56 | 1024.48 |
| 10 | Gular | Ficus racemosa | 36.75 | 11.70 | 26.36 | 541.57 | 649.88 | 471.16 | 235.58 | 863.71 | 3.00 | 2591.14 | 1176.38 |
| 11 | Gulmohar | Delonix regia | 78.74 | 25.08 | 48.65 | 4588.98 | 5506.77 | 3992.41 | 1996.20 | 7318.69 | 2.00 | 14637.37 | 6645.37 |
| 12 | Horsetail She-oak | casuarina equisetifolia | 41.34 | 13.17 | 32.77 | 852.00 | 1022.41 | 741.24 | 370.62 | 1358.81 | 5.00 | 6794.06 | 3084.50 |
| 13 | Indian cork | Millingtonia hortensis | 69.34 | 22.08 | 39.74 | 2906.62 | 3487.94 | 2528.76 | 1264.38 | 4635.59 | 6.00 | 27813.56 | 12627.36 |
| 14 | Jackfruit | Artocarpusheteroph yllus | 34.25 | 10.91 | 29.63 | 881.48 | 1057.77 | 766.89 | 383.44 | 1405.82 | 1.00 | 1405.82 | 638.24 |
| 15 | Jamun | Syzygium cumini L | 74.02 | 23.57 | 43.88 | 3657.23 | 4388.67 | 3181.79 | 1590.89 | 5832.69 | 6.00 | 34996.17 | 15888.26 |
| 16 | Kanji | Millettia pinnata | 92.13 | 29.34 | 50.29 | 6493.66 | 7792.39 | 5649.48 | 2824.74 | 10356.35 | 4.00 | 41425.41 | 18807.14 |
| 17 | Kedondong | Garuga floribunda | 38.99 | 12.42 | 33.85 | 782.98 | 939.57 | 681.19 | 340.59 | 1248.72 | 4.00 | 4994.89 | 2267.68 |
| 18 | Maharukh | Ailanthus altissima | 90.55 | 28.84 | 46.58 | 5811.08 | 6973.30 | 5055.64 | 2527.82 | 9267.75 | 8.00 | 74142.00 | 33660.47 |
| 19 | Mango | Mangiferaindica | 68.90 | 21.94 | 43.93 | 3172.72 | 3807.26 | 2760.27 | 1380.13 | 5059.98 | 5.00 | 25299.91 | 11486.16 |
| 20 | Neem | Azadirachta indica | 74.13 | 23.61 | 32.20 | 2692.35 | 3230.82 | 2342.35 | 1171.17 | 4293.87 | 4.00 | 17175.48 | 7797.67 |
| 21 | Palash | Butea monosperma | 31.50 | 10.03 | 44.29 | 1114.07 | 1336.89 | 969.24 | 484.62 | 1776.77 | 15.00 | 26651.50 | 12099.78 |
| 22 | Palm | Arecacene | 9.84 | 3.13 | 13.36 | 32.81 | 39.37 | 28.54 | 14.27 | 52.33 | 13.00 | 680.24 | 308.83 |
| 23 | Paras pipal | Thespesia populnea | 38.98 | 12.41 | 21.55 | 498.09 | 597.71 | 433.34 | 216.67 | 794.37 | 3.00 | 2383.11 | 1081.93 |
| 24 | Pipal | Ficusreligiosa | 78.62 | 25.04 | 36.69 | 3450.30 | 4140.35 | 3001.76 | 1500.88 | 5502.67 | 11.00 | 60529.37 | 27480.34 |
| 25 | Pseudo Ashoka | Polyathialongifolia | 46.30 | 14.74 | 29.56 | 964.01 | 1156.81 | 838.68 | 419.34 | 1537.43 | 25.00 | 38435.86 | 17449.88 |
| 26 | Red cotton tree | Bombax ceiba | 53.15 | 16.93 | 38.56 | 1657.04 | 1988.44 | 1441.62 | 720.81 | 2642.71 | 5.00 | 13213.54 | 5998.95 |
| 27 | Royal Palm | Roystoncaregia | 49.21 | 15.67 | 55.21 | 2034.34 | 2441.20 | 1769.87 | 884.94 | 3244.44 | 3.00 | 9733.33 | 4418.93 |
| 28 | Safed Siris | Albiziaprocera | 78.74 | 25.08 | 71.62 | 6755.23 | 8106.27 | 5877.05 | 2938.52 | 10773.51 | 4.00 | 43094.03 | 19564.69 |
| 29 | Shahtoot | Morus alba | 15.75 | 5.02 | 17.25 | 108.45 | 130.14 | 94.35 | 47.17 | 172.96 | 1.00 | 172.96 | 78.52 |
| 30 | Shisham | Dalbergia sissoo | 98.97 | 31.52 | 35.53 | 5294.04 | 6352.85 | 4605.82 | 2302.91 | 8443.15 | 6.00 | 50658.92 | 22999.15 |
| 31 | Silver Oak | Grevillea robusta | 57.48 | 18.31 | 34.44 | 1731.06 | 2077.27 | 1506.02 | 753.01 | 2760.76 | 1.00 | 2760.76 | 1253.39 |
| 32 | Subabul | Leucaena leucocephala | 35.68 | 11.36 | 29.49 | 951.81 | 1142.17 | 828.08 | 414.04 | 1517.99 | 2.00 | 3035.98 | 1378.33 |
| 33 | Tamarind | Tamarindus indica L | 54.33 | 17.30 | 19.47 | 874.41 | 1049.29 | 760.73 | 380.37 | 1394.54 | 11.00 | 15339.94 | 6964.33 |
| 34 | Tasmanian blue gum | Eucalyptus | 72.05 | 22.94 | 62.36 | 4925.00 | 5910.00 | 4284.75 | 2142.37 | 7854.58 | 18.00 | 141382.51 | 64187.66 |
| | | | | | TOTAL | | | | | | 250.00 | 963688.19 | 437514.44 |

*Table 10.2* $CO_2$ sequestered by Chinar Park

| *Description* | *Abbr.* | *Amount (kg)* | *±SE* | *Value (INR)* | *±SE (INR)* |
|---|---|---|---|---|---|
| Carbon monoxide removed annually | CO | 3.81 | ±0.30 | ₹ 0.00 | ±0 |
| Nitrogen dioxide removed annually | $NO_2$ | 20.77 | ±1.66 | ₹ 82.52 | ±0 |
| Ozone removed annually | $O_3$ | 206.89 | ±16.50 | ₹ 2,640.64 | ±247.56 |
| Sulfur dioxide removed annually | $SO_2$ | 13.09 | ±1.04 | ₹ 0.00 | ±0 |
| Particulate matter <2.5 µm removed annually | PM2.5 | 10.05 | ±0.80 | ₹ 5,446.20 | ±412.56 |
| Particulate matter >2.5 µm and <10 µm removed annually | PM10* | 69.3 | ±5.53 | ₹ 1,897.96 | ±165.04 |
| Total | | 323.91 | ±25.84 | ₹ 10,149.96 | ±852.2 |

*Source*: i-Tree canopy.

*Table 10.3* Number of gases removed annually

| *Description* | *Carbon (t)* | *±SE* | *$CO_2$ equiv. (t)* | *$±SE_2$* | *Value (INR)* | *±SE (INR)* |
|---|---|---|---|---|---|---|
| Sequestered annually in trees | 11.53 | ±0.92 | 42.29 | ±3.37 | ₹ 1,78,904 | ±14,276 |
| Stored in trees (note: this benefit is not an annual rate) | 289.67 | ±23.11 | 1,062.13 | ±84.73 | ₹ 4,493,901 | ±3,58,469 |

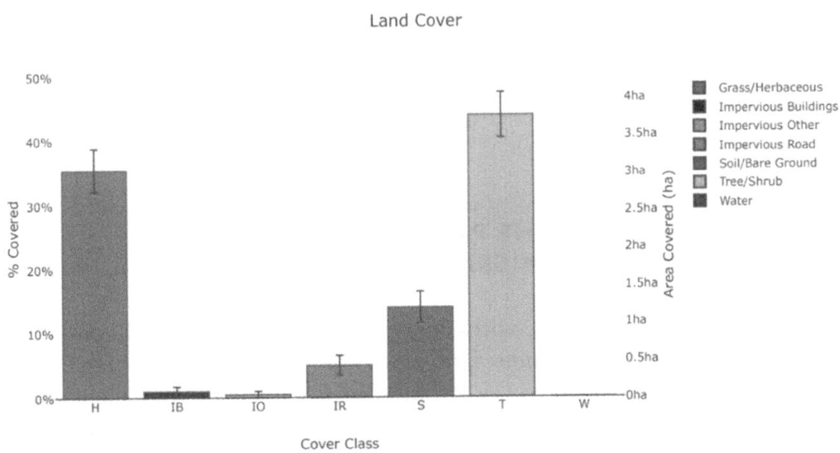


*Figure 10.10* Land cover percentage of Chinar Park.

*Table 10.4* Difference in $CO_2$ sequestered

| *Particulars* | *$CO_2$ amount (in tons)* |
|---|---|
| Total $CO_2$ sequestered (i-Tree) | 1,062.13 |
| $CO_2$ sequestered (survey) | 437.51 |
| Difference | **624.62** |

The total financial benefit from removing other pollutants amounts to INR 10,000. The total monetary benefits of the park per year are estimated to be around 2 lakhs. The results of the i-Tree include the sequestration benefits of soil, grass, and shrubs, leading to a higher estimate than the survey conducted on a sample size of 250 identified trees. The analysis concludes the monetary and environmental benefits of the park are significantly offsetting carbon emissions overall.

### 10.5.3 Validation of i-Tree

There was a variation in $CO_2$ sequestration levels between the i-Tree tool results and manual surveying. The i-Tree tool's estimate was almost 143% higher than the survey outcomes (Table 10.4).

Several reasons contribute to this variation. The i-Tree Canopy tool works on the general average model, governed by the user choices and shows carbon stored in various land covers such as grass, trees, shrubs, roads, and buildings. In contrast, the manual survey was focused solely on fully grown trees only analyzed identified species. A major drawback of i-Tree is that they rely on averages, without considering factors such as height, girth, age, and species of trees. However, its advantage is that it covers large green areas, various vegetation types, and other aspects such as soil and water. which would be very time-consuming to complete manually. When considering scale, the inclusion of major aspects, and time efficiency, i-Tree canopy is a highly efficient tool for estimating carbon sequestration.

## 10.6 POTENTIAL OF THE PARK TO SEQUESTER THE CURRENT EMISSIONS

Bhopal lies in hilly terrain, with slopes toward the north and southeast. Hillocks of different altitudes are along the southwest and northwest portions of the city. As per the Urban and Regional Development Plans Formulation and Implementation guidelines, a city like Bhopal should have 10–12 $m^2$ of green space per capita., but the current green space per capita in Bhopal is only 7–8 $m^2$. The green cover of the city has been declining, decreasing from 92% in 1977 to 18% in 2014, according to researchers from the Indian Institute of Science Bangalore. Green cover is projected to degrade further to 4% in 2030 if no measures are taken. Urban vegetation plays a significant role in reducing atmospheric carbon dioxide by sequestering it within the system.

The total green area of Bhopal is 63.23 $km^2$, while the total area of Bhopal is 410.39 $km^2$ (based on 2016 data). Chinar Park has a total area of 94,264 $m^2$ (0.1 $km^2$).

$$1\,\text{tone} = 1{,}000\,\text{kg}$$

If a green patch of 0.1 $km^2$ of Chinar Park (an urban park) can sequester 4,37,514.44 kgs (437.51 tones) of $CO_2$, then a total green area of Bhopal of 63.23 $km^2$ can sequester an estimated 27,66,40,380.412 kg (2,76,640.38 tones) of $CO_2$.

Carbon sequestration potential can be improved by increasing the green cover and improving water bodies, which also sequester carbon. Improving public transport for better connectivity and using sustainable solutions will help reduce the city's carbon emissions.

## 10.7 DISCUSSIONS

Using the above method, carbon sequestration estimation of an urban area was carried out. With freely available, easy-to-use tools like i-Tree and geospatial techniques, we could analyze and validate the estimated results. Further carbon sequestration comparisons could be made ward-wise or at the city level using this methodology. Carbon sequestration potential can be improved by increasing green cover and improving water bodies, which also sequester carbon. Improving public transport for better connectivity and using sustainable solutions will help reduce the city's carbon emissions. However, there are limitations in the estimation process as the age of trees and other parameters were unknown. An in-depth study of the sequestration process is needed for each tree species. Using the current methodology, an assessment of carbon reduction by trees could guide decision-makers in shaping policies to improve the quality of life in urban areas. Policies should mandate the inclusion of green spaces along roads with heavy vehicular traffic, in industrial areas, and in other regions with high carbon emissions. Tree species with high sequestration potential should be identified and promoted to act as carbon sinks, naturally sequestering carbon released due to human activities.

## 10.8 CONCLUSION AND FUTURE SCOPE

The study successfully estimated the carbon storage capacity of an urban park, validated through primary surveys. Urban parks not only reduce urban heat but also serve as small biodiversity habitats, creating a bridge between carbon emissions and carbon sequestration. They offer a promising avenue for carbon sequestration and climate mitigation, but their full potential can only be realized through careful planning, management, and coordination with other sustainability initiatives. The use of i-Tree software can help cities measure and manage the carbon stored in their parks and other urban green spaces. The i-Tree tool provides valuable data on carbon storage potential of trees, as well as their ecological and economic benefits. The sequestration of a fully grown tree represents the amount of carbon stored over the years. As discussed in this study, various aspects of carbon sequestration were noticed, and the potential of the park for sequestration was assessed through surveys and various tools. The validation of surveys was conducted, determining the accuracy levels and providing insights into the precision of tools that use remotely sensed images. Species with higher carbon sequestration potential can be identified and promoted for use in green belts and urban forestry. Further research could be estimate the potential of trees at the city level and their ability to mitigate emissions from various sector.

## REFERENCES

Baldocchi, D. (2008). Breathing of the terrestrial biosphere: Lessons learned from a global network of carbon dioxide flux measurement systems. *Australian Journal of Botany*, *56*, 1–26.

Behera, S. K., Mishra, S., Sahu, N., Manika, N., Singh, S. N., Anto, S., Kumar, R., Husain, R., Verma, A. K., Pandey, N. (2022). Assessment of carbon sequestration potential of tropical tree species for urban forestry in India. *Ecological Engineering*, *181*, 106692.

Falkowski, P. G. (1997). Evolution of the nitrogen cycle and its influence on biological sequestration of $CO_2$ in the ocean. *Nature*, *387*(6630), 272–275.

Huston, M. A., Marland, G. (2003). Carbon management and biodiversity. *Journal of Environmental Management*, *67*(1), 77–86.

Intergovernmental Panel on Climate Change. (2005). Carbon Dioxide Capture and Storage.

Lal, R. (2008). Carbon sequestration. *Philosophical Transactions of the Royal Society B: Biological Sciences*, *363*(1492), 815–830.

Nowak, D. (1994). Air pollution removal by Chicago's urban forest. In *Chicago's Urban Forest Ecosystem: Results of the Chicago Urban Forest Climate Project*, pp. 63–81. Radnor, PA: U.S. Department of Agriculture, Forest Service, Northeastern Forest Experiment Station.

Ramachandra, T. V., Aithal, B. H., Sreejith, K. (2015). GHG footprint of major cities in India. *Renewable and Sustainable Energy Reviews*, *44*, 473–495.
School of Planning and Architecture. (n.d.). *Climate Informed Environmental Planning for the Smart Cities of Madhya Pradesh*. Prepared by School of Planning and Architecture, Bhopal for Centre for Climate Change & Sustainable Development, October 2019.
Velasco, E., Roth, M., Norford, L., Molina, L. T. (2016). Does urban vegetation enhance carbon sequestration? *Landscape and Urban Planning*, *148*, 99–107.
Zhang, D., Song, J. (2014). Mechanisms for geological carbon sequestration. *Procedia IUTAM*, *10*, 319–327.

Chapter 11

# Evolution of carbon capture, utilization, and storage technology

## Past, present, and future developments

*Ashwani Kumar*

### 11.1 INTRODUCTION

Carbon capture, utilization, and storage (CCUS) has emerged as a critical solution in the global effort to combat climate change. As the world grapples with the escalating challenges posed by rising carbon dioxide ($CO_2$) levels, technological innovations in CCUS have become pivotal in mitigating the impact of anthropogenic emissions (Figures 11.1 and 11.2). This comprehensive review explores the recent advancements in CCUS technologies, focusing on their effectiveness, scalability, and potential impact on sustainable development [1]. Table 11.1 assesses the key parameters typically considered in the assessment of $CO_2$ purity, with the importance of each parameter varying depending on the end use of $CO_2$.

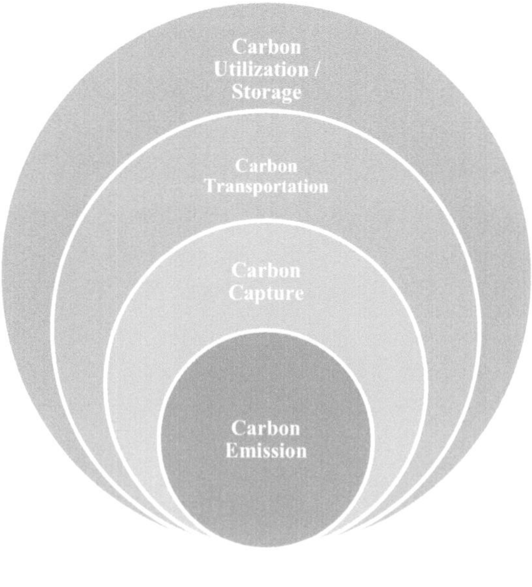


*Figure 11.1* Carbon emission to carbon utilization and storage.


DOI: 10.1201/9781003521341-11

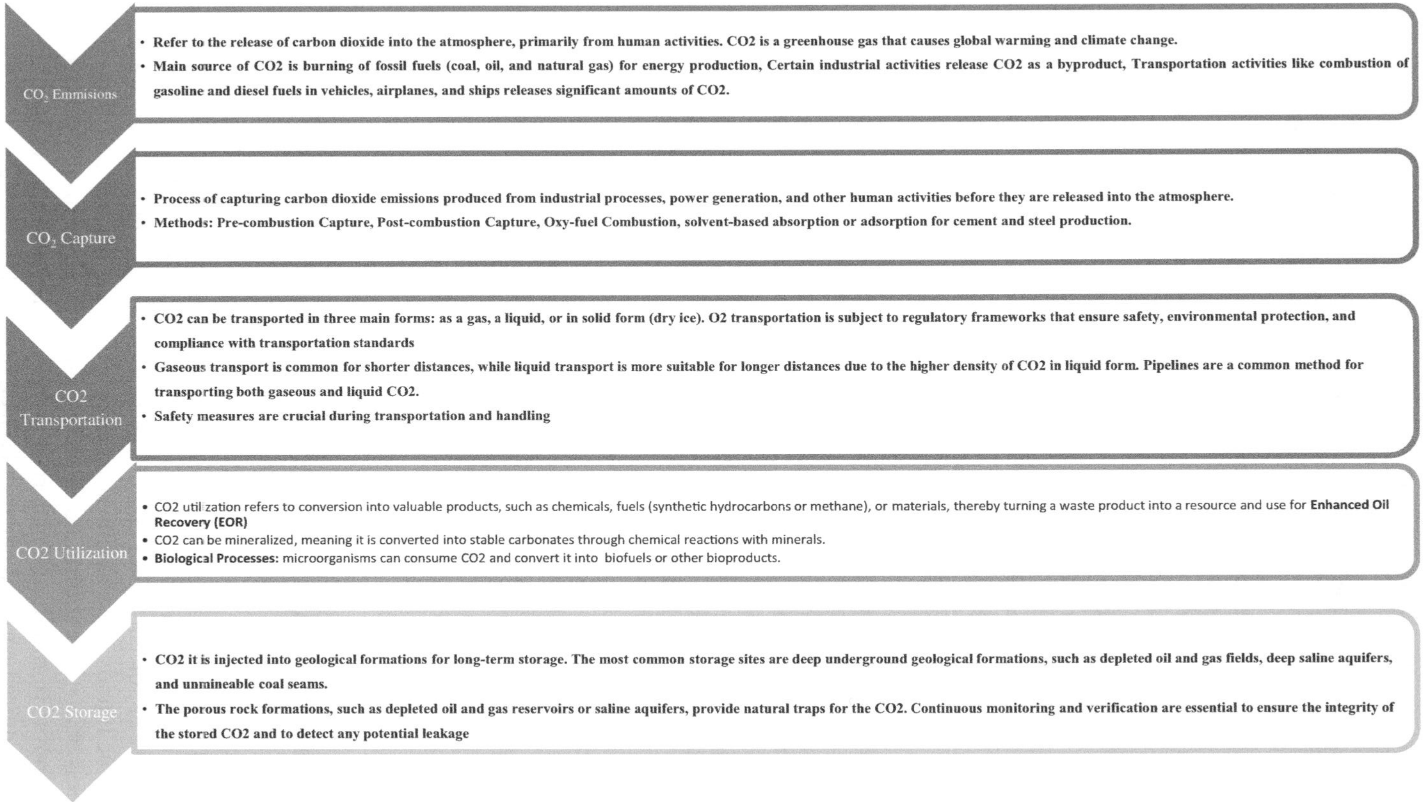


*Figure 11.2* Key points of CCUS [2,3].

*Table 11.1* Assessment of the different parameters affecting $CO_2$ purity [4–9]

| *Key parameter* | *Details* | *Importance* | *Measurement method* | *Acceptable range* |
|---|---|---|---|---|
| Source of $CO_2$ | Origin of the $CO_2$ (e.g., combustion and fermentation) | High | Traceability to source | N/A |
| Impurities | Presence of contaminants (e.g., $H_2O$ and hydrocarbons) | High | Gas chromatography and FTIR spectroscopy | <0.1% |
| Moisture content | Water vapor content | High | Moisture analyzers | <10ppm |
| Pressure | Operating pressure of $CO_2$ | Medium | Pressure gauges | Specified range |
| Temperature | Operating temperature of $CO_2$ | Medium | Temperature sensors | Specified range |
| Purity level | Level of $CO_2$ purity required | High | Gas analyzers and chromatography | Specified level |
| Storage and transportation | Conditions during storage and transportation | Medium | Monitoring during transit/storage | Adherence to standards |
| Certification and compliance | Adherence to industry standards and regulations | High | Compliance certificates | Compliance with standards |
| Contamination prevention | Measures taken to prevent contamination | Medium | Filtration, purification processes | Adherence to protocols |
| Handling procedures | Protocols for handling, filling, and distribution | Medium | Standard operating procedures | Adherence to protocols |
| Quality control | Quality control processes and checks | High | Regular testing and analysis | Adherence to standards |

## 11.2 CARBON CAPTURE TECHNOLOGIES

This section delves into various carbon capture technologies that have gained prominence in recent years. Among these, chemical absorption, adsorption, and membrane separation are widely employed methods. Advances in solvent-based capture systems, such as amine-based solvents and ionic liquids, have significantly improved the efficiency and cost-effectiveness of carbon capture. Additionally, breakthroughs in solid adsorbents and innovative membrane materials have provided alternatives with lower energy requirements and enhanced selectivity [10,11].

*Table 11.2* Explores technological advancements in DAC [12,13]

| *Category* | *Advancements* |
|---|---|
| **Sorbent materials** | **Novel materials:** Introduction of advanced sorbent materials with enhanced $CO_2$ adsorption capacity and selectivity. Materials such as metal–organic frameworks and amine-functionalized polymers are under exploration |
| | **Regeneration techniques:** Development of more efficient and cost-effective methods for regenerating sorbent materials, improving their reusability and overall performance |
| | **Tuning properties:** Research into the modification of sorbent material properties to optimize performance under different environmental conditions and concentrations of $CO_2$ |
| **Air contactors** | **Design optimization:** Advancements in the design of air contactors to improve mass transfer efficiency and reduce energy consumption in the capture process |
| | **Modular systems:** Exploration of modular air contactor systems that can be scaled up or down based on the specific requirements of the DAC facility |
| | **Heat management:** Implementation of innovative heat management techniques to enhance the overall efficiency of the air contactor, reducing energy requirements |
| **Integration of renewable energy** | **Renewable-powered DAC:** Integration of renewable energy sources such as solar and wind to power DAC facilities, reducing the carbon footprint of the overall capture process |
| | **Hybrid systems:** Development of hybrid systems that combine DAC with renewable energy storage technologies to ensure a continuous and stable power supply |
| | **Energy-efficient processes:** Optimization of DAC processes to operate efficiently with intermittent renewable energy sources, allowing for greater flexibility in deployment |

The following are carbon capture technologies currently being developed or implemented:

1. **Direct Air Capture (DAC):** DAC technology involves capturing carbon dioxide directly from the ambient air. Companies like Carbon Engineering and Climeworks have been working on DAC systems. DAC has garnered attention as a technology capable of extracting $CO_2$ directly from the ambient air. Table 11.2 highlights recent innovations in DAC, including advancements in sorbent materials, air contactors, and the integration of renewable energy sources [12].
2. **Bioenergy with Carbon Capture and Storage (BECCS):** BECCS combines bioenergy (usually from biomass) with carbon capture and storage (CCS) to remove carbon dioxide from the atmosphere. This technology is still in the early stages of development.

3. **Post-Combustion Capture:** This technology involves capturing carbon dioxide emissions after the combustion of fossil fuels. Amine-based solvents are commonly used in post-combustion capture processes.
4. **Pre-Combustion Capture:** This approach involves capturing carbon dioxide before the combustion of fossil fuels. Gasification of coal or biomass is one method used in pre-combustion capture.
5. **Oxy-Fuel Combustion:** In oxy-fuel combustion, fossil fuels are burned in an atmosphere of pure oxygen, resulting in a flue gas primarily composed of carbon dioxide and water vapor, which makes carbon capture more feasible.
6. **Mineralization:** Some technologies focus on the mineralization of carbon dioxide, where $CO_2$ is captured and converted into stable mineral forms. This process is often referred to as mineral carbonation.
7. **Enhanced Weathering:** This concept involves accelerating the natural process of weathering to capture and store carbon dioxide. The idea is to expose minerals that react with $CO_2$ and convert it into stable carbonates.
8. **Membrane-based Capture:** Membrane separation technologies are being explored for capturing $CO_2$ from industrial processes. Membranes selectively allow certain gases to pass through while capturing $CO_2$.

Table 11.3 shows the effectiveness and applicability of CCUS technologies vary based on specific project requirements, site conditions, and regulatory considerations.

## 11.3 CARBON UTILIZATION TECHNOLOGIES

Carbon capture is only one part of the solution; utilizing captured carbon for value-added products is equally crucial. Table 11.4 highlights recent innovations in carbon utilization technologies (CUT), including the conversion of $CO_2$ into fuels, chemicals, and materials. Table 11.5 shows the environmental implications of electrochemical and biological processes for converting $CO_2$ into valuable products. The environmental impact can vary depending on the specific technology, energy source used, and scale of implementation [17–19].

## 11.4 UNDERGROUND STORAGE AND ENHANCED OIL RECOVERY

The storage of captured carbon in geological formations is a key component of CCUS. CCUS can be integrated with enhanced oil recovery (EOR) techniques to improve oil production while mitigating greenhouse gas

*Table 11.3* Different CCUS technologies and industrial applications [14–16]

| *Carbon capture technology* | *Key parameters* | *Purpose* | *Technology used* | *Targeted industry* |
|---|---|---|---|---|
| **Post-combustion capture** | Absorption efficiency, solvent regeneration, and energy consumption | Capture $CO_2$ emissions from flue gas after combustion | Solvent-based absorption processes (e.g., amine scrubbing) | Power plants, industrial facilities |
| **Pre-combustion capture** | Gasification efficiency, hydrogen production, and overall plant efficiency | Capture $CO_2$ before combustion by converting fuel into syngas | Gasification technology with a water–gas shift reaction | Integrated gasification combined cycle plants |
| **Oxy-fuel combustion** | Oxygen purity, combustion temperature, and flue gas treatment | Burn fuel in oxygen-rich environment, producing a concentrated $CO_2$ stream | Oxygen-fired combustion with subsequent $CO_2$ separation | Power plants, industrial furnaces |
| **DAC** | Air contact time, sorbent capacity, and energy consumption | Extract $CO_2$ directly from ambient air | Chemical sorbents or solid materials with high affinity for $CO_2$ | General applications for atmospheric carbon removal |
| **BECCS** | Biomass feedstock, carbon sequestration in biomass and storage | Combine bioenergy production with carbon capture from flue gases | Biomass combustion or gasification, followed by $CO_2$ capture | Power generation from biomass and industrial processes |
| **Mineralization** | Mineral reactivity, $CO_2$ conversion rate, and by-products | Convert $CO_2$ into stable minerals through chemical reactions | Injection of $CO_2$ into geological formations or reaction with alkaline minerals | Geological sequestration, waste utilization |
| **EOR** | Oil recovery efficiency, $CO_2$ injection rate, and reservoir characteristics | Use captured $CO_2$ to enhance the recovery of oil from depleted fields | Injection of $CO_2$ into oil reservoirs to reduce viscosity and increase oil mobility | Oil and gas industry |

*Table 11.4* Carbon utilization technologies and recent developments [20–25]

| *Technology* | *Application* | *Key points* | *Key industry* | *Current status* |
|---|---|---|---|---|
| Direct air capture | Fuel production | Captures $CO_2$ directly from the atmosphere for conversion | Climeworks and Carbon Engineering | Ongoing commercialization |
| Electrochemical conversion | Fuels and chemicals | Uses electrochemical processes to convert $CO_2$ into fuels or chemicals | Carbon Clean Solutions and Electrochaea | Ongoing development |
| **Carbon-to-fuel technologies** | | | | |
| Synthetic fuels | Liquid fuels | Produces synthetic fuels using renewable energy and captured $CO_2$ | Sunfire and INERATEC | Pilot projects (2020s) |
| Power-to-gas | Gaseous fuels | Converts $CO_2$ into methane or hydrogen for energy storage | Siemens and Audi | Pilot projects (2020s) |
| **Carbon-to-chemicals technologies** | | | | |
| $CO_2$-based polymers | Materials | Converts $CO_2$ into polymers and materials | Covestro and Novomer | Research and development (2020s) |
| Carbonate mineralization | Building materials | Converts $CO_2$ into carbonates for use in construction materials | CarbonCure and Solidia Technologies | Commercial deployment (2020s) |
| **Biological conversion** | | | | |
| Algae-based systems | Biomass production | Utilizes algae to capture $CO_2$ and produce biomass for various applications | AlgaeCytes and Pond Technologies | Research and pilot projects (2020s) |
| **Advanced technologies** | | | | |
| Photocatalytic conversion | Solar-driven processes | Uses sunlight to convert $CO_2$ into fuels or chemicals | LanzaTech and Joint Center for Artificial Photosynthesis | Research stage (2020s) |
| Artificial Intelligence (AI) | Process optimization | Applies AI to optimize carbon utilization processes for efficiency | C-Capture and Climeworks | Integration into existing technologies (2020s) |

*Table 11.5* Integration of CCUS and EOR

| *EOR technique* | *Key features* |
|---|---|
| **EOR** | Various EOR techniques can be employed, including water flooding, gas injection, and chemical injection |
| **Water flooding with $CO_2$** | $CO_2$ is injected into the oil reservoir along with water. The injected $CO_2$ dissolves in oil, reducing its viscosity and improving sweep efficiency, leading to enhanced oil recovery |
| **Gas injection ($CO_2$ or $N_2$)** | Injection of $CO_2$ or nitrogen ($N_2$) into the reservoir to maintain reservoir pressure and displace oil toward production wells. $CO_2$ can also react with the oil, improving recovery |
| **Chemical EOR with $CO_2$** | $CO_2$ can be used as a solvent in combination with chemicals to improve the mobility of oil. This method is known as $CO_2$-assisted chemical flooding |
| **Storage of captured $CO_2$** | Captured $CO_2$ is injected and stored in geological formations (such as depleted oil reservoirs or deep saline aquifers) to prevent its release into the atmosphere |
| **Benefits** | (i) Increased oil recovery. (ii) Significant reduction in greenhouse gas emissions. (iii) Geological storage of $CO_2$ contributes to long-term carbon sequestration |
| **Challenges** | (i) High upfront costs. (ii) Technological and logistical challenges in capturing and transporting $CO_2$. (iii) Monitoring and verifying the stored $CO_2$ for environmental and regulatory compliance |

emissions. Table 11.5 evaluates cutting-edge underground storage technologies, examining their safety, longevity, and storage capacity. It also explores the potential of combining carbon capture, utilization, and storage (CCUS) with enhanced oil recovery (EOR) techniques. This integration can offer economic advantages by capturing carbon while simultaneously increasing oil production.

## 11.5 TECHNOLOGICAL CHALLENGES AND OPPORTUNITIES

Despite significant progress, CCUS technologies still face various challenges. Table 11.6 critically evaluates the technological barriers, including high costs, energy requirements, and the need for infrastructure development. Table 11.6 also identifies emerging opportunities, such as policy support, international collaboration, and the role of public–private partnerships in advancing CCUS deployment on a global scale.

*Table 11.6* Technological challenges and opportunities related to CCUS [26–28]

| *Category* | *Technological challenges* | *Technological opportunities* |
|---|---|---|
| **Capture technologies [29]** | High energy costs and efficiency losses in capture processes | Advancements in solvent-based and solid adsorbent technologies |
| | Availability of suitable solvents for various flue gases | Integration of capture with industrial processes |
| | Scale-up challenges for new capture technologies | Modular and scalable capture systems |
| | Reduction of environmental impact of capture processes | Development of low-energy and environmentally friendly solvents |
| **Transportation** | Infrastructure for $CO_2$ transportation and pipeline networks | Innovations in pipeline materials and monitoring technologies |
| | Energy-efficient compression and transportation methods | Advanced sensors for leak detection |
| | Regulatory challenges related to transportation and storage | Streamlined permitting processes for pipeline construction |
| **Storage technologies [30–32]** | Long-term storage integrity and leakage risks | Enhanced monitoring and verification technologies |
| | Identification and characterization of suitable storage sites | Research on alternative storage methods (e.g., mineralization) |
| | Public acceptance and regulatory frameworks for storage | Development of risk assessment and mitigation strategies |
| **Carbon utilization** | Limited economic viability of CUT | Exploration of new and improved carbon utilization pathways |
| | Scalability and integration with existing industries | Collaboration between industries for integrated solutions |
| | Development of markets and incentives for carbon products | Policies promoting the use of carbon-based products |
| **Cross-cutting issues [33–35]** | High upfront costs for CCUS implementation | Financial incentives, subsidies, and carbon pricing mechanisms |
| | Lack of standardized protocols and regulations | Development of international standards and agreements |
| | Research and development funding for CCUS innovations | Increased public and private investment in CCUS research [36] |

## 11.6 ENVIRONMENTAL AND SOCIAL IMPACTS

As CCUS technologies become more prevalent, it is essential to assess their environmental and social implications. Table 11.7 examines the potential environmental impacts of CCUS, such as leakage risks and induced

*Table 11.7* The environmental effect of electrochemical and biological processes based on CUT [28–32]

| *Environmental impact* | *Electrochemical processes* | *Biological processes* |
|---|---|---|
| **Greenhouse gas emissions** | Typically low emissions, especially if renewable energy is used in the process | Minimal to low emissions, as biological processes often use microorganisms that do not produce significant greenhouse gases |
| **Energy consumption** | Energy-intensive, especially if electricity is used for electrochemical conversion | Energy consumption varies, but biological processes can be less energy-intensive than electrochemical methods. They may utilize natural processes and organisms [46–51] |
| **Waste generation** | Some waste generation, depending on the specific electrochemical process and the materials used | Biological processes can generate by-products, but they are often biodegradable and may be less of a concern than some electrochemical waste products |
| **Resource utilization** | Requires materials for electrodes, catalysts, and other components. The extraction and processing of these materials may have environmental impacts | Relies on biological systems, which may have lower resource requirements than some electrochemical processes |
| **Land use** | Typically requires less land than some biological processes | Biological processes may require larger land areas for cultivation of microorganisms or plants used in the conversion process |
| **Chemical use** | The use of chemicals for catalysts and electrode materials may have environmental implications, especially if not handled properly | Relies on biological agents that may have lower environmental impact, but the use of growth media and nutrients should be considered |
| **Scalability** | Electrochemical processes can be more easily scaled up and integrated into existing industrial infrastructure | Biological processes may face challenges in scalability, and large-scale implementation may require careful optimization and control |
| **Economic viability** | Can be economically viable, especially as technology advances and becomes more efficient | Economic viability may vary and is often influenced by factors such as the type of biological process and the products generated |
| **Environmental benefits** | Can contribute to reducing $CO_2$ emissions by converting $CO_2$ into valuable products | Offers a potential way to sequester carbon and produce valuable compounds using natural biological processes [52–58] |

seismicity, while also considering the social acceptance and equity aspects of large-scale deployment. Table 11.7 provides a balanced discussion on the benefits and concerns associated with CCUS is provided, emphasizing the importance of a holistic approach to technology deployment [37–45].

## 11.7 POLICY AND REGULATORY LANDSCAPE

The success of CCUS relies heavily on a supportive policy and regulatory framework. Table 11.8 highlights recent developments in global policies related to CCUS. Table 11.8's analysis also includes incentives, regulations,

*Table 11.8* Developments in global policies related to CCUS [32–36]

| *Country* | *CCUS policy* |
|---|---|
| United States | **45Q Tax Credit:** Provides tax credits for CCS projects.<br>**USE IT Act:** Encourages the development of carbon capture technologies and infrastructure. |
| European Union | **EU Emission Trading System:** Sets limits on greenhouse gas emissions and includes provisions for carbon capture projects.<br>**Innovation Fund:** Supports projects including CCUS to accelerate the transition to a low-carbon economy. |
| Canada | **Output-Based Pricing System:** Imposes a carbon price on industrial facilities and provides incentives for emission reduction technologies, including CCUS.<br>**CCS Regulatory Framework:** Provides guidelines for the development and operation of CCUS projects. |
| Norway | **Full-Scale CCS Project at Sleipner and Snøhvit:** Regulations and policies supporting the development of large-scale CCUS projects in collaboration with industry. |
| United Kingdom | **CCS Infrastructure Fund:** Provides funding for the development of CCUS infrastructure, including CCS facilities.<br>**CCUS Deployment Pathway:** Outlines the government's strategy for the deployment of CCUS in the United Kingdom. |
| Australia | **Emissions Reduction Fund:** Supports projects that reduce emissions, including CCUS projects.<br>**National Carbon Offset Standard:** Provides guidelines for achieving carbon neutrality, which may involve CCUS projects. |
| International Energy Agency | **CCS Guidelines and Best Practices:** Develops guidelines and best practices for the implementation of CCS technologies globally.<br>**Clean Energy Ministerial:** Encourages international collaboration on clean energy technologies, including CCUS. |

and international agreements aimed at promoting the widespread adoption of CCUS technologies. It also explores the role of carbon pricing mechanisms in driving investments and innovation in the CCUS sector.

## 11.8 APPLICATION OF CCUS

The global application of CCUS projects presents a complex landscape with both promises and challenges. On the positive side, CCUS has gained attention as a potential mitigation strategy to address carbon emissions from industrial processes and power generation. Table 11.9 shows notable projects underway globally, reflecting a growing recognition of the need for innovative solutions to combat climate change. However, the success of CCUS hinges on addressing key issues such as high costs, regulatory frameworks, and public acceptance [59–67]. The geographic distribution of CCUS projects is uneven, with some regions demonstrating stronger commitments

*Table 11.9* Notable CCUS projects around the globe

| *Project name* | *Location* | *Status* | *Capacity/scale* | *Technology* |
|---|---|---|---|---|
| Sleipner | North Sea, Norway | Operational | 1 million tons of $CO_2$ per year | CCS |
| Quest | Alberta, Canada | Operational | 1 million tons of $CO_2$ per year | CCS |
| Petra Nova | Texas, USA | Operational | 1.4 million tons of $CO_2$ per year | CCS |
| Gorgon | Western Australia | Operational | About 4 million tons of $CO_2$ per year | CCS |
| Boundary Dam | Saskatchewan, Canada | Operational | 1 million tons of $CO_2$ per year | CCS |
| Kemper County | Mississippi, USA | Operational (as natural gas plant) | 3 million tons of $CO_2$ per year (planned for CCS) | Integrated Gasification Combined Cycle with CCS |
| Northern Lights | Norway | Under development | Planned capacity of 1.5 million tons of $CO_2$ per year | CCS |
| Port of Rotterdam CCS | Netherlands | Under development | Planned capacity of 2.5 million tons of $CO_2$ per year | CCS |
| Acorn CCS | Scotland, UK | Under development | Planned capacity of 5 million tons of $CO_2$ per year | CCS |
| Project Tundra | North Dakota, USA | Under development | Planned capacity of 1.5 million tons of $CO_2$ per year | CCS |

than others. While CCUS holds significant potential to reduce carbon dioxide emissions, its widespread adoption requires concerted international efforts to overcome technical, economic, and social barriers, ensuring a balanced and equitable deployment across the globe [68–76].

Table 11.9 lists some notable CCUS projects that are either in operation or under development globally.

## 11.9 CONCLUSION

In conclusion, the comprehensive review of CCUS underscores its pivotal role as a potential mitigation strategy in the global fight against climate change.

- The research illuminates the landscape of CCUS technologies, encapsulating the latest advancements in capture methods, utilization strategies, and storage techniques. The study emphasizes the importance of a supportive policy and regulatory framework while also addressing the economic feasibility and global regulatory landscape surrounding these innovations.
- The positive aspects of CCUS, as highlighted in the analysis of recent global policies and ongoing projects, reveal a growing recognition of the need for innovative solutions to combat climate change.
- However, the success of CCUS depends on overcoming significant challenges, including high costs, regulatory hurdles, and public acceptance. The uneven geographic distribution of CCUS projects indicates the necessity for concerted international efforts to address technical, economic, and social barriers, ensuring a balanced and equitable deployment across the globe.
- As the world navigates the complex landscape of climate change mitigation, CCUS emerges as a testament to the ongoing pursuit of sustainable solutions, with the potential to reshape our approach to carbon management.
- Despite its potential, the critical analysis serves as a valuable resource for researchers, policymakers, and industry stakeholders.
- It highlights the need for a holistic approach that takes into account not only technological advancements but also the socioeconomic and regulatory factors that influence the success and widespread adoption of CCUS technologies.

In essence, while CCUS holds significant promise to reducing carbon dioxide emissions, its journey to widespread adoption and success is a multifaceted challenge that requires global collaboration and a steadfast commitment to overcoming barriers, contributing to a more resilient and environmentally conscious future.

## REFERENCES


1. Kramer, G. J., Haigh, M. (2009). No quick switch to low-carbon energy. *Nature* 462, 568–569.
2. Belhaj, H., Bera, A. (2017). A brief review of mechanisms for carbon dioxide sequestration into aquifer reservoirs. *International Journal of Petroleum Engineering* 3, 49–66.
3. Aminu, M. D., Nabavi, S. A., Rochelle, C. A., Manovic, V. (2017). A review of developments in carbon dioxide storage. *Applied Energy* 208, 1389–1419.
4. Szima, S., Cormos, C. C. (2021). $CO_2$ utilization technologies: A techno-economic analysis for synthetic natural gas production. *Energies* 14, 1258.
5. Dincer, I. (2000). Renewable energy and sustainable development: A crucial review. *Renewable and Sustainable Energy Reviews* 4(2), 157–175.
6. IPCC, Masson-Delmotte, V., Zhai, P., Potner, H.-O., Roberts, D., Skea, J., Shukla, P. R., Pirani, A., Moufouma-Okia, W., Péan, C., Pidcock, R., et al., (eds.) (2018). *Global Warming of 1.5°C*. Cambridge University Press: Cambridge, UK; New York.
7. https://carbonengineering.com/, accessed on 22/12/2023.
8. https://climeworks.com/direct-air-capture, accessed on 22/12/2023.
9. IPCC, Metz, B., Davidson, O., de Coninck, H. C., Loos, M., Meyer, L. A., (eds.) (2005). *IPCC Special Report on Carbon Dioxide Capture and Storage*, Prepared by Working Group III of the intergovernmental panel on climate change. Cambridge University Press: Cambridge, UK; New York.
10. Nam, K. M., Waugh, C. J., Paltsev, S., Reilly, J. M., Karplus, V. J. (2014). Synergy between pollution and carbon emissions control: Comparing China and the United States. *Energy Economics* 46, 186–201.
11. Nasrollahi, Z., Hashemi, M., Bameri, S., Mohamad Taghvaee, V. (2018). Environmental pollution, economic growth, population, industrialization, and technology in weak and strong sustainability: Using STIRPAT model. *Environment Development and Sustainability* 22(2), 1105–1122.
12. Wan, Y. H. (2022). A techno-economic review on carbon capture, utilization and storage systems for achieving a net–zero $CO_2$ emissions future. *Carbon Capture Science & Technology* 3, 100044.
13. Zhao, Y., Itakura, K. (2023). A state-of-the-art review on technology for carbon utilization and storage. *Energies* 16(10), 3992. https://doi.org/10.3390/en16103992.
14. Saboori, B., Rasoulinezhad, E., Sung, J. (2017). The nexus of oil consumption, $CO_2$ emissions and economic growth in China, Japan and South Korea. *Environmental Science and Pollution Research* 24(8), 7436–7455.
15. He, Y., Wang, Y., Zhang, L., Yang, Z., Yu, L. (2021). Public attitudes toward carbon capture, utilization and storage (CCUS) technology in the United States: A survey-based study. *Journal of Cleaner Production* 304, 127092.
16. Kapila, S., Oni, A. O., Gemechu, E. D., Kumar, A. (2019). Development of net energy ratios and life cycle greenhouse gas emissions of large–scale mechanical energy storage systems. *Energy* 170, 592–603.
17. Vandyck, T., Keramidas, K., Kitous, A., Spadaro, J. V., Van Dingenen, R., Holland, M., Saveyn, B. (2018). Air quality co-benefits for human health and agriculture counterbalance costs to meet Paris Agreement pledges. *Nature Communications* 9(1), 4939.

18. Solomon, S., Plattner, G. K., Knutti, R., Friedlingstein, P. (2009). Irreversible climate change due to carbon dioxide emissions. *Proceedings of the National Academy of Sciences of the United States of America* 106, 1704–1709.
19. Chunbao Charles, X. U., Cang, D. Q. (2010). A brief overview of low CO2 emission technologies for iron and steel making. *Journal of Iron and Steel Research International*, 17, 1–7.
20. Bodor, M., Santos, R. M., Van Gerven, T., Vlad, M. (2013). Recent developments and perspectives on the treatment of industrial wastes by mineral carbonation: A review. *Central European Journal of Engineering* 3, 566–584.
21. Osman, A. I., Hefny, M., Abdel Maksoud, M. I. A., et al. (2021). Recent advances in carbon capture storage and utilisation technologies: A review. *Environmental Chemistry Letters* 19, 797–849. https://doi.org/10.1007/s10311-020-01133-3.
22. Fasihi, M., Efimova, O., Breyer, C. (2019). Techno-economic assessment of CO2 direct air capture plants. *Journal of Cleaner Production* 224, 957–980. https://doi.org/10.1016/j.jclepro.2019.03.086.
23. Chen, C., Tavoni, M. (2013). Direct air capture of $CO_2$ and climate stabilization: A model based assessment. *Climatic Change* 118, 59–72. https://doi.org/10.1007/s10584-013-0714-7.
24. Wise, M., Calvin, K., Thomson, A., Clarke, L., Bond-Lamberty, B., Sands, R., Smith, S. J., Janetos, A., Edmonds, J. (2009). Implications of limiting $CO_2$ concentrations for land use and energy. *Science* 324(5931), 1183–1186.
25. Alvarez-Herranz, A., Balsalobre-Lorente, D., Shahbaz, M., Cantos, J. M. (2017). Energy innovation and renewable energy consumption in the correction of air pollution levels. *Energy Policy* 105, 386–397.
26. Boadway, R., Song, Z., Tremblay, J. F., European Journal of Political Economy (2011). The efficiency of voluntary pollution abatement when countries can commit. *European Journal of Political Economy* 27(2), 352–368.
27. Bond, T. C., Wehner, B., Plewka, A., Wiedensohler, A., Heintzenberg, J., Charlson, R. J. (2006). Climate-relevant properties of primary particulate emissions from oil and natural gas combustion. *Atmospheric Environment* 40(19), 3574–3587.
28. Colvile, R., Hutchinson, E., Mindell, J., Warren, R. (2001). The transport sector as a source of air pollution. *Atmospheric Environment* 35(9), 1537–1565.
29. Seneviratne, S. I., Donat, M. G., Pitman, A. J., Knutti, R., Wilby, R. L. J. N. (2016). Allowable $CO_2$ emissions based on regional and impact-related climate targets. *Nature Communications* 529(7587), 477–483.
30. Roggero, M., Kähler, L., Hagen, A., (2019). Strategic cooperation for transnational adaptation: Lessons from the economics of climate change mitigation. *International Environmental Agreements: Politics Law and Economics* 19(4–5), 395–410.
31. Sawlani, R., Agnihotri, R., Sharma, C. (2020). Chemical and isotopic characteristics of $PM_{2.5}$ over New Delhi from September 2014 to May 2015: Evidences for synergy between air-pollution and meteorological changes. *Science of the Total Environment* 763, 142966.
32. Böhringer, C., Garcia-Muros, X., Cazcarro, I., Arto, I. (2017). The efficiency cost of protective measures in climate policy. *Energy Policy* 104, 446–454.
33. Oh, I., Yoo, W. J., Yoo, Y. (2019). Impact and interactions of policies for mitigation of air pollutants and greenhouse gas emissions in Korea. *International Journal of Environmental Research and Public Health* 16(7), 1161.

34. Baron, R. M., Kenny, D. A. (1986). The moderator–mediator variable distinction in social psychological research: Conceptual, strategic, and statistical considerations. *Journal of Personality and Social Psychology* 51(6), 1173.
35. Potoski, M., Woods, N. D. J. P. S. J. (2002). Dimensions of state environmental policies: Air pollution regulation in the United States. *Policy Studies Journal* 30(2), 208–226.
36. Roggero, M., Kähler, L., Hagen, A. (2019). Strategic cooperation for transnational adaptation: Lessons from the economics of climate change mitigation. *International Environmental Agreements: Politics Law and Economics* 19(4–5), 395–410.
37. Kumar, A., et al. (2021). *Advanced Computational Methods in Mechanical and Materials Engineering.* CRC Press: Boca Raton, FL. https://doi.org/10.1201/9781003202233.
38. Prasad, A., et al. (2022). *Biodegradable Composites for Packaging Application.* CRC Press: Boca Raton, FL. https://doi.org/10.1201/9781003227908.
39. Gajrani, K. K., et al. (2022). *Advances in Sustainable Machining and Manufacturing Processes.* CRC Press: Boca Raton, FL. https://doi.org/10.1201/9781003284574.
40. Singh, S., et al. (2022). *Nanomanufacturing and Nanomaterials Design: Principles and Applications.* Taylor & Francis (CRC Press): Boca Raton, FL. https://doi.org/10.1201/9781003220602.
41. Sharma, V. K., et al. (2022). *Additive Manufacturing in Industry 4.0: Methods, Techniques, Modeling, and Nano Aspects.* CRC Press: Boca Raton, FL. https://doi.org/10.1201/9781003360001.
42. Kumar, A., et al. (2023). *Thermal Energy Systems: Design, Computational Techniques and Applications.* CRC Press: Boca Raton, FL. https://doi.org/10.1201/9781003395768.
43. Kumar, A., et al. (2023). *Laser Based Technologies for Sustainable Manufacturing.* CRC Press: Boca Raton, FL. https://doi.org/10.1201/9781003402398.
44. Kumar, R., et al. (2023). *Sustainable Smart Manufacturing Processes in Industry 4.0.* Taylor & Francis (CRC Press): Boca Raton, FL. https://doi.org/10.1201/9781003436072.
45. Awasthi, M. K., et al. (2024). *Computational Fluid Flow and Heat Transfer: Advances, Design, Control & Applications.* CRC Press: Boca Raton, FL. https://doi.org/10.1201/9781003465171.
46. Kundu, A., et al. (2023). Introduction to thermal energy resources and its smart application. In Kumar, A., et al. (eds.), *Thermal Energy Systems: Design, Computational Techniques, and Applications.* CRC Press: Boca Raton, FL, Chapter 1, pp. 1–15. https://dx.doi.org/10.1201/9781003395768-1.
47. Dewangan, A. K., et al. (2023). Thermal energy storage: Opportunities, challenges and future scope. In Kumar, A., et al. (eds.), *Thermal Energy Systems: Design, Computational Techniques, and Applications.* CRC Press: Boca Raton, FL, Chapter 2, pp. 17–28. https://dx.doi.org/10.1201/9781003395768-2.
48. Kundu, A., et al. (2023). Modelling and simulation of thermal energy system for design optimization. In Kumar, A., et al. (eds.), *Thermal Energy Systems: Design, Computational Techniques, and Applications.* CRC Press: Boca Raton, FL, Chapter 7, pp. 103–140. https://dx.doi.org/10.1201/9781003395768-7.

49. Singh, S., et al. (2022). Challenges and opportunities in nanomanufacturing. In Singh, S., et al. (eds.), *Nanomanufacturing and Nanomaterials Design: Principles and Applications*. CRC Press, Taylor & Francis: Boca Raton, FL. DOI: 10.1201/9781003220602-2.
50. Singh, V. P., et al. (2022). Nanomanufacturing and design of high-performance piezoelectric nanogenerator for energy harvesting. In Singh, S., et al. (eds.), *Nanomanufacturing and Nanomaterials Design*. CRC Press, Taylor & Francis: Boca Raton, FL. DOI: 10.1201/9781003220602-15.
51. Kumar, A., et al. (2021). Thermal contact conductance prediction using FEM based computational techniques. In Kumar, A., et al. (eds.), *Advanced Computational Methods in Mechanical and Materials Engineering*, pp. 183–220. CRC Press: Boca Raton, FL. DOI: 10.1201/9781003202233-13.
52. Singh, V. P., et al. (2024). *Energy Efficient Vehicles: Technologies and Challenges*. CRC Press: Boca Raton, FL. https://doi.org/10.1201/9781003464556.
53. Shah, S., et al. (2024). *Sustainability in Smart Manufacturing: Trends, Scope, and Challenges*. CRC Press: Boca Raton, FL. https://doi.org/10.1201/9781003467496.
54. Verma, V., et al. (2024). *Highly Efficient Thermal Renewable Energy Systems: Design, Optimization and Applications*. CRC Press: Boca Raton, FL. https://doi.org/10.1201/9781003472629.
55. Kumar, A., et al. (2024). *Applications of Biotribology in Biomedical Systems*. Springer Nature. https://link.springer.com/book/9783031583261.
56. Awasthi, M. K., et al. (2024). Viscous correction to the potential flow analysis of Rayleigh–Taylor instability in a Rivlin–Ericksen viscoelastic fluid layer with heat and mass transfer. *Heat Transfer* 1–17. https://onlinelibrary.wiley.com/doi/10.1002/htj.23076.
57. Dutt, N., et al. (2024). Thermohydraulic performance investigation of solar air heater duct having staggered D-shaped ribs: Numerical approach. *Heat Transfer* 53(3), 1501–1531. https://onlinelibrary.wiley.com/doi/10.1002/htj.22998.
58. Awasthi, M. K., et al. (2024). Electrohydrodynamic capillary instability of Rivlin–Ericksen viscoelastic fluid film with mass and heat transfer. *Heat Transfer* 53(1), 115–133. doi: 10.1002/htj.22944.
59. Verma, V., et al. (2023). Ground and solar assisted heat pump systems for space heating and cooling applications in the northern region of India: A study on energy and CO2 saving potential. *SETA* 59, 103405. https://doi.org/10.1016/j.seta.2023.103405.
60. Dutt, N., et al. (2023). Thermo-hydraulic performance of solar air heater having discrete D-shaped ribs as artificial roughness. *ESPR*. https://doi.org/10.1007/s11356-023-28247-9.
61. Kushwaha, P. K., et al. (2023). Recent advancements in augmentation of solar water heaters using nanocomposites with PCM: Past, present, and future. *Buildings* 13, 79. https://doi.org/10.3390/buildings13010079.
62. Meena, C. S., et al. (2022). Utilization of solar energy for water heating application to improve building energy efficiency: An experimental study. *Buildings* 12, 2166. https://doi.org/10.3390/buildings12122166.
63. Meena, C. S., et al. (2022). Innovation in green building sector for sustainable future. *Energies* 15, 6631. https://doi.org/10.3390/en15186631.
64. Meena, C. S., et al. (2022). Review on boiling heat transfer enhancement techniques. *Energies* 15, 5759. https://doi.org/10.3390/en15155759.

65. Bhoi, S., et al. (2022). Performance evaluation of different coating materials in delamination for micro- milling applications on high-speed steel substrate. *Micromachines* 13, 1277. https://doi.org/10.3390/mi13081277.
66. Reddy, P. N., et al. (2023). CFD simulation and thermal performance optimization of flow in a channel with multiple baffles. *Journal of Heat and Mass Transfer Research* 10(2), 257–268. https://doi.org/10.22075/JHMTR.2023.31108.1458.
67. Paul, A. K., et al. (2022). Review on artificial neural network and its application in the field of engineering. *Journal of Mechanical Engineering: PRAKASH* 1(1), 53–61.
68. Verma, V., et al. (2024). Introduction to thermal energy storage: Solar, geothermal and hydrogen energy. In Verma, V., et al. (eds.), *Highly Efficient Thermal Renewable Energy Systems: Design, Optimization and Applications*. CRC Press: Boca Raton, FL, Chapter 01, pp. 01–22. https://doi.org/10.1201/9781003472629-1.
69. Ahmadizadeh, M., et al. (2024). Technological advancements in sustainable and renewable solar energy systems. In Verma, V., et al. (eds.), *Highly Efficient Thermal Renewable Energy Systems: Design, Optimization and Applications*. CRC Press: Boca Raton, FL, Chapter 02, pp. 23–39. https://doi.org/10.1201/9781003472629-2.
70. Chitt, M., et al. (2024). Green hydrogen productions: Methods, design and smart applications. In Verma, V., et al. (eds.), *Highly Efficient Thermal Renewable Energy Systems: Design, Optimization and Applications*. CRC Press: Boca Raton, FL, Chapter 16, pp. 261–276. https://doi.org/10.1201/9781003472629-16.
71. Khashehchi, M., et al. (2024). Solar desalination techniques: Challenges and opportunities. In Verma, V., et al. (eds.), *Highly Efficient Thermal Renewable Energy Systems: Design, Optimization and Applications*. CRC Press: Boca Raton, FL, Chapter 19, pp. 305–329. https://doi.org/10.1201/9781003472629-19.
72. Kumar, A., et al. (2024). Globalization and international issues in sustainable manufacturing. In Shah, S., et al. (eds.), *Sustainability in Smart Manufacturing: Trends, Scope, and Challenges*. CRC Press: Boca Raton, FL, Chapter 01, pp. 01–18. https://doi.org/10.1201/9781003467496-1.
73. Singh, V. P., et al. (2024). Introduction to energy-efficient vehicles for sustainable transportation: Transitions and challenges. In Singh, V. P., et al. (eds.), Energy Efficient Vehicles*: Technologies and Challenges*. CRC Press: Boca Raton, FL. https://doi.org/10.1201/9781003464556-1.
74. Upadhyay, R. K., et al. (2024). Sustainable transportation. In Singh, V. P., et al. (eds.), *Energy Efficient Vehicles: Technologies and Challenges*. CRC Press: Boca Raton, FL, Chapter 05, pp. 74–95. https://doi.org/10.1201/9781003464556-5.
75. Singh, V. P., et al. (2024). Techno-economic and future aspects of the HEV-EV-FCV: Decarbonization, digitalization and sustainability. In Singh, V. P., et al. (eds.), *Energy Efficient Vehicles: Technologies and Challenges*. CRC Press: Boca Raton, FL, Chapter 11, pp. 212–231. https://doi.org/10.1201/9781003464556-11.
76. Pathak, A., et al. (2024). Application of Laser Technology in the Mechanical and Machine Manufacturing Industry. In Shah, S., et al. (eds.), *Sustainability in Smart Manufacturing: Trends, Scope, and Challenges*. CRC Press: Boca Raton, FL, Chapter 06, pp. 107–155. DOI: 10.1201/9781003402398-6.

Chapter 12

# Optimizing solar air heater performance

## A comparative analysis of artificial roughness geometries

*Ashwani Kumar, Arun Kumar Singh Gangwar, Bipasa Bimalendu Patra, and Gaurav Dwivedi*

## 12.1 INTRODUCTION

Fueled by rapid industrialization, urbanization, and population growth, India faces a growing demand for sustainable and scalable energy solutions. Fortunately, the nation boasts abundant sunshine, making solar energy a transformative force. Recognizing this potential, the Indian government has actively invested in solar initiatives. At the heart of any solar thermal system lies the solar collector, which captures sunlight and converts it into heat energy for various applications. Among different collector designs, flat plate collectors play a vital role in solar air heaters (SAHs) (Figure 12.1). SAHs offer a sustainable and economical solution for space and process heating, making them a promising technology for India [1,2].

India's geographical advantage of abundant sunlight ensures widespread accessibility to solar energy across the nation. Furthermore, the government's commitment to sustainable development aligns with international efforts to combat climate change, solidifying India's position as a leader in the solar energy revolution. The country has made significant progress in deploying solar technologies in recent years, focusing on both solar photovoltaic systems for electricity generation and solar thermal applications like SAHs [3].

SAHs have become a key technology for harnessing solar energy for space and process heating in India. Their development has been marked by continuous progress. Early designs were simple, often consisting of flat plates with basic absorbers like blackened metal. While affordable and accessible, these early designs suffered from low efficiency due to limited heat transfer area and uncontrolled heat loss. Recognizing these limitations, research efforts have focused on improving collector design. Innovations like finned absorbers and selective coatings have significantly improved SAH performance by increasing heat transfer surface area, enhancing solar absorption, and minimizing heat loss. The present landscape reflects a more advanced approach. Modern collectors utilize


DOI: 10.1201/9781003521341-12

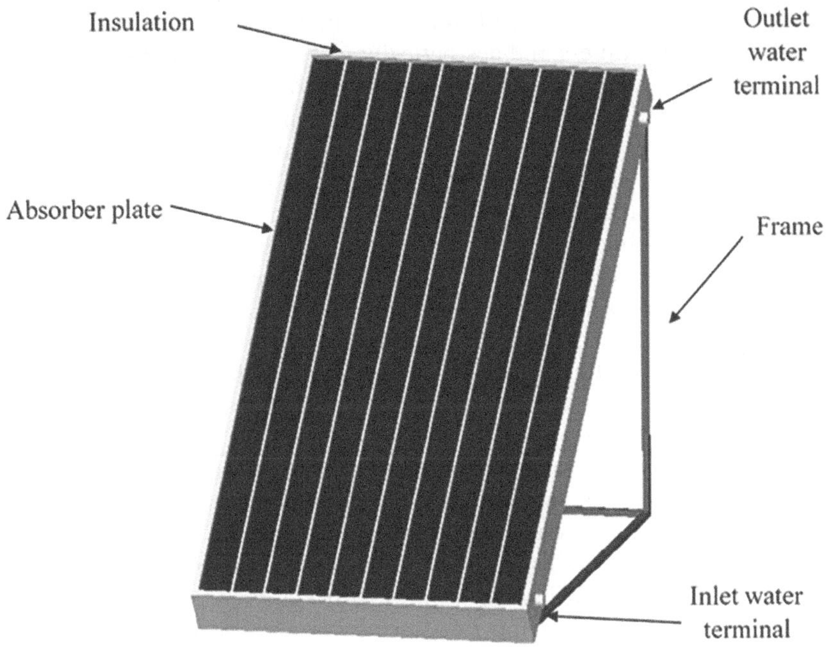


*Figure 12.1* Flat plate solar collectors.

sophisticated materials like transpired absorbers and evacuated tubes for superior heat retention. Additionally, integrated fans and control systems allow for efficient air circulation and targeted heating, optimizing the utilization of captured solar energy. SAH applications have also expanded beyond traditional space heating to include functions like pool heating, ventilation pre-heating, and even crop drying [4,5].

However, challenges remain. The inability to effectively store thermal energy limits SAH operation during non-sunshine hours. Additionally, large-scale integration with buildings necessitates improved architectural considerations. These challenges present exciting opportunities for further innovation in SAH technology.

The future holds immense potential for SAHs, with research actively exploring ways to improve efficiency and broaden their applications. Novel collector designs utilizing nanomaterials are a promising avenue. These materials offer the dual advantage of enhanced solar capture, meaning they absorb more sunlight, and improved thermal conductivity, allowing for better heat transfer within the collector. Another exciting development is the investigation of thermal energy storage (TES) solutions using phase change materials (PCMs). These materials can store excess heat generated during sunny periods, allowing SAHs to extend their heating capabilities beyond daylight hours. Additionally, advancements in building integration are paving the way for seamless and aesthetically pleasing solar heating solutions. For example, incorporating SAHs directly into building envelopes could eliminate the need for bulky external installations [6,7].

SAHs have evolved into versatile and efficient solar thermal systems, with applications extending beyond space heating. They are now used for pool heating, ventilation pre-heating, and even crop drying (Table 12.1). Although addressing limitations in thermal storage and improving building integration are crucial for widespread adoption, SAHs present a compelling option for space and process heating. Their inherent simplicity and lower risk of corrosion compared to solar water heaters make them a strong contender. Recent advancements are pushing the boundaries of their capabilities, opening doors to exciting new applications beyond traditional ones [8–10].

SAHs offer a clean method to harness solar energy for heating purposes. However, their inability to store thermal energy restricts heating to sunshine hours. This challenge is being addressed by integrating PCMs into the design. PCMs absorb heat during sunlight exposure and release it later, extending heating capabilities, especially in regions with fluctuating sunshine. Optimizing PCM selection and placement within the SAH design is crucial for effective heat transfer and minimizing energy losses. Additionally, the initial cost of incorporating PCMs needs careful consideration [11,12]. Another approach to enhance SAH performance involves introducing artificial roughness geometries on the absorber plate. Roughness disrupts the smooth flow of air, promoting turbulence, which leads to better heat transfer. Studies have shown that strategically placed ribs, dimples, or other geometries can significantly improve heat transfer coefficients, translating to capturing more solar energy and heating air more efficiently. However, increased turbulence also leads to a higher pressure drop, requiring more pumping power. Therefore, optimal design involves balancing enhanced heat transfer with minimized pumping power consumption. Researchers are actively investigating various roughness geometries to achieve this balance. In conclusion, advancements like PCM integration and artificial roughness are enhancing the versatility and impact of SAHs. Further research on optimizing these technologies can pave the way for broader adoption of this sustainable heating solution [13,14].

Implementing different roughness geometries on the absorber plate improves the thermo-hydraulic performance of SAHs. Careful consideration of roughness geometry and its impact on both heat transfer and pressure drop is essential for optimal SAH design. Further research on the exploration of novel roughness configurations holds promise for pushing the boundaries of SAH efficiency [15].

*Table 12.1* Types, advantages, and disadvantages of SAH [11]

| *S.No.* | *Types of SAH* | *Key point* | *Advantage* | *Disadvantage* | *Recent progress* |
|---|---|---|---|---|---|
| 1 | Flat plate SAH | Consists of a flat absorber plate – most common type | Simple design and cost-effective | Lower efficiency than some advanced types | Integration of advanced materials to enhance efficiency and durability. Ongoing research on nano-coatings for improved heat absorption |
| 2 | Concentrating SAH | Concentrates sunlight with mirrors or lenses in to smaller surface area – enhances temperature gain | Higher temperature output | Complex design and expensive | Advancements in concentrating optics and tracking systems for improved efficiency. Integration with thermal storage for continuous heat supply. |
| 3 | Unglazed transpired solar collector | Perforated absorber plate without glass cover – allows air to flow through the perforations | Simple construction, suitable for pre-heating ventilation air | Limited efficiency in harsh weather conditions | Research on advanced materials and improved designs for enhanced efficiency, especially in colder climates |
| 4 | Vented SAH | Uses a fan to force air through the collector – improves heat transfer | Better heat exchange, can be used in various climates | Requires additional energy for fan operation | Integration with low-energy consumption fans and smart control systems for improved energy efficiency |
| 5 | Evacuated tube SAH | Consists of tubes containing absorber material and a vacuum, reduces heat loss | High efficiency, good performance in colder climates | More expensive than flat plate collectors | Ongoing research on cost reduction through improved manufacturing processes. Integration with phase-change materials for enhanced heat storage |
| 6 | Hybrid SAH | Combines solar air heating with other renewable technologies (e.g., photovoltaics) – enhances overall energy production | Synergistic benefits from multiple technologies | Increased complexity and cost | Advancements in integration with advanced energy storage systems. Research on hybrid systems that optimize multiple energy sources for maximum efficiency |

## 12.2 RECENT DEVELOPMENTS IN SAHS

In recent years, research on SAHs has focused on enhancing performance and broadening applications. A primary focus has been optimizing the thermo-hydraulic performance of SAHs, particularly through modifications to the roughness geometry of the absorber plate. Additionally, researchers are exploring selective coatings that maximize solar absorption while minimizing heat losses. Another promising area is the integration of TES with SAHs. TES materials like PCMs can store excess heat during periods of high solar irradiance and release it during times of low irradiance, allowing SAHs to provide heated air even after sunset, thus significantly extending their usability. Furthermore, recent studies address application-specific SAH designs tailored for tasks like crop drying or building space heating. These designs consider factors like required air temperature, flow rate, and economic feasibility to ensure that SAHs effectively meet specific needs [16,17]. A comprehensive literature review, summarized in Table 12.2, highlights promising advancements in enhancing thermal performance, integrating TES, and customizing designs. However, for widespread adoption, cost-effective materials and manufacturing processes are crucial. Future research should focus on integrating SAHs with existing heating systems and developing smart control strategies to optimize operation and effectiveness, making SAHs a more practical and widely used renewable energy technology.

The past five years (2019–2024) have witnessed significant advancements in SAH technology, driven by the pursuit of enhanced efficiency and broader applicability. A critical review of recent literature reveals several key trends. One major area of focus is optimizing heat transfer through the use of artificial roughness geometries on the absorber plate. Researchers have explored various geometries like ribs, V-shaped elements, and dimples, analyzing their impact on factors like Nusselt number (heat transfer coefficient) and friction factor (pressure drop), [29]. This optimization approach helps strike a balance between increased heat transfer and maintaining acceptable pressure losses.

Another promising development is the integration of TES with SAHs. TES systems store excess thermal energy collected during peak sunlight hours for later use, extending the operational timeframe of SAHs [30,32]. This integration holds immense potential for applications requiring consistent thermal output. Furthermore, researchers are customizing SAH designs for specific applications, tailoring collector geometries, airflow patterns, and material selection to suit diverse requirements like drying agricultural products or space heating in different climatic zones [28,33]. These advancements highlight the ongoing efforts to make SAHs a more versatile and efficient renewable energy technology. However, challenges remain, particularly regarding cost-effective materials and large-scale manufacturing processes. Future research should address these aspects alongside exploring the seamless integration of SAHs with existing heating systems and developing intelligent control strategies for optimal operation.

*Table 12.2* Types of ribbed surfaces [11]

| *S.No.* | *Types of rib* | *Figure* |
|---|---|---|
| 1 | Transverse ribs | 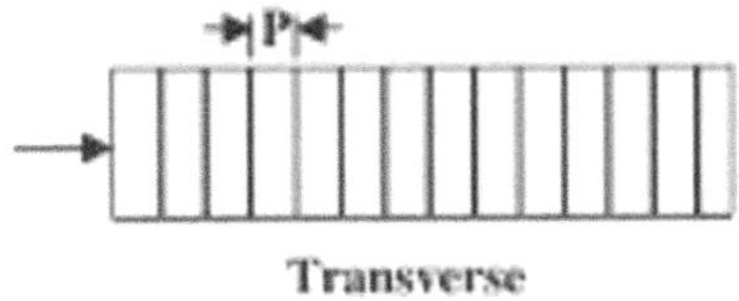  |
| 2 | Inclined ribs | 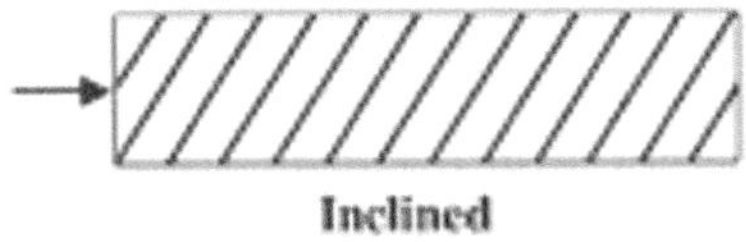  |
| 3 | V-Up continuous ribs | 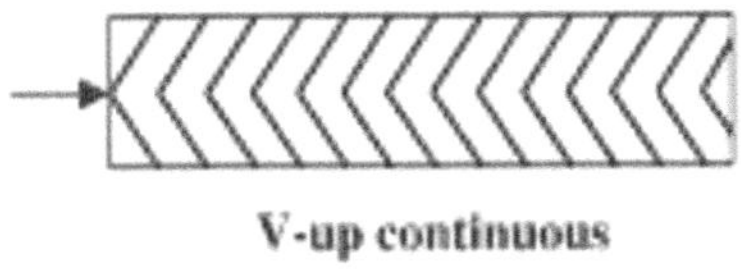  |
| 4 | V-Up discrete ribs | 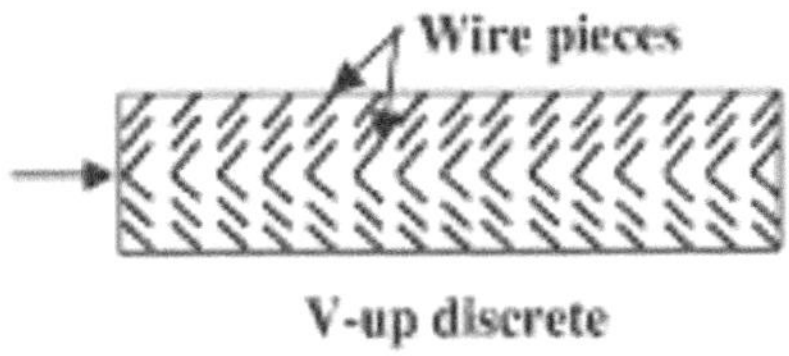  |
| 5 | V-down continuous ribs | 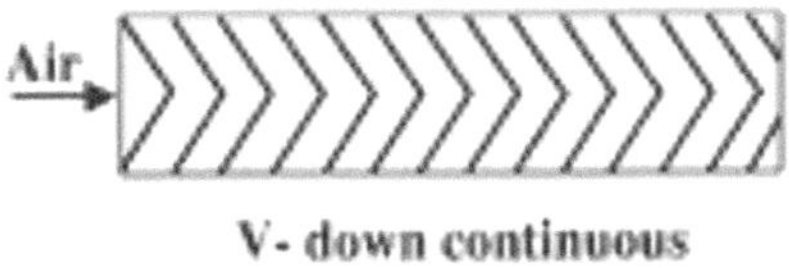  |
| 6 | V-down discrete ribs | 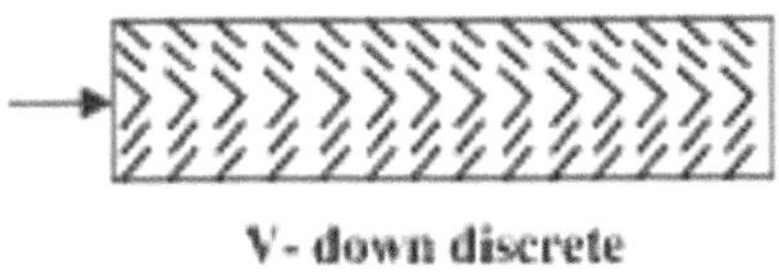  |
| 7 | Scattered ribs | 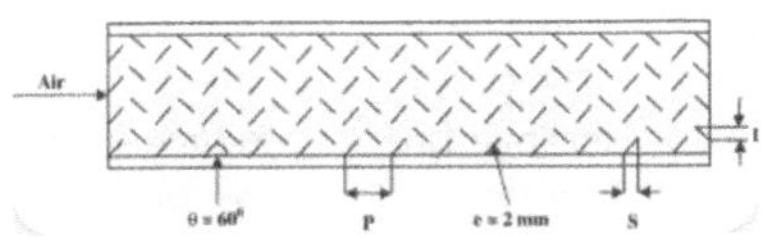 |

## 12.3 EFFECT OF ARTIFICIAL ROUGHNESS IN SAHS

Ribbed surfaces, as detailed in Table 12.2, enhance heat transfer in SAHs through two key mechanisms. First, they disrupt the insulating boundary layer that forms near the surface. By promoting turbulence in the airflow, ribs encourage the mixing of hot fluid near the surface with the cooler bulk fluid. This intensified convection significantly improves heat transfer. Second, ribs increase the overall heat transfer area. Imagine unfolding the ribbed surface – the expanded area allows more fluid to contact the hot surface, further boosting heat transfer efficiency [18,19].

Ribbed surfaces are a recognized method for enhancing heat transfer in SAHs, according to Tables 12.2 and 12.3. This approach impacts airflow and thermal properties in two key ways. First, ribs introduce additional surface area for heat transfer between the absorber plate and the air, providing a clear geometric advantage [20–23]. Second, ribs disrupt the smooth laminar air flow, generating eddies and turbulence. This turbulent flow improves air mixing, bringing cooler air into better contact with the hot absorber plate and promoting heat transfer. Notably, specific rib shapes, like inclined or V-shaped ribs, can induce secondary flows within the air stream, further intensifying this effect.

As illustrated in Figure 12.2 and Table 12.4, ribbed surfaces on the absorber plate demonstrably improve heat transfer in SAHs. However, this

*Table 12.3* Roles of ribbed surfaces in heat transfer enhancement [24–26]

| *S.No* | *Rib type* | *Role in heat transfer enhancement* | *Benefits* | *Drawbacks* |
|---|---|---|---|---|
| 1 | Transverse ribs (perpendicular to airflow) | Increase flow disruption and turbulence | Effective at enhancing heat transfer | May cause significant pressure drop in the air flow |
| 2 | V-shaped ribs | Promote secondary flows along with turbulence. These are more effective than transverse ribs at lower Reynolds numbers | Can offer higher heat transfer compared to straight ribs | Manufacturing complexity can be higher |
| 3 | Artificial roughness (e.g., W-shaped ribs) | Disrupt flow and create turbulence similar to transverse ribs | Good alternative to ribs for achieving enhanced heat transfer | Optimization of roughness parameters might be required |
| 4 | Inclined ribs (angled relative to airflow) | Induce stronger secondary flows compared to straight ribs | Potentially achieve higher heat transfer augmentation | Drop in pressure penalty might be more because of complex flow patterns |
| 5 | Ripple surfaces (sinusoidal waves) | Create a wavy flow path, promoting mixing and heat transfer | Provide better balance between heat transfer improvement and pressure drop | Might require more complex manufacturing techniques |

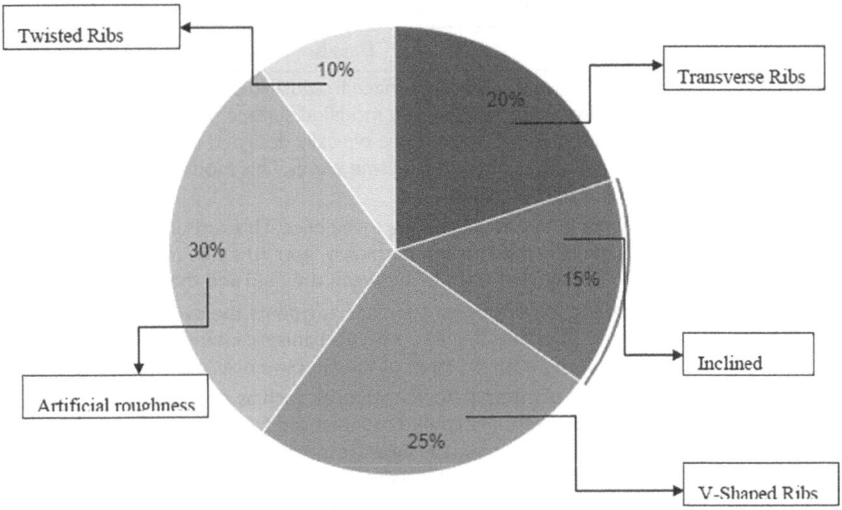


*Figure 12.2* Roles of different ribbed surfaces in heat transfer enhancement.

effectiveness depends on several factors: rib height, spacing (pitch), inclination angle (for angled ribs), and the airflow's Reynolds number. While ribs enhance heat transfer, they also increase pressure drop, which needs careful optimization to maintain sufficient airflow through the collector. Finding the optimal ribbed surface geometry requires balancing high heat transfer with minimal pressure loss and will vary depending on specific design requirements [27–29].

### 12.3.1 Effect of artificial roughness-based turbulence

Artificial roughness isn't a magic bullet for boosting heat transfer in SAHs. Its effectiveness depends on a delicate interplay between various factors. These include the specific roughness geometry itself (ribs and dimples), the flow regime (represented by the Reynolds number), and the size and spacing of the roughness elements (pitch and height). Table 12.5 summarizes the intricate relationship between artificial roughness and turbulence generation within the heater [33].

### 12.3.2 Parameters affecting the performance of SAHs

SAHs hold promise for sustainable space and water heating, but their success relies heavily on a well-tuned balance between design and usage. This section explores three critical factors that significantly impact SAH performance, as detailed in references [34,35].

*Table 12.4* Key parameters effecting heat transfer enhancement in SAH [30–32]

| *Parameter* | *Details* |
|---|---|
| Geometry | Ribbed surfaces incorporate surface irregularities in the form of raised or recessed features, resulting in a modified surface profile. These protrusions and depressions are typically designed in geometric patterns such as rectangles, triangles, or sine waves. This modification enhances the area of heat transfer. |
| Turbulence | Ribs disrupt fluid flow, triggering turbulence. This turbulence increases mixing, thinning the thermal boundary layer (the hot zone near the rib) and promoting heat transfer between the fluid and the rib. |
| Heat transfer coefficient | The introduction of ribs on a surface augments the overall heat transfer coefficient by promoting two key mechanisms: intensification of flow turbulence and augmentation of the effective heat transfer area. |
| Pressure drop | The inclusion of internal flow structures, such as ribs, can induce significant pressure losses due to increased frictional and form drag on the flowing fluid. In practical applications, it is essential to optimize the design by balancing the enhancement of heat transfer achieved by these structures with the resulting pressure drop penalty. |
| Reynolds number | The intensification of convective heat transfer becomes more significant with increasing Reynolds numbers. However, at low Reynolds numbers (laminar flow regime), the advantages of employing ribbed surfaces for heat transfer augmentation might be negligible. |
| Thermal performance factor | The thermal performance factor quantifies the trade-off between augmented heat transfer and increased pressure losses in a system. A thermal performance factor greater than one signifies a net improvement in heat transfer efficiency. |
| Applications | In heat transfer applications encountered in Industry 4.0, ribbed surfaces are strategically introduced within heat exchangers, cooling channels, and electronic devices to promote augmented heat dissipation. The selection of the optimal rib geometry is contingent upon the specific application under consideration and the prevailing flow conditions, as comprehensively documented in references. |
| Challenges | Challenges associated with ribbed surfaces include manufacturing complexities, increased pressure drop, and the potential for fouling in certain applications. |
| Optimization | Designing ribs for best performance requires balancing how well they improve heat transfer against the increased resistance to flow (pressure drop) they cause. Engineers use computer simulations (CFD) and real-world testing to find the ideal rib shape and size for a specific application. |
| Materials | The ability of the ribbed surface material to conduct heat (thermal conductivity) affects how well it transfers heat. Materials that conduct heat more easily (high thermal conductivity) will transfer heat more efficiently. |
| Research trends | Scientists are looking at new designs for ribs (like fins), special materials, and different surface textures to improve heat transfer and avoid any problems that might come with these improvements. Engineers are developing innovative rib shapes, advanced materials, and special surface treatments to boost heat transfer efficiency while minimizing negative side effects. Research is underway to optimize heat transfer by exploring novel rib geometries, cutting-edge materials, and advanced surface modifications. This aims to maximize heat transfer while reducing any downsides. |

*Table 12.5* Effect of artificial roughness-based turbulence on heat transfer [25–30]

| *Parameters* | *Details* |
|---|---|
| Type of roughness | Heat transfer varied on different types of ribs (e.g., ribbed, dimpled, and vortex generators) |
| Reynolds number | It has been observed better mixing and turbulence increases at higher Reynolds numbers |
| Roughness height | Heat transfer increased with the increase in roughness height but it should be up to optimal limit |
| Pitch | Pitch affects the flow between roughness elements; optimal spacing is critical for maximizing heat transfer enhancement |
| Orientation of roughness | The orientation of roughness elements influences the flow patterns and heat transfer; optimal orientations vary |
| Flow inlet conditions | Inlet velocity and temperature conditions affect the onset and intensity of heat transfer enhancement |
| Fluid properties | Viscosity and thermal conductivity of the working fluid influence the heat transfer performance |
| Surface material | Manufacturing industry, the thermal conductivity, and surface characteristics of the material used for roughness elements impact heat transfer |
| Surface roughness distribution | Uniform or non-uniform distribution of roughness elements affects the overall heat transfer enhancement |
| Surface curvature | Curved surfaces with artificial roughness may exhibit different heat transfer characteristics compared to flat surfaces |
| Aspect ratio roughness | The ratio of the height to the characteristic length influences the heat transfer enhancement; optimal ratios exist |
| Turbulence intensity | Higher turbulence levels generally enhance heat transfer, and artificial roughness can contribute to increased turbulence |

- **Absorber Plate Design:** The absorber plate, the core component of an SAH, significantly impacts its performance. Material properties like thermal conductivity and solar absorptivity determine how efficiently the plate captures and transfers solar energy. Researchers have explored introducing artificial roughness on the absorber surface using fins or ribs to disrupt airflow, thereby enhancing heat transfer between air and the plate. However, the geometry of this roughness, including fin spacing, height, and angle, plays a critical role. It affects both heat transfer efficiency and pressure drop across the system. Striking a balance between maximizing heat gain and minimizing the energy needed to circulate air (pumping power). Therefore, optimizing these parameters is essential for achieving optimal SAH performance [27–29].
- **Air Mass Flow Rate:** SAH performance is heavily influenced by the air mass flow rate. Higher flow rates can heat a larger volume of air, potentially boosting overall heat output. However, this comes at the cost of a lower temperature increase per unit of air due to the shorter

time it spends within the collector. Conversely, a slower flow rate allows for a more significant temperature rise for each air unit but reduces the total heat output. Striking the optimal balance between these factors is crucial, involving carefully consideration of the desired heating requirements and ensuring sufficient air circulation to minimize pressure drop across the SAH.

- **External Factors:** Beyond the design of the SAH itself, external factors also play a crucial role in its effectiveness. Ambient temperature significantly impacts performance. As the surrounding air heats up, the temperature difference between the heated air and its environment shrinks, reducing the efficiency of heat transfer. Similarly, fluctuations in solar irradiance, the amount of sunlight hitting the collector, directly affect the available solar energy for air heating. Understanding these external conditions and their influence is critical for designing and operating SAHs effectively in real-world scenarios [30–32].

## 12.4 ECONOMIC ANALYSIS OF SAHS

Evaluating an SAH's economic viability requires a techno-economic analysis that considers both upfront costs and operational savings. Here are key factors influencing economic attractiveness:

- **Payback Period (PBP):** This measures how long it takes for energy cost savings to recover the initial investment. Factors influencing PBP include collector efficiency, system cost, local energy prices, and solar radiation. A shorter PBP signifies faster returns, making the SAH more appealing.
- Life Cycle Cost (LCC): LCC considers all costs throughout the SAH's lifespan, including the initial investment, installation, maintenance, repairs, and eventual replacement. A lower LCC indicates a more cost-effective solution. Material selection for durability and ease of maintenance is crucial here.
- **System Design and Efficiency:** SAH design significantly impacts economics. The absorber plate material, glazing type, and flow channel geometry influence the collector's thermal efficiency. Higher efficiency translates to more captured heat and lower operational costs, leading to faster PBPs.
- **Government Incentives:** Subsidies, tax breaks, and net metering programs can significantly improve SAH economics. These incentives can help offset initial investment costs, making SAHs more competitive with conventional heating systems. Analyzing available financial and policy support is crucial for a complete assessment.

Evaluating an SAH's technical merits goes hand in hand with a thorough economic analysis. This financial assessment determines the feasibility of installing an SAH for a particular application. By considering factors like installation costs, operational expenses, and potential energy savings, the analysis empowers individuals and businesses to make informed decisions, ensuring they can leverage the cost-effectiveness and sustainability benefits of SAH technology [33–39].

## 12.5 CONCLUSION

In conclusion, the field of heat transfer enhancement is undergoing a remarkable transformation. A convergence of diverse technologies, like artificial roughness and novel materials, alongside methodologies like machine learning. This creates a new era of "intelligent thermal management" that adapts to changing conditions and optimizes efficiency. Research is actively exploring these areas, including innovative surface modifications, advanced simulations, and smart control strategies. The combined impact is undeniable, promising significant advancements in industrial processes, electronics cooling, and renewable energy applications. This exploration signifies not just current achievements but a future filled with heightened efficiency and continuous innovation in heat transfer.

## REFERENCES

1. B.N. Prasad, J.S. Saini (1988). Effect of artificial roughness on heat transfer and friction factor in a solar air heater. *Solar Energy* 41, 555–560.
2. D. Gupta, S.C.C. Solanki, J.S.S. Saini (1997). Thermohydraueic performance of solar air heaters with roughened absorber plates. *Solar Energy* 61, 33–42.
3. N. Dutt, A. Hedau, A. Kumar, M.K. Awasthi (2024). Thermo-hydraulic performance investigation of solar air heater duct having staggered D-shaped ribs: Numerical approach. *Heat Transfer.* https://onlinelibrary.wiley.com/doi/10.1002/htj.22998.
4. P. N. Reddy, V. Verma, A. Kumar, M. K. Awasthi (2023). CFD simulation and thermal performance optimization of channel flow with multiple baffles. *Journal of Heat and Mass Transfer Research* 10(2), 257–268. https://doi.org/10.22075/JHMTR.2023.31108.1458.
5. N. Dutt, A. J. Hedau, A. Kumar, M. K. Awasthi, V. P. Singh, G. Dwivedi (2023). Thermo-hydraulic performance of solar air heater having discrete D-shaped ribs as artificial roughness. *ESPR* 1–22. https://doi.org/10.1007/s11356-023-28247-9.
6. V. Verma, C.S. Meena, S. Thangavel, A. Kumar (2023). Ground and solar assisted heat pump systems for space heating and cooling applications in the northern region of India: A study on energy and $CO_2$ saving potential. *SETA* 59, 103405. https://doi.org/10.1016/j.seta.2023.103405.

7. P. K. Kushwaha, N. K. Sharma, A. Kumar, C. S. Meena (2023). Recent advancements in augmentation of solar water heaters using nanocomposites with PCM: Past, present, and future. *Buildings* 13, 79. https://doi.org/10.3390/buildings13010079.
8. V. P. Singh, S. Jain, A. Karn, A. Kumar (2022). Experimental assessment of variation in open area ratio on thermohydraulic performance of parallel flow solar air heater. *Arabian Journal for Science and Engineering*. https://doi.org/10.1007/s13369-022-07525-7.
9. C. S. Meena, A.N. Prajapati, A. Kumar, M. Kumar (2022). Utilization of solar energy for water heating application to improve building energy efficiency: An experimental study. *Buildings* 12, 2166. https://doi.org/10.3390/buildings12122166.
10. V. P. Singh, S. Jain, A. Kumar, S. Mishra, N. K. Sharma (2022). Heat transfer and friction factor correlations development for double pass solar air heater artificially roughened with perforated multi-V ribs. *Case Studies in Thermal Engineering* 39,102461. https://doi.org/10.1016/j.csite.2022.102461.
11. V. P. Singh, S. Jain, A. Karn, A. Kumar, G. Dwivedi, C. S. Meena (2022). Recent developments and advancements in solar air heaters: A detailed review. *Sustainability* 14, 12149. https://doi.org/10.3390/su141912149.
12. C. S. Meena, A. Kumar, S. Jain (2022). Innovation in green building sector for sustainable future. *Energies* 15, 6631. https://doi.org/10.3390/en15186631.
13. V. P. Singh, S. Jain, A. Karn, A. Kumar, G. Dwivedi, C.S. Meena, R. Cozzolino (2022). Mathematical modeling of efficiency evaluation of double-pass parallel flow solar air heater. *Sustainability* 14, 10535. https://www.mdpi.com/2071-1050/14/17/10535.
14. V. P. Singh, C. S. Meena, A. Kumar, N. Dutt (2022). Double pass solar air heater: A review. *International Journal of Energy Resources Applications* 1(2), 22–43, https://doi.org/10.56896/IJERA.2022.1.2.009.
15. C. S. Meena, A. Kumar, S. Roy, A. Cannavale, A. Ghosh (2022). Review on boiling heat transfer enhancement techniques. *Energies* 15, 5759. https://doi.org/10.3390/en15155759.
16. N. Dutt, A. Binjola, A. J. Hedau, A. Kumar, V. P. Singh, C.S. Meena (2022). Comparison of CFD results of smooth air duct with experimental and available equations in literature. *International Journal of Energy Resources Applications (IJERA)* 1(1), 40–47. https://doi.org/10.56896/IJERA.2022.1.1.006.
17. A. Kumar, Y. Gori, N. Dutt, Y. K. Singla, A. Maurya (eds.) (2021). *Advanced Computational Methods in Mechanical and Materials Engineering*. CRC Press: Boca Raton, FL. https://doi.org/10.1201/9781003202233.
18. A. Kumar, V. P. Singh, C. S. Meena, N. Dutt (eds.) (2023). *Thermal Energy Systems: Design, Computational Techniques and Applications*. CRC Press: Boca Raton, FL. https://doi.org/10.1201/9781003395768.
19. A.K. Dewangan, S.Q. Moinuddin, M. Cheepu, S.K. Sajjan, A. Kumar (2023). Thermal energy storage: Opportunities, challenges and future scope. In A. Kumar, V. P. Singh, C. S. Meena, N. Dutt (eds.), *Thermal Energy Systems: Design, Computational Techniques and Applications*. CRC Press: Boca Raton, FL, Chapter 2, pp. 17–28. https://dx.doi.org/10.1201/9781003395768-2.

20. G. Pant, C. S. Meena, A. Saxena, A. Kumar, V. P. Singh, N. Dutt (2023). Study the temperature variation in alternate coils of insulated condenser cum storage tank: Experimental study. In Sikarwar, B.S., et al. (eds.), *Advances in Fluid and Thermal Engineering. FLAME 2022. Lecture Notes in Mechanical Engineering.* Springer: Singapore. https://doi.org/10.1007/978-981-99-2382-3_52.
21. A. Kundu, A. Kumar, N. Dutt, C.S. Meena, V.P. Singh (2023). Introduction to thermal energy resources and its smart application. In A. Kumar, V. P. Singh, C. S. Meena, N. Dutt (eds.), *Thermal Energy Systems: Design, Computational Techniques and Applications.* CRC Press: Boca Raton, FL, Chapter 1, pp. 1–15. https://dx.doi.org/10.1201/9781003395768-1.
22. A. Kundu, A. Kumar, N. Dutt, V.P. Singh, C.S. Meena (2023). Modelling and simulation of thermal energy system for design optimization. In A. Kumar, V. P. Singh, C. S. Meena, N. Dutt (eds.), *Thermal Energy Systems: Design, Computational Techniques and Applications.* CRC Press: Boca Raton, FL, Chapter 7, pp. 103–140. https://dx.doi.org/10.1201/9781003395768-7.
23. V. Verma, S. Thangavel, N. Dutt, A. Kumar, R. Weerasighe (eds.) (2024). *Highly Efficient Thermal Renewable Energy Systems: Design, Optimization and Applications.* CRC Press: Boca Raton, FL. https://doi.org/10.1201/9781003472629.
24. M. K. Awasthi, A. Kumar, N. Dutt, S. Singh (eds.) (2024). *Computational Fluid Flow and Heat Transfer: Advances, Design, Control and Applications.* CRC Press: Boca Raton, FL. https://doi.org/10.1201/9781003465171.
25. V. Verma, S. Thangavel, N. Dutt, A. Kumar, R. Weerasinghe (eds.) (2024). Recent development of thermal energy storage: Solar, geothermal and hydrogen energy. In *Highly Efficient Thermal Renewable Energy Systems: Design, Optimization and Applications.* CRC Press: Boca Raton, FL, Chapter 01. https://doi.org/10.1201/9781003472629-1.
26. M. Ahmadizadeh, M. Heidari, S. Thangavel, E.A. Naamani, M. Khashehchi, V. Verma, A. Kumar (2024). Technological advancements in sustainable and renewable solar energy systems. In V. Verma, S. Thangavel, N. Dutt, A. Kumar, R. Weerasighe (eds.), *Highly Efficient Thermal Renewable Energy Systems: Design, Optimization and Applications.* CRC Press: Boca Raton, FL, Chapter 02. https://doi.org/10.1201/9781003472629-2.
27. S.P. Shetty, A. Paineni, M. Kande, N. Madhwesh, N.Y. Sharma, K.V. Karanth (2020). Experimental investigations on a cross flow solar air heater having perforated circular absorber plate for thermal performance augmentation. *Solar Energy* 197, 254–265.
28. V. Ahmed, A. Tiwari, Z. A. Said (2020). Design and performance analysis of a double pass solar air heater integrated with latent heat storage for space heating applications. *Solar Energy* 207, 1142–1153.
29. S. Kumar, J. S. Saini, M. Singh (2020). Performance evaluation of solar air heater integrated with V-shaped artificial roughness. *Materials Today: Proceedings* 47(2), 1722–1727.
30. C. Li, J. Yang, H. Xu, Y. Li (2023). Numerical investigation on thermal performance of a novel solar air heater integrated with phase change materials for space heating. *Applied Thermal Engineering* 219, 116783.

31. H. K. Singh, J. S. Saini, V. K. Verma (2022). Heat transfer and friction factor analysis of solar air heater with integrated discrete ribs on absorber plate. *International Journal of Thermal Technologies* 14(1), 121–132.
32. Y. Tian, C. Zhao, H. Xu, Y. Li (2021). Numerical investigation on the unsteady thermal performance of a solar air heater integrated with latent heat storage for space heating. *Energy* 239, 122222.
33. Z. Xu, J. Ji, Y. Li, H. Li (2022). Performance analysis of a novel compound parabolic concentrator solar air heater for drying agricultural products. *Journal of Food Engineering* 322, 111342.
34. M.K. Awasthi, A. Kumar, N. Dutt (2024). Modeling Rayleigh-Taylor Instability in Nanofluid Layers. In M. K. Awasthi, A. Kumar, N. Dutt, and S. Singh (eds.), *Computational Fluid Flow and Heat Transfer: Advances, Design, Control and Applications*. CRC Press: Boca Raton, FL, Chapter 13. https://doi.org/10.1201/9781003465171-13.
35. M. K. Awasthi, N. Dutt, A. Kumar, S. Kumar (2024). Electrohydrodynamic capillary instability of Rivlin–Ericksen viscoelastic fluid film with mass and heat transfer. *Heat Transfer* 53(1), 115–133. doi:10.1002/htj.22944.
36. M. K. Awasthi, A. K. Shukla, A. Kumar, D. Yadav, N. Dutt (2024). Viscous correction to the potential flow analysis of Rayleigh-Taylor instability in a Rivlin-Ericksen viscoelastic fluid layer with heat and mass transfer. *Heat Transfer*. https://onlinelibrary.wiley.com/doi/10.1002/htj.23076.
37. M. Chitt, S. Thangavel, V. Verma, A. Kumar (2024). Green hydrogen productions: Methods, designs and smart applications. In V. Verma, S. Thangavel, N. Dutt, A. Kumar, R. Weerasighe (eds.), *Highly Efficient Thermal Renewable Energy Systems: Design, Optimization and Applications*. CRC Press: Boca Raton, FL, Chapter 16, pp. 261–276. https://doi.org/10.1201/9781003472629-16.
38. M. Khashehchi, S. Thangavel, P. Rahmanivahid, M. Heidari, T. Moazzeni, V. Verma, A. Kumar (2024). Solar desalination techniques: Challenges and opportunities. In V. Verma, S. Thangavel, N. Dutt, A. Kumar, R. Weerasighe (eds.), *Highly Efficient Thermal Renewable Energy Systems: Design, Optimization and Applications*. CRC Press: Boca Raton, FL, Chapter 19, pp. 305–329. https://doi.org/10.1201/9781003472629-19.
39. A. Kumar, Y. Singla, T. Namboodri (2024). Globalization and international issues in sustainable manufacturing. In S. Shah, et al. (eds.), *Sustainability in Smart Manufacturing: Trends, Scope, and Challenges*. CRC Press: Boca Raton, FL, Chapter 01, pp. 1–18. https://dx.doi.org/10.1201/9781003467496-1.

Chapter 13

# Competing alternative fuel technologies for commercial vehicles in India

*Deeptam Trivedi, Sujeet Kesharvani, Raj Suman, Gaurav Dwivedi, and Olusegun David Samuel*

## 13.1 INTRODUCTION

Carbon dioxide ($CO_2$), methane, and nitrous oxide are the three main greenhouse gases (GHGs) responsible for global warming, which causes climate change, impacting people, ecosystems, and the environment. In India, carbon emissions increased by 7.5% from January to May 2022 as compared to the same period last year [1]. Simultaneously, a sustained decline in indigenous crude oil production over the years has increase the country's oil crude oil import dependency [2]. Domestic crude oil production in 2021–2022 was 18% lower than in 2016–2017. India's crude oil imports were reported at 4,243.758 barrels/day in December 2021, an increase from the previous number of 4,033.050 barrels/day for December 2020. The top import sources include Iraq, Saudi Arabia, Russia, and the United States. As the population is increasing day by day, the demand and usage of petroleum products have increased the emissions level and the cost of oil imports [3]. Hence, focusing on the alternative sources of fuels is essential to improving the air quality and energy security by reducing imports.

## NOMENCLATURE

| | | | |
|---|---|---|---|
| $CO_2$: | Carbon di oxide | MT: | Metric tonnes |
| LNG: | Liquefied natural gas | ERS: | Electric road system |
| CNG: | Compressed natural gas | FC: | Fuel Cell |
| EV: | Electric vehicle | HDT: | Heavy-duty truck |
| MMT: | Million metric tonnes | FFA: | Free fatty acid |
| PM: | Particulate matter | B20: | Binary blend |
| CAGR: | Compound annual growth rate | UCO: | Used Cooking Oil |
| GHG: | Greenhouse gas | | |

DOI: 10.1201/9781003521341-13

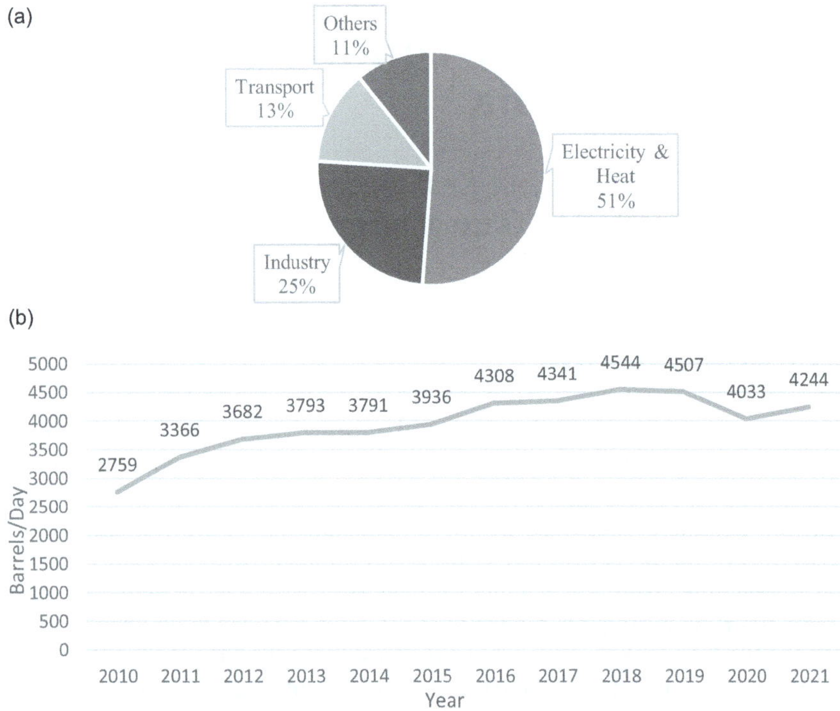


*Figure 13.1* (a) Sector-wise carbon emission. (b) Year-wise crude oil imports in barrels per day.

Petrol and diesel together account for 93% of the transportation fuel basket, and the government's effort to control emissions will impact the energy transition matrix. Figure 13a and 13b shows sector-wise emissions generation in India and year-wise crude oil imports in barrels per day.

As an alternative to petrol, blending of ethanol is seen as a feasible option. In June 2022, the blending of 10% ethanol with 90% petrol was implemented 5 months ahead of schedule, resulting in less carbon emissions, more savings for the country, and better income for farmers. Blending 10% ethanol reduced GHG emissions by 27 lakh tonnes and had a forex impact of approximately Rs. 41,500 crore. By April 2023, India plans to begin delivering gasoline containing 20% ethanol at a few gas stations, which is expected to save an estimated $4 billion yearly [4]. The other alternative for petrol engines in compressed natural gas (CNG). CNG has been used as an alternate fuel for quite a few years after its potential as a clean fuel was demonstrated as an exclusive fuel powering the public transport bus system in the National Capital Region of India. CNG gives more mileage than petrol as fuel. Many cities have been experimenting with CNG as a fuel to power buses, three-wheelers, and passenger vehicles [5].

For diesel, a perfect alternative is yet to be determined as diesel is widely used for transportation of heavy loads over long distances. While CNG can be used as an alternative, its range is quite shorter and the torque generated loads is lower than diesel, making is less suitable for heavy-duty applications. Biodiesel blending is another beneficial option, but the main problem lies in continuous perennial supply of feedstock for large-scale production. Converting diesel engines into electric drivetrain is also an option, but the weight of the battery pack reduces the load-carrying capacity of commercial vehicles, and the long refuelling or charging times keeps the vehicle out of service for a long period. Hydrogen can be the most perfect substitute for diesel, but the technology is quite expensive and is still in the early stages of development. Hence, given the current scenario, liquefied natural gas (LNG) can appears to be the most viable option alternative for commercial vehicles dependent on diesel [6].

This chapter reviews alternative fuel technologies such as LNG, hydrogen, biodiesel, and electric vehicle (EV) technologies for commercial vehicles in India. It analyses their potential, scope, and government support and policies for their development as a viable alternative in the commercial vehicle segment, which contributes to high crude oil import costs and large amount of emissions in the air.

## 13.2 PRESENT FUEL OPTIONS

India consumes 5 million barrels of oil per day, and its growth rate is 3%, which is higher than the roughly 1% average for the rest of the world. India's total petroleum consumption in 2019–2020 was 194.3 million metric tonnes (MMT), which rose by about 5% to 204.2 MMT in the next year. The annualised consumption growth from April to September 2022 was 107.7 MMT. The consumption of high-speed diesel increased by 16.1% during the half year 2022–2023 and of motor spirit or petrol increased by 18.5%. For vehicle emissions and fuel quality requirements, India has adopted the European model, which are known as Bharat Stages [6].

### 13.2.1 Petrol/gasoline

In recent years, the consumption of petrol/gasoline has been mainly driven by the two-wheeler sector. Two-wheelers consume about 62% of petrol, while four-wheelers consume about 27%. Commercial three- and four-wheelers account for 6%, and other purposes making up the remaining 5%. To reduce the cost of imports and emissions, the Government of India plans to introduce 20% ethanol blending with petrol in some parts of the country from April 2023, with a nationwide rollout is expected by 2025–2026. India has achieved 10% ethanol blending with gasoline as of June 2022. According to a 2021 NITI Aayog report, the country could

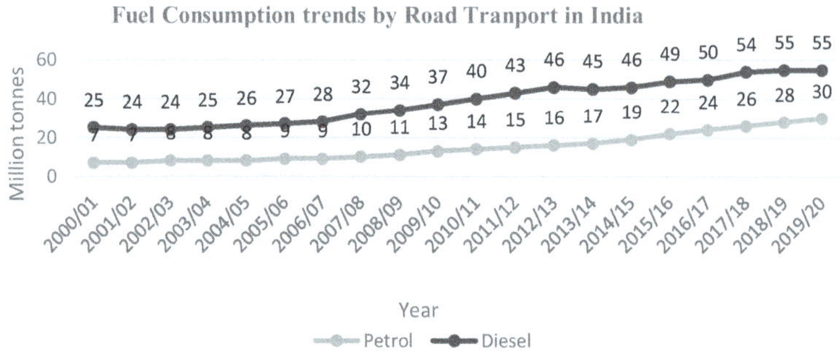


*Figure 13.2* Fuel consumption trends by road transport in India.

experience "immense benefits" from 20% ethanol blending by 2025, including annual savings of 30,000 crore rupees, increased energy security, decreased carbon emissions, improved air quality, self-reliance, better use of ruined food grains, higher farmer incomes, and more investment opportunities. While a 10% ethanol blended petrol does not necessitate significant engine modifications, a 20% ethanol blended petrol may necessitate considerable modifications and could significantly increase the cost of automobiles. Additionally, a higher blending ration might lead to more land being diverted for water-demanding crops like sugarcane, which are currently subsidised by the government finances, might also result from [7] (Figure 13.2).

While new cars will be E20 compliant, older cars will have issues with the higher corrosive nature of the fuel. While there is a study underway at ARAI measuring the impact of E20 and higher ethanol blends, there is certainly going to be an issue with older cars running E20. Thus, one possible workaround is to continue the supply of E10 blends. Distribution and storage would pose a challenge, but not so much for refineries as the blending of ethanol, and petrol is quite simply a mixing that happens after distillation, and downstream in the overall production process [8] (Figure 13.3).

### 13.2.2 Diesel

In recent years, diesel has gained a reputation for being more environmentally harmful than gasoline throughout the world. This is why there are tighter restrictions on diesel vehicles in India too. In 2015, the National Green Tribunal ruled that diesel vehicles older than 10 years would not be permitted to operate in the Delhi–NCR zone; for petrol vehicles, this window was maintained at 15 years. It is not only that $CO_2$ emissions, in general, are not the problem but also that the production of soot and other hazardous emissions are also injurious to people's health. This is where

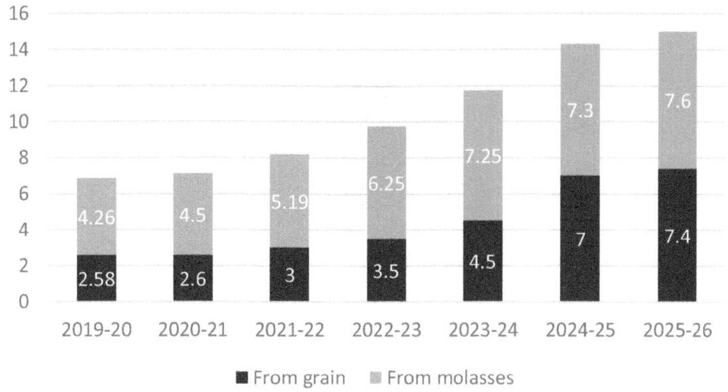


*Figure 13.3* Ethanol capacity requirements for blending with petrol.

diesel is worse off than petrol. Nitrogen oxides (NOx) are created when air is heated in a vehicles engine. Among these are the hazardous nitrogen dioxide ($NO_2$), nitrous oxide ($N_2O$), and nitric oxide (NO), which combine with oxygen to generate poisonous nitrogen dioxide ($NO_2$). Long-term risk of respiratory issues is also increased by nitric oxide alone. Diesel is primarily used in heavier vehicles like trucks and buses, which are typically more likely to be less well-maintained and thus more polluting in India. Additionally, a lot of trucks are outdated and use older emission control technology. In this way, tougher laws for diesel also mean tougher rules for these commercial vehicles. Additionally, the price of a litre of diesel was much lower than the price of petrol before the decontrol of automobile fuel prices started in late 2014, which encouraged a very large number of people to buy diesel cars [9] (Figure 13.4).

### 13.2.3 Compressed natural gas

Compared to diesel and petrol vehicles, the operational costs of CNG vehicles are much lower. However, the cost of CNG has increased in India over the past few years. India's demand for alternative fuels like CNG, LPG, and EVs is growing as people become more conscious of the negative environmental effects of diesel and petrol automobiles. Among different fossil fuels, CNG has advantages in terms of reduced emissions of CO, $CO_2$, $NO_x$, and particulate matter (PM). The Indian government also aims to increase the sales of EVs in commercial vehicles by 70%, in passenger vehicles by 30%, in buses by 40%, and in two- and three-wheelers by 80%. This strategy supports the Indian government's objective of achieving net zero carbon emissions by 2070. This development and increased EV adoption in India is anticipated to hinder the growth of CNG vehicles, which is expected to affect the CNG market in the forecast period negatively.

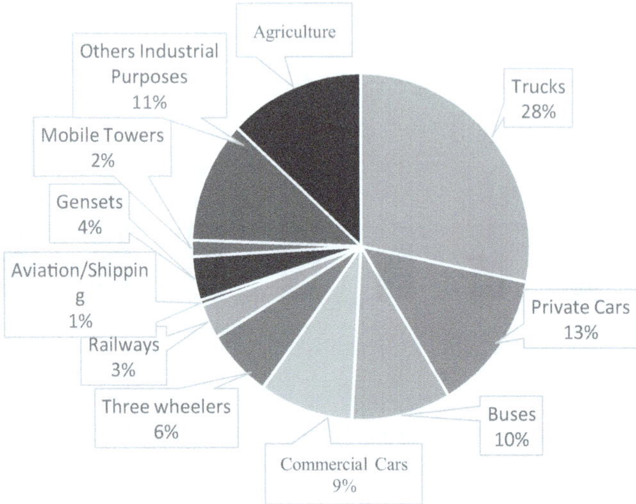


*Figure 13.4* India's diesel consumption pattern.

By 2023–2024, 5,000 bio-CNG plants will be installed nationwide as part of the effort to accomplish this. As the government aims to invest over INR 200 billion in bio CNG plants over the next few years, rising investment in bio CNG is anticipated to provide enormous potential for the CNG market. Currently, India has nearly 4,500 CNG stations, with aggressive plans to set up another 3,500 CNG stations in the next 2 years and extend the network to 10,000 by 2030. With CNG looking to enter a sweet spot of growth, LNG is the next logical step for India to reduce heavy-duty transportation emissions. LNG takes up to three times less space than CNG, providing higher energy density and vehicle range with a similar size of fuel tank that makes LNG a viable alternative fuel for India's long-haul, heavy-duty truck (HDT) fleets that are currently powered by diesel [10] (Figure 13.5).

## 13.3 OPPORTUNITIES AND OPTIONS FOR FUTURE

Commercial vehicles are the backbone of the Indian logistics industry, but they also contribute significantly to pollution. Over 3 trillion ton/km of goods are moved as a result of commercial activity in India, which yearly produces 4.6 billion tonnes of freight. Of the total petroleum consumed by the country's transport sector, nearly 60% is used by freight trucks. These trucks are responsible for 71% of the carbon dioxide emissions, 74% of the PM emissions, and 55% of the nitrogen oxide emissions from road vehicles.

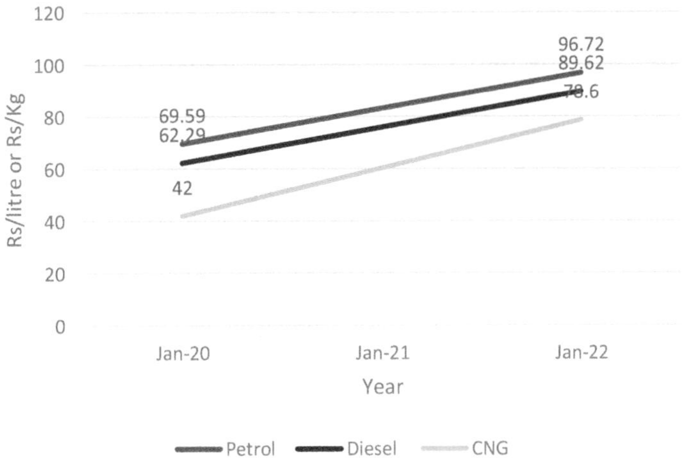


*Figure 13.5* Inflation of fuel prices in India (Delhi) 2020–2022.

The demand for goods is anticipated to grow at a compound annual growth rate (CAGR) of 7% due to rising income levels, increased exports, a quickly expanding e-commerce industry, and an expanding retail sales market (CAGR). This would result in a 16 trillion ton/km overall demand for goods transportation in 2050. Every diesel truck in India emits 1,300 tonnes of $CO_2$ over its lifetime; hence, with the increase in the number of commercial vehicles, the emissions levels will increase, so there is strong requirement for finding an alternative fuel for diesel vehicles.

### 13.3.1 Liquefied natural gas

Natural gas when cooled to its liquid state is called LNG. LNG is odourless, colourless, non-toxic, and non-corrosive mixture of hydrocarbons such as methane and ethane. In comparison, CNG takes about three times more space than LNG at certain temperature, which makes the usage of CNG restricted to intra-state services. LNG in the same storage capacity offers a much better driving range that is suitable for long-distance transportation. Additionally, LNG has an emission factor that is 25% lower than diesel, making it quieter and cleaner. LNG operations have the potential to reduce $CO_2$ emissions by 20% while also substantially reducing nitrogen oxide and PM emissions. Compared to diesel, which has an energy density of 10,900 kcal/kg, LNG has a calorific value of 12,950 kcal/kg, providing a longer driving range with a fuel tank that is the same size. While theft problems are frequent with diesel, LNG can be regarded as a theft-proof fuel because it is cryogenic at a temperature of –162°C [11] (Table 13.1).

*Table 13.1* Thermo-physical properties of LNG

| *Parameter* | *Value* |
| --- | --- |
| Boiling point | −160°C to −162°C |
| Molecular weight | 16–19 g/mol |
| Density | 425–485 kg/m3 |
| Specific heat capacity | 2.2–3.7 kJ/kg/°C |
| Viscosity | 0.11–0.18 mPa·s |
| Higher heat value | 38–44 MJ/m3 |

LNG is a cost-effective and clean fuel for heavy commercial vehicles in the transportation sector. Long-haul vehicles such as trucks and buses benefit from the longer range offered by LNG owing to its comparatively greater energy density per volume. LNG trucks are unique as they use the diesel cycle to run the LNG engine as opposed to the industry practice of using the Otto or petrol cycle using spark plugs for combustion. One fill of a full-fill LNG truck can go around 900 km, which makes it a good alternative for diesel in terms of range.

In a wide range of aspects, LNG is advantageous. LNG is a practical substitute because its price per unit is currently less than that of diesel. An LNG truck costs around 22% more than a comparable diesel vehicle, but its operational costs are roughly 40% less expensive, resulting in a 32% cheaper total cost of ownership.

Natural gas in the form of LNG offers remarkable flexibility in storage and transportation. On land, natural gas is conveniently delivered to consumption centres in its gaseous form through pipelines. But if pipeline transportation is not an option available, natural gas is converted and stored as LNG at minus 162°C, by cooling it down using a refrigeration cycle. At this low temperature, natural gas converts into its liquid form. The resulting LNG occupies only 1/600th of its original gaseous volume and is thus transported efficiently in large specialised tanker ships that have the requisite onboard cryogenic storage facilities (thermally insulated tanks) to keep LNG in liquid form. Upon arrival at an LNG import terminal, LNG is degasified by heating it to convert it back into its gaseous form, which is called R-LNG (degasified LNG). This natural gas regenerated from LNG is transported swiftly via pipeline networks stretching thousands of kilometres. LNG can also be delivered employing LNG trucks with double-walled insulated cryogenic tanks to store LNG at ultra-low temperatures. Similarly, on-ground LNG storage tanks are specially designed cryogenic tanks that maintain frigid conditions to keep LNG in the liquid state. These tank systems are specially engineered to use auto-refrigeration to keep the pressure and temperature in the tank constant [12] (Figure 13.6).

The most crucial gas quality issues are the calorific value, sulphur, and mercury contents. The liquefaction process must be precisely improved and

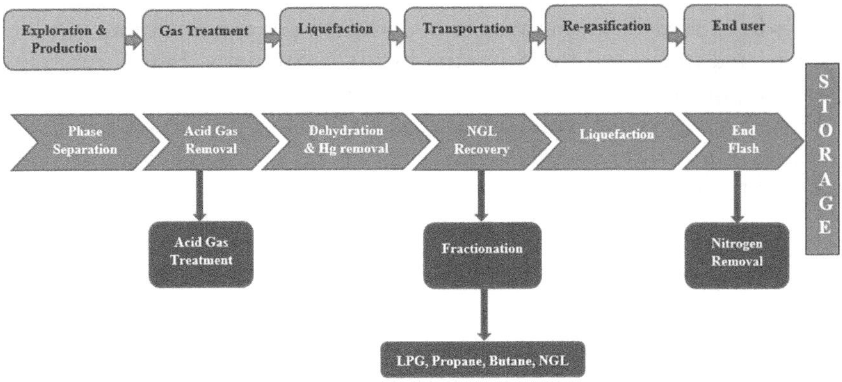


*Figure 13.6* LNG treatment and liquefaction

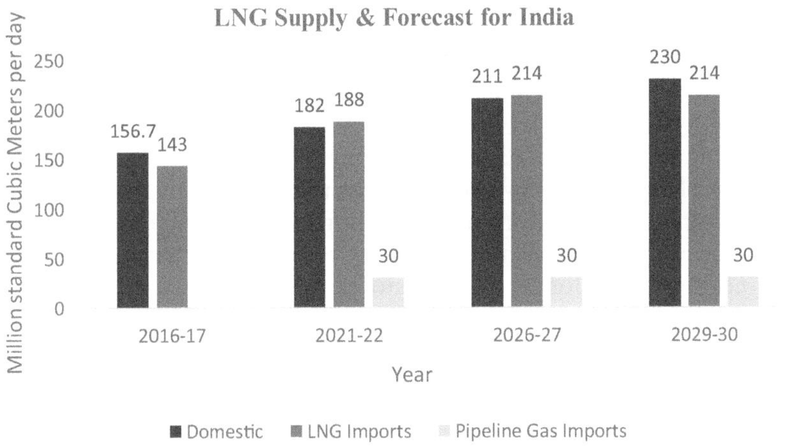


*Figure 13.7* India's LNG supply and forecast.

verified to guarantee the lowest concentration of sulphur and mercury in LNG. LNG cannot ignite when it is in the liquid state and is not explosive. LNG must first evaporate and combine with air in the right amounts to burn. LNG rapidly vaporises and transforms into a gas that readily combines with the air during leaking. There is a chance of ignition starting a fire and thermal radiation dangers in such a situation (Figure 13.7).

By 2030, India aims to raise its natural gas consumption from 6.2% to 15%. In addition to being a more ecologically friendly fuel, LNG can cut fleet operators' fuel costs by 25% and the nation's import costs by 30%–40%. India is preparing a roadmap for setting up floating storage facilities for LNG at all its major ports. The project is estimated to cost ₹20,000 crores in total and will be open for private sector participation. India currently has a regasification capacity of 30 MMT per annum (mmtpa), which is expected to expand to 55 mmtpa by 2025. Similarly, India's gas transmission pipeline of 16,200 km is also witnessing huge capacity augmentation and is expected to reach around 27,430 km by 2025. India has enough natural gas, with recoverable reserves of about 5 trillion cubic meters, on the front of energy security as well [13].

### 13.3.2 EV commercial vehicles

Electric mobility is the most common issue in the transportation sector. EVs draw power from the grid and store it in batteries through an EV charger, which recharges an energy storage device by converting alternating current energy to regulated direct current (DC). There are two types of EV battery chargers: on-board (for use inside an EV) and off-board (at a fixed location) [14].

The three primary methods of charging are typically conductive charging, inductive charging, and battery swapping. With the conductive approach, direct contact is made between the EV connector and the charge inlet; a cable is used to connect the battery to the electrical source. While there is no physical touch between the EV and the charging infrastructure when using the inductive approach, it operates by electromagnetic transfer. Induction chargers work by using an induction coil to generate an alternating electromagnetic field inside a charging base, and a second induction coil in the portable device absorbs energy from the alternating electromagnetic field and transforms it back into electric current to charge the battery. When the inductive charging system makes use of resonant inductive coupling, greater separations between the sender and receiver coils can be made. Recent advancements in this resonant system include the use of a moveable transmission coil and reception coils made of silver-plated copper or occasionally aluminium to reduce weight and reduce skin effect resistance [15] (Table 13.2).

Since conductive charging is less expensive and more effective, it is currently preferred by both the car industry and owners of charging stations. However, a number of R&D initiatives concentrate on inductive charging's

*Table 13.2* Levels of conductive charging

| *Charging level* | *Charging power* | *Nature of supply current* | *Location* |
|---|---|---|---|
| Level 1 | Up to 3.3 kW | Alternating current | Home |
| Level 2 | 7 kW (single phase) and 21 kW (three phase) | Alternating Current | Home, workplace, and public charging facilities |
| Level 3 | Up to 50 kW | DC | Public locations |

ongoing development because it offers a better technique to promote user comfort and may be a crucial component of electric mobility (Figure 13.8).

The third option considers the process of exchanging used batteries for new ones at a swapping station. In battery swapping, EV owners visit the switch stations and swap out their exhausted batteries for charged ones. Multiple batteries are constantly being charged as part of a swapping station that is set at a specific area. A user of an EV can find a swapping station, switch out their declining battery for a charged one, recharge their empty battery, and then head to work. Additionally, it eases drivers' range concern while assisting in the solution of the charging station installation issue. However, for this to be feasible, the battery's internal connections and dimensions must be standardised. Every EV from every manufacturer would need to have batteries that are essentially the same size, shape, and kind. The majority of manufacturers are considering rejecting this strategy since it greatly limits the OEM's freedom of design and because there would be far fewer options for where to put the battery [16] (Figure 13.9).

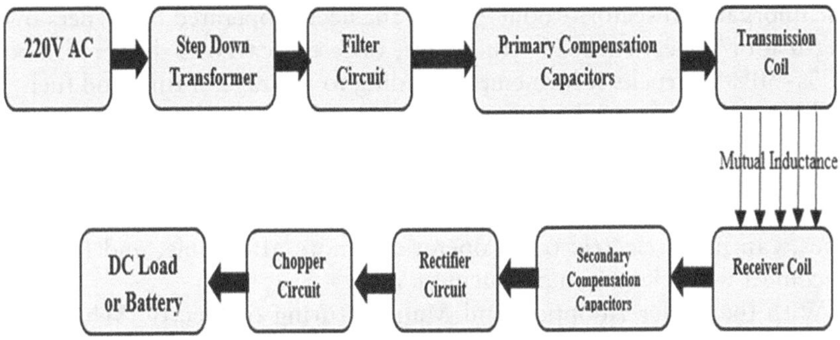


*Figure 13.8* Block diagram for wireless inductive charging system.

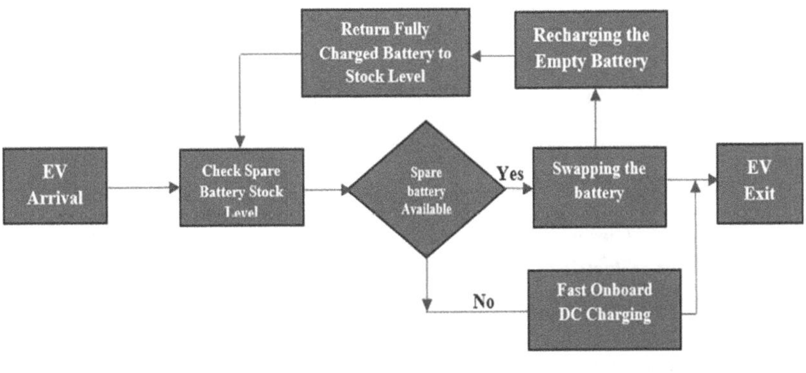


*Figure 13.9* Flow chart for EV battery swapping process.

Electrification of road transport will definitely go a long way in reduction of country's overall GHG emissions. While electrification of two- and three-wheelers, cars, and buses has already gained a momentum, the heavy-duty logistics sector electrification is still lagging behind in India. Over 70% of all road freight movement in India is carried by diesel trucks, and this percentage has been gradually increasing for more than two decades. Additionally, 57% of the petroleum used for transportation in India comes from diesel trucks, despite the country importing 16% of all imports and 88% of its crude oil needs. Therefore, issues about air pollution, GHG emissions, freight costs, the balance of trade, and energy security are greatly influenced by diesel-based transportation. Opportunities for truck electrification have emerged recently, thanks to remarkable reductions in battery costs and energy density, which were hardly anticipated just a few years ago [17] (Table 13.3).

There are multiple reasons why the sector has not yet gained the importance it needs. Some major reasons include lack of clear financial incentive that may propel the consumer toward new EVs, especially heavy-duty vehicles like trucks. Over 80% of the overall logistics, cost in India goes to the unorganised sector. About 75% of the fleet is operated by owners owning five or fewer trucks. Furthermore, once the cargo is delivered, about 30%–50% of trucks return empty, leading to wastage of time and fuel.

Though electric trucks face some unique challenges like harsh environmental constraints on weight and volume of battery packs so as to not compromise on payload capacity that generates revenues, advancements in battery technology and costs, combined with the right policies and incentives, can make electric truck operations more affordable, and India can become a world leader in producing EVs.

With the Faster Adoption and Manufacturing of Electric Vehicles and several other incentives—including the low 5% Goods and Services Tax as opposed to 28% on conventional vehicles, as well as special electricity tariffs for EV charging, India has already made great progress toward electrifying light vehicles, including cars, buses for public transportation, and two- and three-wheelers. The next natural step is therefore for India to start

*Table 13.3* Comparison of fuel economy and emissions per kilometre for diesel and electric

| | | *Units (metric tonnes)* | *7.5MT* | *12MT* | *25MT* | *40MT* |
|---|---|---|---|---|---|---|
| Fuel economy | Diesel | km/litre | 8 | 5.5 | 3.45 | 2.2 |
| | Electric | km/kWh | 2.5 | 1.69 | 0.88 | 0.49 |
| Emissions | Diesel | $gCO_2$/km | 446.9 | 650 | 1036.3 | 1625.1 |
| | Electric | $gCO_2$/km | 288.3 | 427.3 | 820.2 | 1485.1 |
| GHG reduction by electric truck | | | 35% | 34% | 21% | 9% |

*Table 13.4* Specifications of some commercial EVs

| *Type* | *Gross vehicle weight in tonnes* | *Battery capacity (kWh)* | *Approximate range (km)* | *Energy consumption (kWh/km)* |
|---|---|---|---|---|
| E buses | 10–17 tonnes | 120–180 | 175–250 | 0.6–0.7 |
| Light-duty trucks | 5–8 tonnes | 50–80 | 100–150 | 0.5–0.6 |
| Medium-duty trucks | 8–15 tonnes | 80–200 | 120–300 | 0.6–0.9 |
| Heavy-duty trucks | 18–36 tonnes | 200–500 | 250–500 | 1.0–1.5 |

electrifying trucks. Along with this, further regulations will be required to encourage investments in the value chain's most capital-intensive sectors—battery manufacturing, vehicle manufacture, and charging infrastructure [18] (Table 13.4).

#### 13.3.2.1 EV charging infrastructure and key technologies for future

##### *13.3.2.1.1 Megawatt charging system for big trucks*

The pace of battery technology advancement is happening faster than expected, and battery costs are also dropping faster than anticipated. As a result, newly available zero-emission trucks, generally, and heavy-duty zero-emission trucks, in particular, are shifting toward commercialisation at an accelerating pace. The key to increasing the popularity of e-mobility in this region is user-friendly and speedy high power charging due to the size and weight of battery packs in these transport options. A new high-power charging solution is required to meet the market need of the truck and bus industry to charge electric heavy-duty vehicles in an acceptable amount of time. Powerful megawatt charging system (MCS) systems can deliver a massive 3.75 MW DC, recharging a large vehicle, rather in minutes and not hours. The system provides 3,000 A at 1,250 V; its energy transfer far faster than current DC charging systems that deliver 350 kW (CCS Combo connector). MCS will become the universal standard, developed specifically for class 6, 7, and 8 commercial vehicles, which require over 1 MW charging rates. The MCS will not only be used for road vehicles but also for vessels at sea and in the air [19].

##### *13.3.2.1.2 Electric road systems*

Over the past 10 years, the idea of an electric road system (ERS) has been more widely recognised. ERS is described as a system that offers dynamic EV charging for many types of automobiles on roads and highways using either conductive or inductive (wireless) methods. These ERS models each employ a different type of technology to perform the same basic task, such as automatically transferring power to EVs on demand while moving at low and typical traffic speeds (quasi-dynamic and dynamic). Either the vehicle's onboard battery system receives the electricity or its propulsion system can receive direct power [20].

Both the overhead and conductive in-road rail ERS rely on direct contact (through a mechanical arm or pantograph) between the power source and the vehicle to transfer energy. However, it makes use of segmented electric rails that are buried in or placed above the road's surface. The parts of the rail track are electrified by the road side units once a vehicle is recognised travelling over it. A mechanical arm automatically extends from the vehicle's back or underbelly to connect with the rail once it is lined up with the track. The propulsion system or the battery is then directly or indirectly given power. These types of ERS systems are in testing phase in Germany, Sweden, and South Korea [21].

### 13.3.3 Hydrogen as a fuel

Despite being the most prevalent element on the planet, hydrogen is rarely encountered in its solitary state. It usually occurs in combination with other elements, such as water ($H_2O$), hydrocarbons (like methane and $CH_4$), and other organic materials. To isolate hydrogen, chemical, solar-driven, and biological techniques must be used. Hydrogen is considered one of the most sustainable fuels of the future. Combustion of hydrogen produces water vapor, with no residue or climate-harming impact [22] (Figure 13.10).

Today, natural gas is used in the chemical process of steam reforming to produce the majority (over 95%) of the hydrogen used in the world The most intriguing part is that renewable energy sources can also be used to manufacture hydrogen, such as biogas, a gaseous form of methane acquired from biomass, or electricity produced by solar or wind power for electrolysis As a result, the $CO_2$ emissions produced during the manufacturing of

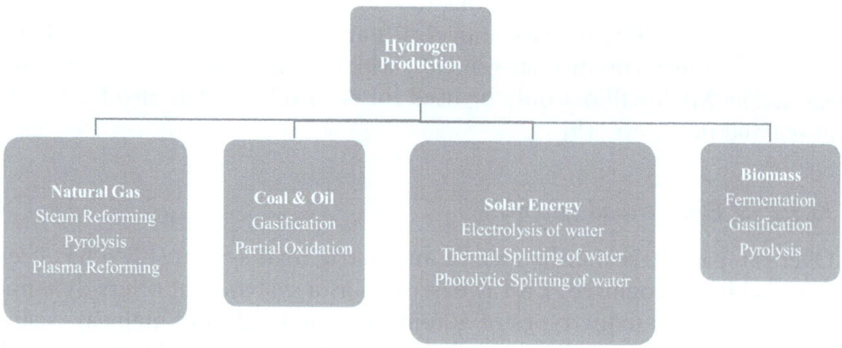


*Figure 13.10* Hydrogen production processes with energy sources.

hydrogen will be significantly reduced. The goal of developing these diverse methods is to compete with natural gas-based hydrogen generation by offering cheaper prices and bigger quantities. Although hydrogen is a colourless gas, it can be distinguished by a number of colours, including green, yellow, white, black, brown, pink, turquoise, grey, and blue. The source or method utilised to create hydrogen is indicated by the colour codes for hydrogen [23] (Table 13.5).

The increasing popularity of EVs (EVs) as the preferred green alternative is arguably the biggest obstacle blocking the adoption of hydrogen fuel cell (FC) vehicles. Most of the funding is going into electric automobiles in comparison to hydrogen being used as a fuel. Hence, the production costs remains high, and infrastructure required remains on hold, preventing hydrogen becoming a more viable option. But this is not only the prime reason, the current conditions under which the hydrogen used in FCs is produced also responsible. There are still hazardous emissions produced during the process of creating hydrogen from natural gas, which raises the carbon footprint and reverses the progress achieved toward having no tailpipe emissions. However, there would be no net increase in GHG emissions and a neutral carbon footprint if the electrolysis step of the hydrogen manufacturing process uses renewable energy sources like solar and wind [24].

As heavy-duty vehicles running on diesel require large power to pull heavy loads over vast distances, the time it takes to replenish an FCEV over the time to charge an EV makes hydrogen FCs in this case more economically viable than EVs. The EV charging downtime results in trucks being out of usage for long hours while they charge; on the other hand, refuelling of hydrogen can be done in minutes. Hydrogen storage takes up much less space than batteries, restoring the cargo capacity for transportation vehicles and small site space for stationary power. Another advantage is hydrogen's energy density. Diesel has an energy density of 45.5 mega joules per kilogram (MJ/kg), whereas hydrogen has an energy density of approximately 120 MJ/kg, almost three times more than diesel [25] (Table 13.6).

*Table 13.5* Hydrogen colour code description

| *Colour code* | *Production process* | *Source* |
|---|---|---|
| Grey | Steam reforming | Natural gas |
| Blue | Steam reforming with carbon capture | Natural gas |
| Green | Renewable electrolysis | Water |
| Yellow | Solar electrolysis | Water |
| Pink | Nuclear electrolysis | Water |
| Turquoise | Methane pyrolysis | Natural gas |
| Black | Gasification | Coal |

*Table 13.6* Physiochemical properties of hydrogen

| *Property* | *Unit* | *Value* |
|---|---|---|
| Molecular weight | g/g-mole | 2.016 |
| Freezing point | °C | −259.20 |
| Boiling point | °C | −253 |
| Density | gm/cm³ | $0.0899 \times 10^{-3}$ |
| Critical temperature | °C | −240 |
| High heating value | MJ/kg | 141.7 |
| Fuel/air mass ratio | – | 0.029 |
| Flame velocity | m/s | 2.65–3.25 |

#### *13.3.3.1 Hydrogen fuel cells*

An FC uses the chemical energy of hydrogen or other fuels to cleanly and efficiently produce electricity. FCs can be used in a wide range of applications, providing power for applications across multiple sectors, including transportation, industries, commercial and residential buildings, and long-term energy storage for the grid in reversible systems. Compared to combustion engines, FCs run more efficiently and are capable of directly converting the chemical energy in fuel into electrical energy at efficiencies of more than 60%. Hence, compared to combustion engines, FCs emit either no emissions or very little. FCs are quiet during operation as they have few moving parts. The use of FCs in equipment and commercial vehicles could become crucial for the future of transportation.

The pressurised fuel tank holds hydrogen onboard. A positively charged electrode (anode), a negatively charged electrode, and liquid (polymer electrolyte) make up an FC (cathode). The FC is filled with hydrogen fuel to start turning hydrogen into energy. Hydrogen enters the FC and undergoes an electrochemical process to produce electricity. The FC also adds oxygen to help the chemical reaction along with hydrogen. Hydrogen atoms split during the reaction to produce protons and electrons. Here, protons decay into water-based exhaust, while electrons create an electric current. The electric motor can be directly powered by the FC or the electricity can be stored onboard using a battery [26] (Figure 13.11).

On-board energy is produced using hydrogen fuel and an FC, and the only exhaust coming out of the tailpipe is clean water. As a result, FC automobiles are environmentally benign, much like EVs. The net chemical reaction of the FC is given below (Table 13.7):

$$2H_2(g) \rightarrow 4H^+ + 4e^- \left[\text{reaction at anode}\right]$$

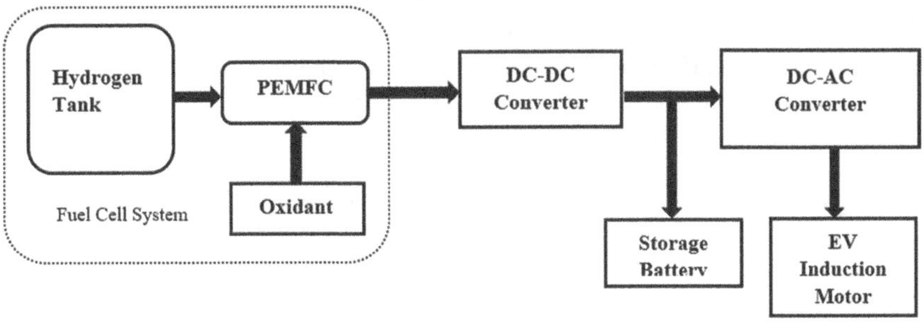


*Figure 13.11* Block diagram for fuel cell electric vehicle.

*Table 13.7* Various types of fuel cells and their characteristics

| *Fuel cell type* | *Fuel mixture* | *Electrolyte* | *Operating temp (°C)* | *Efficiency (%)* |
|---|---|---|---|---|
| PEMFC | Hydrogen | Proton exchange membrane | 30–100 | 40–60 |
| DMFC | Methanol-air | Proton exchange membrane | 20–90 | 20–30 |
| SOFC | Hydrocarbon | Oxide ion conducting ceramic | 500–1,000 | 30–50 |
| AFC | Hydrogen | Potassium hydroxide | 150–200 | 60–70 |

$$O_2(g) + 4H^+ + 4e^- \rightarrow 2H_2O(l) \quad [\text{reaction at cathode}$$

$$2H_2(g) + O_2(g) \rightarrow 2H_2O(l) \quad [\text{Overall reaction}]$$

Hydrogen has the promise of transforming India from an energy-deficient country to an energy-rich country. It can even make India a net exporter of energy. As of now, India is a developing nation when it comes to hydrogen fuel technology. The technology of FC automobiles is still in its nascent stage when compared to that of EVs.

In India, the commercial vehicle market is expanding at a pace of 15% annually, and more than 65% of the country's freight is being transported by trucks. Thus, the switch to hydrogen FC vehicles presents a potential opportunity for the Indian transportation industry. FC medium-duty trucks and HDTs may replace battery-powered commercial EVs as early as 2030 at a cheaper cost. Even with high-capacity fast chargers, larger batteries are expensive, heavy, and lower the vehicle's payload and also take a long time to charge. The price of hydrogen fuel at the pump and the cost of the vehicles in terms of the engine and on-board hydrogen tanks are the two primary cost components that must be focused on for FC trucks (FCT) to be priced similarly to rivals. In addition, the high cost of hydrogen FCs and hydrogen tanks drives up the cost of the powertrain, making FCTs around three times more expensive than equivalent diesel trucks. With just 10,000 trucks produced annually, manufacturers might achieve a large powertrain cost reduction (about 60%), thanks to the contribution of mass manufacturing [27].

To reduce its reliance on energy imports and transition the economy away from fossil fuels to meet climate goals, the Indian government is proposing a significant increase in the generation of green hydrogen. Launched in August 2021, India's National Hydrogen Mission aims to increase the generation of "green" hydrogen through renewable electrolysis and use it in a variety of industries, including transportation. Green hydrogen has a substantial decarbonisation potential, but its widespread adoption in India's road transport industry will depend in part on consumer prices. Production costs, transportation costs, fuelling costs, and potential tariffs throughout the supply chain make up the four main cost factors that go into the price of hydrogen fuel at the pump [28] (Figure 13.12).

According to the findings, while green hydrogen is around 1.5 times more expensive than blue hydrogen in 2030, its cost at the pump will be comparable in 2050. The difference gets smaller primarily as a result of rising natural gas prices and falling solar electricity and electrolysis expenses. In addition, increased capacity utilisation and economies of scale will result in a per-kilogram cost of pipeline transport that is about six times lower in 2050 than in 2030. Creating enough demand through policies and programmes that encourage early market development of pipeline infrastructure and FC electric car production is essential to realising these per-kilogram cost reductions. To ensure that investments in green hydrogen are not hampered by both new and existing regulations, governments and businesses must collaborate. A unified international standard will help the trade of hydrogen by tracking environmental effects of various hydrogen supply as well as for shipping and storing huge volumes [29].

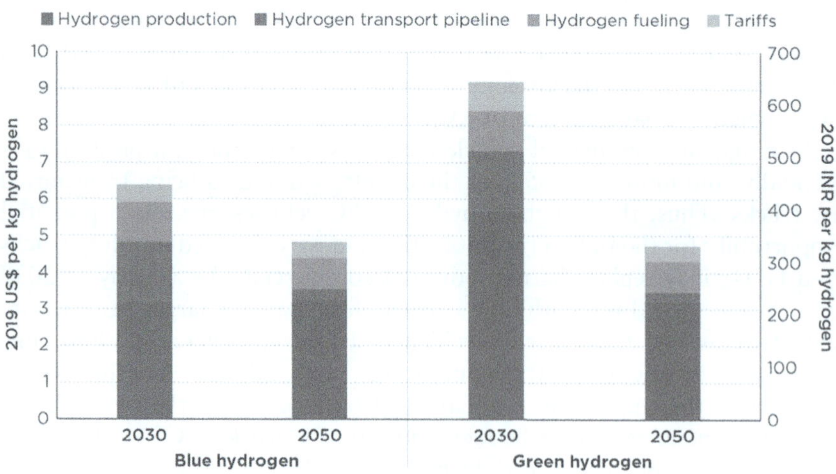


*Figure 13.12* Estimated cost variation of hydrogen as fuel in 2030 and 2050.

### 13.3.4 Biodiesel

Alternative diesel fuel made from renewable sources is known as biodiesel. Biodiesel is an alkyl ester made from renewable feedstocks such as vegetable oils, animal fat wastes, and algae. It is also a non-toxic, biodegradable, and renewable diesel fuel. Although biodiesel does not include oil, it can still be used as a fuel in either its pure form or when combined in any amount with diesel oil.

Vegetable oils (sunflower, soybean, rapeseed, safflower, cotton, and palm oils), non-edible oils (neem, karanja, jatropha, pongemia, and rubberseed), used cooking oil (UCO), animal oils (fish oils and poultry oils), algae, and waste cooking oil are all sources of oil that can be used to make biodiesel. The kind and quantity of vegetable oils utilised as raw materials sources in the synthesis of biodiesel are crucial. The types and ratios of the fatty acids present in the oil used to make the biodiesel reveal the fuel quality of the substance [30] (Figure 13.13).

There are numerous concentrations and blends of biodiesel that can be employed. Both B5 (up to 5% biodiesel) and binary blend (B20) (6%–20% biodiesel) are the most often used. Rarely utilised as a transportation fuel, B100 (pure biodiesel) is primarily used as a blend stock to create lower blends. B20 is a popular blend because it strikes a fair mix between price, emissions, cold-weather performance, material compatibility, and solvent

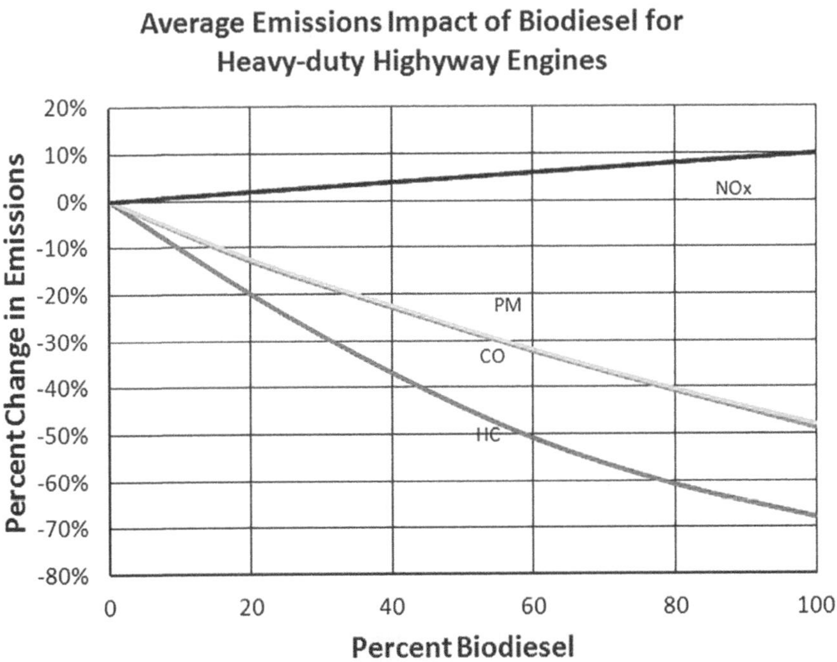


*Figure 13.13* Variation of emissions with biodiesel blending for heavy-duty engines.

*Table 13.8* Comparison of properties of diesel and biodiesel as per the ASTM standard

| *Property* | *Units* | *Diesel fuel ASTM D975* | *Biodiesel ASTM D6751* |
|---|---|---|---|
| Kinematic viscosity @40°C | C.st | 1.3–4.1 | 4.0–6.0 |
| Density @15°C | g/cm$^3$ | 0.85 | 0.86–0.90 |
| Calorific value | MJ/kg | 45 | 37.3 |
| Hydrogen | wt % | 13 | 123 |
| Oxygen, by dif. | wt % | 0 | 11 |
| Sulphur | ppm | <10 | <5 |
| Flash point | °C | 60–80 | 100–170 |
| Cloud point | °C | −15 to 5 | −3 to 12 |
| Pour point | °C | −35 to −15 | −15 to 10 |
| Cetane number | --- | 40–50 | 48–65 |

properties. It has been found that neat vegetable oil can be used as diesel fuel in conventional diesel engines and can lead to the problems such as incomplete combustion, leading to heavy smoke emission, poor fuel atomisation due to high viscosity, high flash point attributes to lower volatility characteristics, high carbon deposits, etc. The various processes for production of biodiesel are dilution, microemulsion creation method, pyrolysis, and transesterification. The process used to create biodiesel most frequently is transesterification [31] (Table 13.8).

Oil is esterified with a monohydric alcohol (ethanol, methanol) in the transesterification reaction, producing fatty acid esters and glycerine as the major products in the presence of a catalyst (acidic, basic catalysts, and enzymes). Diglycerides and monoglycerides are also produced during the esterification reaction, along with extra reactants and free fatty acids. Transesterification reaction can be given as

$$\text{Triglyceride} + \text{Alcohol} \rightarrow \text{Alkyl Ester}(\text{Biodiesel}) + \text{Glycerol}$$

The degree of biodiesel purity is the most crucial factor in production. As a result, the refining process becomes crucial. The two major processes for making biodiesel from vegetable oils or fats are transesterification with a base catalyst and transesterification with an acid catalyst. The most popular method for producing biodiesel is base catalyst transesterification. Low temperature, low pressure, minimal side effects, a high conversion rate of 98%, direct conversion to biodiesel without intermediary molecules, and the absence of foreign, extraneous elements in the structure are a few of the causes [32].

Due to their widespread availability, refined vegetable oils are frequently used to make biofuel; however, this practice has the potential to lead to food insecurity. This worrying scenario has led to the investigation of numerous non-edible oils with high free fatty acid levels. The esterification process is influenced by a number of factors, including the free fatty acid (FFA) content, catalyst concentration, methanol-to-oil molar ratio, reaction temperature, and reaction time.

The FFA content reduction is the most difficult and important of these factors. The high FFA content (>2% w/w) in the oil is undesirable because it reduces the biodiesel yield due to saponification with the alkali catalyst, which makes it more difficult to separate the biofuel from the glycerine fraction and makes ester purification challenging, both of which lead to a significant yield loss. By directly turning FFA and oil into biofuel, acid-catalysed transesterification could solve the drawbacks. However, the lengthy reaction time and subpar yields prevent this procedure from being used. Due to other economic factors, a pre-treatment procedure to lower the FFA level to under 2% w/w is required because it will increase product yields during base transesterification [33] (Table 13.9).

Long-term biodiesel storage is problematic, particularly during the winter. The amount of time that biodiesel can be stored depends on a number of crucial factors, one of which is its stability. Biodiesel is typically more prone to microbial deterioration than petroleum diesel. Fuel dispensers, vehicle fuel filters, and stalling automobiles can all be caused by microbial contamination of fuel storage tanks. If the biodiesel is going to be stored before being blended, the induction time and acid number need to be checked periodically to make sure it is not oxidising [34].

While authorising the sale of E-20 mixed gasoline starting in April 2023, India's national average aim of 20% ethanol blended with gasoline was moved to 2025 under the National Biofuel Policy of 2018. India has not changed its goal of achieving a 5% blend of biodiesel and regular diesel

*Table 13.9* Comparison of physiochemical properties of biodiesel from various non-edible oil sources

| *Properties* | *Regular diesel* | *Jatropha biodiesel* | *Neem biodiesel* | *Mahua biodiesel* | *Karanj (Pongamia) biodiesel* | *Palm oil biodiesel* |
|---|---|---|---|---|---|---|
| Density 15°C g/m³ | 0.8116 | 0.865 | 0.898 | 0.882 | 0.89 | 0.855 |
| Kinematic viscosity cst@40°C | 4.68 | 5.2 | 5.81 | 4.2 | 5.12 | 4.5 |
| Flash point °C | 98 | 191 | 175 | 208 | 166 | 174 |
| Pour point °C | 15 | NA | 8 | 6 | 5 | 16 |
| Cloud point °C | 18 | 13 | 4 | 13 | 3 | 16 |
| Cetane number | 53 | 51 | 48–65 | 51 | 36 | 65 |
| Sulphur ppm or wt% | 0.1 | NA | 0.03 | 180 ppm | 0.02 | 0.04 |
| Carbon residue wt% | 0.14 | 0.02 | 0.08 | 0.2 | 0 | 0.02 |
| Water content% | 0.001 | 0.16 | 0.1 | 0.04 | 0.005 | 0.1 |

by 2030. The Indian government believes that the aims can be achieved by increasing domestic biofuel production, using a variety of feedstocks, and promoting biofuel blending to replace gasoline and diesel in stationary and portable power applications as well as in automobiles and machinery.

By 2030, India hopes to have a B5 biodiesel blend for on-road use. In 2021, the national average blend rate was estimated at 0.09%. UCO, imported palm stearin, tiny amounts of non-edible oils, and animal fats from domestic sources are used to make biodiesel. Due to restricted feedstock availability, a lack of an integrated and committed supply chain, and import limitations, biodiesel utilisation is still very low. India would need to significantly invest in new plants to increase annual production capacity from its current effective capacity of 520 million litres, as well as build a supply chain infrastructure for UCO while enforcing necessary collection mechanisms, to comply with its blending mandate. According to estimations, India may save up to Rs 7 per litre by increasing the biodiesel blend ratio.

In contrast to India, where entrepreneurs running micro, small, and medium-sized enterprises predominate the market, many countries that produce biodiesel rely heavily on manufacturing facilities established by large oil companies or vegetable oil refineries. These entrepreneurs are typically fuel traders with relatively better access to the domestic fuels market. However, because the biodiesel industry is volume-driven, most traders choose to forgo capital expenditures by setting up low-cost biodiesel production facilities that primarily process low free fatty acids feedstock like palm stearin and palm oil, among other substances that are frequently subject to the fluctuations of vegetable oil markets. The operating profitability of these units is impacted by market price swings, which reduces the commercial appeal of biodiesel production compared to traditional diesel. Since the cost of the feedstock is crucial to the production of biodiesel, industrial facilities must be updated for flexibility and to have the necessary infrastructure to process many kinds of feedstocks, including UCO [35].

India must continue to pursue a workable plan for creating a domestic sector with a strong financial foundation and access to markets for imports. Comparatively few domestic producers generate biodiesel, and the majority of their production capacity is underutilised. Additionally, there are few reliable feedstock sources and few government assistance channels (like subsidised feedstock, tax credits, or loan guarantees). Due to the lack of virgin (first use) animal fat or vegetable oils and the failure of jatropha and other non-arable oilseeds to produce the required yields to make cultivation economically viable, India's biodiesel sector is still in its infancy. India will need to raise its production capacity to more than 3.45 billion litres to reach B-5 by 2030, assuming that the size of the estimated diesel pool for on-road consumption will be 69 billion litres in that year (Table 13.10).

*Table 13.10* Biodiesel production analysis from multiple feedstock (million litres), India

| *Year* | *2011* | *2012* | *2013* | *2014* | *2015* | *2016* | *2017* | *2018* | *2019* | *2020* | *2021* |
|---|---|---|---|---|---|---|---|---|---|---|---|
| Production | 111 | 126 | 132 | 138 | 152 | 158 | 170 | 185 | 230 | 200 | 180 |
| Imports | 0.0 | 0.0 | 0.3 | 1.7 | 0.8 | 2.7 | 7.1 | 25.2 | 7.0 | 1 | 1 |
| Exports | 0 | 0 | 3.9 | 41.5 | 33.1 | 41.7 | 7.6 | 23.1 | 54.0 | 68 | 50 |
| Consumption | 113 | 125 | 128 | 102 | 118 | 119 | 165 | 180 | 185 | 140 | 140 |
| **Feedstock use for fuel (1,000MT)** | | | | | | | | | | | |
| Non-edible industrial | 58 | 65 | 70 | 75 | 85 | 90 | 100 | 110 | 140 | 145 | 90 |
| Used cooking oil | 42 | 48 | 49 | 50 | 55 | 55 | 55 | 60 | 65 | 50 | 55 |
| Animal fats/tallows | 6 | 7 | 7 | 6 | 5 | 6 | 6 | 8 | 10 | 9 | 9 |
| Total | 106 | 120 | 126 | 131 | 145 | 151 | 161 | 178 | 215 | 204 | 154 |

## 13.4 CONCLUSION

The transport industry in India is currently the third largest emitter of $CO_2$ and will continue to play a significant role in the talks and initiatives aimed at reducing emissions. By 2050, the transportation industry might be in charge of 1.5 gigatonnes of yearly carbon emissions under a business-as-usual scenario. Without considerable action, the transportation industry could present a barrier to meeting the global net zero emissions target of 2050 and the climate goals of 2030. The industry urgently needs to change, and all interested parties—including governments and businesses—must work together to meet the challenging objective of decarbonising it. The desire to switch to alternative fuels has been constant, but there has not been much progress in the implementation on the ground. To help consumers embrace new technology automobiles and at the same time give the sector certainty to create vehicles, the Indian government needs to put out infrastructure for the spread of alternative fuels. To introduce alternative fuels across the nation, a rolling plan must be created. It is necessary to draught and frequently review a firm plan for the next 5 years and a tentative plan for the following 10 years. One agency under the Government of India needs to assess how each phase of the implementation of alternative fuels is going.

The current mix for biodiesel is around 0.001%. Additional efforts would be needed to supply at least 5% biodiesel blend at a national level, given the significant proportion of fuel used by buses and trucks. In addition to switching to B5 blend for diesel-powered vehicles, LNG and hydrogen technologies would also show promise if fuel is available when it comes to duty cycles, long-distance hauling, and travelling along highways. The industry may choose to concentrate on both potential fuels and just one in the future depending on how the fuel infrastructure develops. Depending on the geographical locations and the availability of LNG and other different fuels in areas where such fuels are available easily and energy suppliers also have sufficient dispensing units in these areas, it should be ensured that 5%–7% by 2025 and 10%–15% by 2030 of the new commercial vehicle fleet will be operating on LNG or any other alternative fuel.

Adoption of efficient and cleaner vehicle technology will result from a hastened implementation of a scrappage policy and green tax and from state governments joining forces to carry it out in full. The auto industry needs to collaborate with efforts to implement efficiency measures and acknowledge the gravity of the situation. For the commercial vehicle sector, these steps are essential. All commercial vehicle categories should have immediate access to emission standards.

The solution to clean mobility does not lie with a singular tool but will come about from a carefully thought-out and targeted multi-fuel approach by efficient development of all alternative fuel technologies such as LNG, hydrogen, biodiesels, and EVs (Table 13.11).

*Table 13.11* Summary of alternative fuel technologies feasibility for commercial vehicles in India

| *Alternative fuels* | | | | | | |
|---|---|---|---|---|---|---|
| *Vehicle type/technology* | *CNG* | *LNG* | *E10 and E20* | *Biodiesel B5* | *Hydrogen* | *Electric* |
| Three-wheelers – SI | Technology already in application | Not a feasible fuel for vehicle type | For E10, fuel already in use. For E20, technology needs to be developed | Not a feasible fuel for vehicle type | Not a feasible fuel for vehicle type | Technology already in application |
| Three-wheelers – CI | Not a feasible fuel for vehicle type | Not a feasible fuel for vehicle type | Not a feasible fuel for vehicle type | Technology available, fuel feasible | Not a feasible fuel for vehicle type | Technology already in application |
| Buses | Technology already in application | Technology to be developed, feasibility to be established | Not a feasible fuel for vehicle type | Technology available, fuel feasible | Technology to be developed, feasibility to be established | Technology already in application |
| LCVs | Technology already in application | Technology to be developed, feasibility to be established | Not a feasible fuel for vehicle type | Technology available, fuel feasible | Can be developed but economically not feasible | Technology in starting phase of application |
| Trucks | Technology in starting phase of application | Technology to be developed, feasibility to be established | Not a feasible fuel for vehicle type | Technology available, fuel feasible | Technology to be developed, feasibility to be established | Technology to be developed, feasibility to be established |

## ACKNOWLEDGMENTS

We are thankful to those who provided their insights and expertise for this research. We would like to thank MANIT Bhopal for providing all the necessary assistance needed and support for the research.

## REFERENCES

1. Singh, N., Mishra, T., and Banerjee, R. (2019). Greenhouse gas emissions in India's road transport sector. In: Venkataraman, C., Mishra, T., Ghosh, S., and Karmakar, S. (eds.), *Climate Change Signals and Response*. Springer: Singapore. https://doi.org/10.1007/978-981-13-0280-0_12.
2. Dalei, N. N., Roy, H., and Gupta, A. (2017). Crude oil import of India from its major oil trade partner countries: An empirical evidence using panel data analysis. *International Journal of Advanced Research and Development* 2(6): 726–735.
3. Ghosh, S. (2009). Import demand of crude oil and economic growth: Evidence from India. *Energy Policy* 37, 699–702. https://dx.doi.org/10.1016/j.enpol.2008.10.021.
4. Sakthivel, P., Subramanian, K. A., and Mathai, R. (2018). Indian scenario of ethanol fuel and its utilization in automotive transportation sector. *Resources, Conservation and Recycling* 132, 102–120. https://doi.org/10.1016/j.resconrec.2018.01.012.
5. Trivedi, S., Prasad, R., Mishra, A., Kalam, A., and Yadav, P. (2020). Current scenario of CNG vehicular pollution and their possible abatement technologies: An overview. *Environmental Science and Pollution Research* 27(32), 39977–40000. doi: 10.1007/s11356-020-10361-7.
6. Nouni, M. R., Jha, P., Sarkhel, R., Banerjee, C., & Tripathi, A. K. (2021). Alternative fuels for decarbonisation of road transport sector in India : Options, present status, opportunities, and challenges. *Fuel* 305(August), 121583. https://doi.org/10.1016/j.fuel.2021.121583.
7. Dey, B., Roy, B., Datta, S., and Ustun, T. S. (2023). Forecasting ethanol demand in India to meet future blending targets: A comparison of ARIMA and various regression models. *Energy Reports* 9(1), 411–418. https://doi.org/10.1016/j.egyr.2022.11.038.
8. Kumbhar, S. V., and Khot, S. A. (2022). Experimental investigations of ethanol-gasoline blends on the performance, combustion, and emission characteristics of spark ignition engine spark ignition (S.I) engine with partial addition of n-pentane. *Materials Today: Proceedings*. https://doi.org/10.1016/j.matpr.2022.11.284.
9. Ghosh, P. P. (2022). Impact of India's diesel subsidy reforms and pricing policy on growth and inflation. *Energy Economics* 113, 106195. https://doi.org/10.1016/j.eneco.2022.106195.
10. Kaushal, A. V., and Verma, A. S. (2022). Fuel shifting towards fuel of future for Transportation in India. *International Research Journal of Engineering and Technology (IRJET)* 9(6), 2408–2414.

11. Arefin, M.A, Nabi, M. N., Akram, M. W., Islam, M. T., and Chowdhury, M. W. (2020). A review on liquefied natural gas as fuels for dual fuel engines: Opportunities, challenges and responses. *Energies* 13(22), 6127.
12. Locke, K. (2019). Indian LNG demand & the global LNG market. *The Journal of World Energy Law & Business* 12(3), 246–259. https://doi.org/10.1093/jwelb/jwz010.
13. Sinha, A. S. K., Kar, S. K., Ojha, U., and Balathanigaimani, M. S. (2022). *Role of Natural Gas in India: Recent Developments and Future Perspectives.* IntechOpen.
14. Bahrami, A. (2022). EV charging definitions, modes, levels, communication protocols and applied EV charging definitions, modes, levels. *Communication Protocols and Applied Standards.* https://doi.org/10.13140/RG.2.2.15844.53123/11.
15. Unal, C., Yirik, E., Ünal, E., Cuma, M., Onur, B., and Tümay, M. (2018). A review of charging technologies for commercial electric vehicles. *International Journal of Advances on Automotive and Technology.* DOI: 10.15659/ijaat.18.01.892.
16. Ahmad, F., Alam, M., Alsaidan, I., and Shariff, S. (2020). Battery swapping station for electric vehicles: Opportunities and challenges. *IET Smart Grid* 3, 280–286. DOI: 10.1049/iet-stg.2019.0059.
17. Al-hanahi, B., and Member, G. S. (2021). Charging infrastructure for commercial electric vehicles : Challenges and future works. *IEEE Access* 9, 121476–121492. https://doi.org/10.1109/ACCESS.2021.3108817.
18. Nikit, A., Gopinathan, N., Khandekar, A., Karali, N., Phadke, A. A., and Rajagopal, D. (2022). Freight trucks in India are primed for electrification.
19. Mishra, P., Miller, E., Santhanagopalan, S., Bennion, K., and Meintz, A. (2022). A framework to analyze the requirements of a multiport megawatt-level charging station for heavy-duty electric vehicles. *Energies* 15, 3788. DOI: 10.3390/en15103788.
20. Tongur, S., and Sundelin, H. (2016). The electric road system transition from a system to a system-of-systems. *2016 Asian Conference on Energy, Power and Transportation Electrification (ACEPT)*, pp. 1–8. DOI: 10.1109/ACEPT.2016.7811529.
21. Gustavsson, M., and Börjesson, C. (2018). User perspectives on electric roads. *Conference: EVS 31 & EVTeC 2018*, Kobe, Japan.
22. Al-Baghdadi, M. (2020). An overview of hydrogen as an alternative fuel. *Encyclopedia* 6, 1–22. DOI: 10.32545/encyclopedia202006.0003.v1.
23. Panić, I., Cuculić, A., and Ćelić, J. (2022). Color-coded hydrogen: Production and storage in maritime sector. *Journal of Marine Science and Engineering* 10(12), 1995.
24. Offer, G., Howey, D., Contestabile, M., Clague, R., and Brandon, N. P. (2010). Comparative analysis of battery electric, hydrogen fuel cell and hybrid vehicles in a future sustainable road transport system. *Energy Policy* 38, 24–29. DOI: 10.1016/j.enpol.2009.08.040.
25. Dash, S. K, Chakraborty, S., Roccotelli, M., and Sahu, U. K. (2022). Hydrogen fuel for future mobility: Challenges and future aspects. *Sustainability* 14(14), 8285.

26. Ball, M., and Wietschel, M. (2009). The future of hydrogen: Opportunities and challenges. *International Journal of Hydrogen Energy* 34, 615–627. DOI: 10.1016/j.ijhydene.2008.11.014.
27. Harichandan, S., Kar, S. K., Bansal, R., and Mishra, S. K. (2022). Achieving sustainable development goals through adoption of hydrogen fuel cell vehicles in India: An empirical analysis. *International Journal of Hydrogen Energy*. https://doi.org/10.1016/j.ijhydene.2022.11.024.
28. Ajanovic, A., Sayer, M., and Haas, R. (2022). The economics and the environmental benignity of different colors of hydrogen. *International Journal of Hydrogen Energy* 47(57), 24136–24154. https://doi.org/10.1016/j.ijhydene.2022.02.094.
29. Sontakke, U., and Jaju, S. (2021). Green hydrogen economy and opportunities for India. *IOP Conference Series: Materials Science and Engineering* 1206, 012005. DOI: 10.1088/1757-899X/1206/1/012005.
30. Aktaş, E., Demir, Ö., and Deniz, U. Ç. A. R. (2020). Biodiesel production methods. *International Journal of Energy and Smart Grid* 5(1–2), 1–10.
31. Bateni, H., Saraeian, A., and Able, C. (2017). A comprehensive review on biodiesel purification and upgrading. *Biofuel Research Journal* 4(3), 668–690.
32. Dewangan, A., Yadav, A. K., and Mallick, A. (2018). Environmental effects current scenario of biodiesel development in India : prospects and challenges. *Energy Sources, Part A: Recovery, Utilization, and Environmental Effects*, 00(00), 1–8. https://doi.org/10.1080/15567036.2018.1502849.
33. Elgharbawy, A. S., Sadik, W., Sadek, O. M., and Kasaby, M. A. (2021). A review on biodiesel feedstocks and production technologies. *Journal of the Chilean Chemical Society* 66(1), 5098–5109.
34. Halder, S., Dhawane, S. H., Kumar, T., and Halder, G. (2015). Acid-catalyzed esterification of castor (Ricinus communis) oil: optimization through a central composite design approach. *Biofuels* 6(3–4), 191–201.
35. Hosseinzadeh-Bandbafha, H., Kumar, D., Singh, B., Shahbeig, H., Lam, S. S., Aghbashlo, M., Tabatabaei, M. (2022). Biodiesel antioxidants and their impact on the behavior of diesel engines: A comprehensive review, *Fuel Processing Technology*, 232, 107264, ISSN 0378-3820, https://doi.org/10.1016/j.fuproc.2022.107264.

Chapter 14

# Hydrogen production technologies

## Current scenario and future evolution

*Rahul Kumar Rajak, Kaustubh Singh, Gaurav Dwivedi, and Tikendra Nath Verma*

### 14.1 INTRODUCTION

Clean energy transition has gathered significant attention to address the issues the decline in conventional fuel resources and alarming rise in pollution levels. The recent World Climate Action Summit COP 28, held at Dubai, UAE in December 2023 emphasised limiting global temperature rise to about 1.5° and increasing the share of renewable energy with higher energy efficiency (Younas et al., 2022). Thus, it is imperative to identify cleaner alternatives to conventional sources. Among the various clean energy options, renewable hydrogen is considered the most favourable due to its high calorific value, small flame quenching distance, high flammability limits, and clean-burning nature. Hydrogen can be produced from a variety of sources, such as fossil fuel sources (coal and natural gas) and renewable sources (wind, solar, and biomass). Currently, hydrogen production is primarily derived from non-renewable sources (approximately 96%), with major contributions from natural gas steam reforming (about 48%), naphtha reforming (about 30%), and coal gasification (about 18%). Hydrogen has numerous applications in industries such as ammonia production, petroleum refining, steel industries (as a reducing agent), combined heat and power (for generating heat and electricity), spacecraft (as a propellent), and the transportation sector (fuel cell vehicles and internal combustion engine vehicles) (Teoh et al., 2023). Based on the 2023 report of WHA International Inc., the present consumption of hydrogen is primarily for ammonia production (about 55%), followed by petroleum refining (about 25%), methanol production (~10%), and other uses, such as in transportation industry (~10%). To increase the inclusion of hydrogen energy as a fuel in the transportation sector to decarbonise the harmful tailpipe emissions, several governments have implemented clean energy initiatives. China developed its electrolyser capacity from 220 MW in 2022 to 750 MW (under construction). The European Union approved hydrogen-related Important Projects of Common European Interest (2022) and announced auctions for

DOI: 10.1201/9781003521341-14

the European Hydrogen Bank (2023). India approved the National Green Hydrogen Mission, aiming to produce about 5 Mt of renewable hydrogen (2030) to become a leading manufacturer of electrolysers. These policies have encouraged the scientific community to conduct research that makes hydrogen production more sustainable and cleaner for end users (Mission Innovation, n.d.). Conventional methods of hydrogen production offer several advantages such as mature technology, ease of operation, low production costs. However, they also have downsides such as high emissions and environmental damage. To mitigate these effects, efforts are being made to make the hydrogen production more sustainable and environmentally friendly by incorporating carbon capture and sequestration units. These units capture harmful $CO_2$ emissions, a major contributor to global warming, and store them underground in geological formations. To compare hydrogen fuel with conventional fuel sources, several comparative indicators have been developed, such as the environmental impact factor (EIF), greenisation factor (GF), and hydrogen content factor (HCF), which are described in Eqs. (14.1)–(14.3):

$$\mathrm{EIF} = \frac{\mathrm{kg\,CO_2\ product\ of\ combustion\ reactior}}{\mathrm{kg\,fuel}} \quad (14.1)$$

$$\mathrm{GF} = \frac{\mathrm{EIF_{max}} - \mathrm{EIF}}{\mathrm{EIF_{max}}} \quad (14.2)$$

$$\mathrm{HCF} = \frac{\mathrm{kg\ of\ H_2\ in\ the\ fuel}}{\mathrm{kg\,fuel}} \quad (14.3)$$

Figure 14.1 shows a comparative assessment of different fuels based on EIF, GF, and HCF criteria. This clearly demonstrates that hydrogen fuel has the highest GF and HCF values compared to all other fuels, while exhibiting the least EIF value amongst the selected fuels (Dincer & Acar, 2014).

## 14.2 METHODS OF HYDROGEN PRODUCTION

Hydrogen production can be classified based on several criteria, such as the primary energy source or material, colour coding (which indicates the carbon footprint generated during hydrogen production), and the nature of resources, that is, renewable sources or non-renewable sources. From the available literature, two primary types of classification have been identified for hydrogen production: one based on the primary energy source/material, and another based on colour coding of the hydrogen production process.

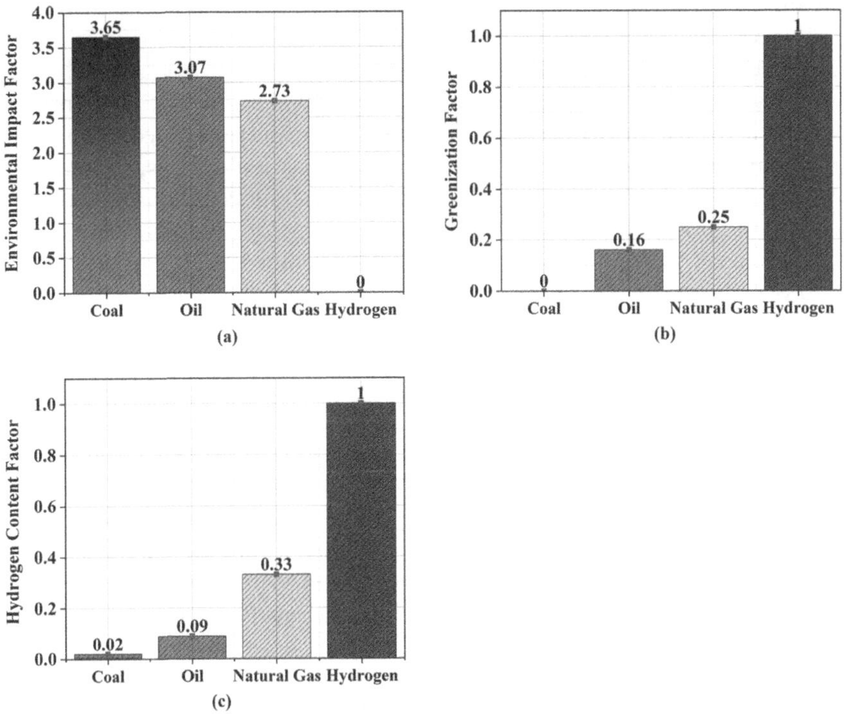


*Figure 14.1* Variation of (a) environmental impact factor, (b) greenisation factor, and (c) hydrogen content factor for coal, oil, natural gas, and hydrogen fuels.

### 14.2.1 Based on primary energy source/material

The primary energy source or material governs the principle of hydrogen production and the raw materials used in the production process. Some preliminary sources of energy commonly used for hydrogen production include thermal energy, nic energy, electrical energy, and biochemical energy.

#### *14.2.1.1 Thermal energy-based methods*

These methods utilise the energy stored in different feedstocks such as coal, natural gas, and biomass to generate hydrogen present in their chemical structure. The different types of thermal energy-based methods for hydrogen production are described as follows:

##### *14.2.1.1.1 Coal gasification*

The coal gasification process utilises carbonaceous materials such as coal to convert it into synthetic gas through the application of heat using air and steam. This technique can also efficiently convert various waste materials, such car tires, sewage sludge, wood, and plastic waste, into value-added products. The output of the gasification process is synthetic gas, which is composed of CO, $CH_4$, $H_2$, ash tar, $H_2S$, $NH_3$, HCl, and HCN. The synthetic gas is further processed to separate hydrogen from harmful contaminants. Different ranks of coal can be used in the gasification process, including (i) lignite (low-rank coal), (ii) bituminous (medium-rank coal), and (iii) anthracite (high-rank coal). Table 14.1 presents the proximate and ultimate analysis of these coal ranks. The quality of coal is determined by its carbon content and volatile matter; high-rank coals have higher carbon percentage and lower volatile matter (Chai et al., 2019; Coppola et al., 2015).

Coal gasification can be carried out at temperatures ranging from 900°C to 1,700°C based on the gasification technique used. The various gasification techniques include (i) fixed-bed gasification, (ii) moving-bed gasification, (iii) fluidised-bed gasification, (iv) entrained-flow gasification, and (v) plasma gasification (Afzanizam Samiran et al., 2014). Figure 14.2 shows the different kinds of gasifiers available for gasifying carbonaceous substances. Fixed-bed gasifiers are further classified as updraft and downdraft gasifiers. In updraft gasifiers, fuel loading and synthesis gas is extracted from the upper part of the gasifier, while the gasifying agent is supplied from the bottom. On the contrary, downdraft gasifiers have fuel loading and synthesis gas output at the bottom of the gasifier, while the gasifying agent is supplied from the middle portion through the channels to the reactor. In fluidised-bed reactors, fuel loading takes place from the side of the reactor, which gets mixed with the bed material and reaches the bed temperature within a short period of time. The lower part of the gasifier is used to supply the gasifying agent and remove the slag, while the upper part of the reactor receives the output synthesis gas. Fluidised-bed gasifiers are also classified as bubbling-bed, circulating-bed, internally circulating bed, spouted-bed, and dual-bed gasifiers. For high-power applications (up to 50 MW), entrained-flow gasifiers are used, in which the top portion supplies the gasifying agent, the middle portion provides the output synthesis gas, while the lower portion removes the ash formed during the gasification process. Plasma gasification, a novel technique, uses plasma torches to achieve high temperatures, decomposing raw materials into syngas and slag. This offers several advantages over conventional gasification techniques, including higher efficiency, lower tar formation, higher quality of syngas, higher hydrogen concentration in syngas, and cleaner exhaust emissions (Midilli et al., 2021).

Table 14.2 provides a comparative assessment of different types of gasifiers based on the type of coal, syngas temperature, scale of operation, reaction zone temperature, and gas efficiency. The integration of carbon capture and sequestration technologies with coal gasification units can be employed to trap harmful pollutants such as carbon oxides, prevent their direct release into the atmosphere, which can be further used for other applications.

*Table 14.1* Proximate and ultimate analysis of different ranks of coal (Habibi & Mehrpooya, 2021)

| | *Proximate analysis of coal* | | | | | *Ultimate analysis of coal* | | | |
|---|---|---|---|---|---|---|---|---|---|
| *Type of coal* | *Moisture (%)* | *Ash (%)* | *Sulphur (%)* | *Phosphorus (%)* | *Volatile matter (%)* | *Carbon (%)* | *Hydrogen (%)* | *Nitrogen (%)* | *Oxygen (%)* |
| Lignite | 10.3 | 3.5–7.5 | 0.5–1.2 | 0.001–0.002 | 52–60 | 70–73 | 4.6–5.5 | 0.6–1 | 22–26 |
| Bituminous | 7.4 | 12.9 | 0.6 | - | 21.8 | 66.5 | 4.4 | 1.2 | 13.3 |
| Anthracite | - | 7.54–9.65 | 0.24–0.62 | - | 9.31–13.63 | 79.32–85.42 | 3.06–3.70 | 0.7–0.8 | 1.82–5.80 |

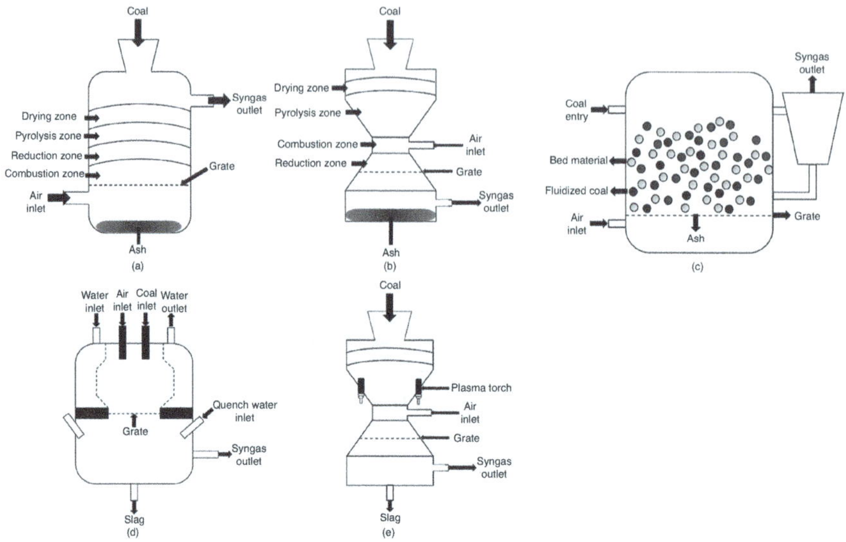


*Figure 14.2* Schematic diagrams of different types of gasifier: (a) updraft gasifier (fixed-bed type), (b) downdraft gasifier (fixed-bed type), (c) fluidised-bed gasifier, (d) entrained-flow gasifier, and (e) plasma gasifier.

*Table 14.2* Comparative assessment of different types of gasifier designs (Midilli et al., 2021)

| | *Fixed-bed gasifier* | *Fluidised-bed gasifier* | *Entrained-bed gasifier* | *Plasma gasifier* |
|---|---|---|---|---|
| *Type of coal* | *Low-rank coal only (lignite)* | *Low-rank coal only (lignite)* | *All types of coal* | *All types of coal* |
| Syngas temperature (°C) | 460–650 | 800–1,000 | 900–1,600 | 1,250 |
| Scale of operation | Large | Large | Large | Small |
| Temperature of reaction zone (°C) | 1,000–1,100 | 800–1,000 | 1,990 | 1,500–5,500 |
| Gas efficiency (%) | 80 | 78–81 | 80 | 80–90 |

### *14.2.1.1.2 Fossil fuel reforming*

Fossil fuel reforming is a commercially accepted technique of hydrogen production. It involves the reaction of fossil fuels with steam or oxygen, known as steam reforming or partial oxidation, respectively. Another method of fossil fuel reforming is autothermal reforming, which has a net zero enthalpy of reaction (Megia et al., 2021).

14.2.1.1.2.1 STEAM REFORMING

In this method, steam reacts with lighter hydrocarbons such as natural gas, liquified petroleum gas, and naphtha at high temperatures to produce hydrogen and carbon dioxide. Methane is commonly used in this method, where it reacts with steam to generate carbon monoxide. Carbon monoxide is further treated using water–gas shift reaction to yield a mixture of hydrogen and carbon dioxide. The overall reaction is shown in Eq. (14.4) as follows:

$$CH_4 + 2H_2O \rightarrow CO_2 + 4H_2 \tag{14.4}$$

The reaction is highly endothermic, requiring an enormous amount of heat energy and is carried out at temperatures between 800°C and 1,000°C. To withstand such high temperatures, reformer material should have high thermal resistance such as nickel–chromium steel. Investigations into high-performance materials have been conducted to ensure stability under high temperatures required in the process (Yoo et al., 2017). Studies on catalysts have established that ceramic oxides are ideal materials for lowering the amount of harmful gases such as carbon monoxides, improving the hydrogen production, and withstanding high temperatures of reaction. The mixture of gases formed after reforming is filtered using a pressure swing adsorption unit to separate pure hydrogen as the final product.

14.2.1.1.2.2 PARTIAL OXIDATION

The partial oxidation method for hydrogen production offers a range of feedstocks such as hydrocarbons, heavy fuel oil. In this method, the feedstock reacts with oxygen following a highly exothermic reaction and produces a mixture of gases including hydrogen, carbon monoxide, and other partially oxidised species. Due to highly exothermic nature of this process, no external source of energy is required. The product of partial oxidation reaction is governed by the C/O ratio of the feedstock and higher temperature limits (above 1,000°C). A suitable heterogenous catalyst is necessary to facilitate the reaction at lower temperatures. Transition metal-based catalysts are used for partial oxidation due to their variable oxidation states and high adsorption capacity for reactants and intermediate products. The partial oxidation for hydrogen production occurs as shown in Eq. (14.5):

$$C_nH_m + \frac{1}{2}nO_2 \rightarrow nCO + \frac{1}{2}mH_2 \tag{14.5}$$

Above 550°C, the reaction primarily yields CO and $H_2$, where the CO is removed by converting into $CO_2$ either by oxidation or by water–gas shift reaction, thereby improving the yield of hydrogen. The partial oxidation method offers several advantages such as flexible feedstocks, lower energy consumption, and simple operation. While the downsides are a smaller life span of the reactor unit due to high exothermicity and catalyst deactivation (Megia et al., 2021).

##### 14.2.1.1.2.3 AUTOTHERMAL REFORMING

This method combines the approaches of both exothermic partial oxidation with oxygen with endothermic steam reforming inside the reformer unit to yield a thermodynamically neutral reaction. Catalyst selection plays a vital role in this process, and usually nickel-based catalysts are preferred due to their high effectiveness and lower cost. This process has higher thermal efficiency than steam reforming and partial oxidation. The rate of hydrogen production is higher than that of partial oxidation but lower than steam reforming (Megia et al., 2021).

#### *14.2.1.1.3 Biomass gasification*

In the biomass gasification process, biological substances undergo a thermochemical process in the presence of a gasifying medium to convert into mixture of gases such as methane ($CH_4$), carbon dioxide ($CO_2$), and hydrogen ($H_2$). The product gases are then further treated to generate pure hydrogen (Dincer & Acar, 2014). The biomass primarily comprises lignin (up to 25%), cellulose (up to 55%), and hemicellulose (up to 40%). Biomass containing a higher ratio of (cellulose+hemicellulose) to lignin produces higher percentage of synthesis gas. Gasifying mediums used for the gasification process can also be of different types such as air, oxygen, steam, or supercritical water (SCW). Air is the most commonly used gasifying medium due to its abundance in nature and low cost. Although air is an economical gasifying agent, it has poor syngas production. Steam, on the other hand, has a higher H/C ratio of syngas, yielding hydrogen with higher concentration and calorific value of syngas. Oxygen is also utilised as a gasifying agent to remove the dilution of air created by excess nitrogen; however, due to being expensive and energy-intensive, it is having limited usage (Cormos, 2023). Recently, SCW has been extensively exploited as a gasifying agent in biomass gasification. SCW is obtained when water reaches a temperature of 647 K and 221.2 bar pressure over the critical point of water. At this stage, the liquid and gaseous phases coexist, losing the surface tension property of water. The advantages of using Supercritical water gasification (SCWG) method is that it allows feedstocks having higher moisture percentage to be utilised in the gasification process, it eliminates the step for drying the feedstock used in gasification and reduces the cost of the process, and SCW provides hydrogen, which acts as a gasifying agent (Mishra & Upadhyay, 2021). Figure 14.3 shows the heating value of syngas produced by using different gasifying agents. From the figure, it can be clearly visualised that SCW

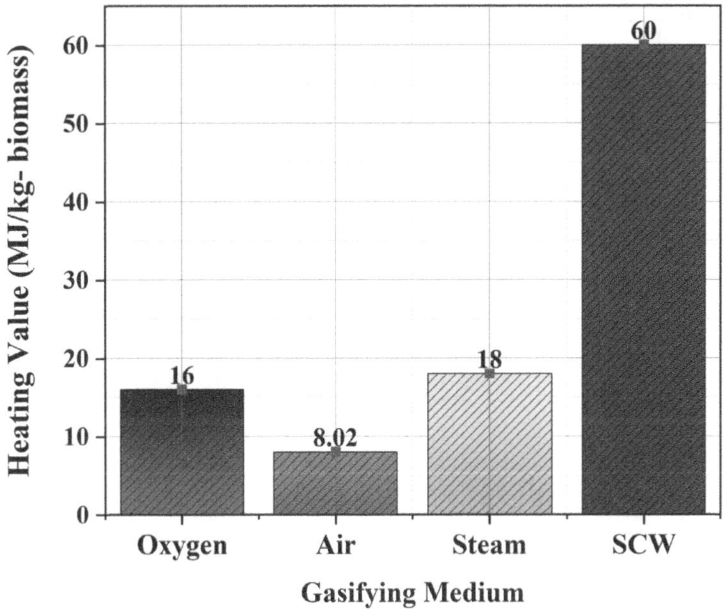


*Figure 14.3* Heating value (MJ/kg biomass) for different gasifying mediums used in biomass gasification.

produces synthesis gas having the highest heating value of up to 60 MJ/kg biomass, followed by steam (18 MJ/kg biomass) and oxygen mediums (16 MJ/kg biomass). Moreover, SCW offers flexibility in choice of feedstocks for the gasification process. The process of gasification is divided into different zones, such as drying zone, pyrolysis zone, combustion zone, and reduction zone. In the drying zone, the feedstock loses its moisture, reaching up to a temperature of 473 K (Afzanizam Samiran et al., 2014). Figure 14.4 shows the moisture content present in different feedstocks such as wheat straw, wood bark, corn stalks, rice straw, and food waste. From the figure, it can be inferred that the highest moisture content is present in food waste (up to 70%), while wheat straw has the least amount of moisture content (up to 15%). Higher the amount of moisture present in the feedstock, higher would be the energy consumption in the drying zone and poor would be the yield of syngas. After the drying zone, the next zone is the pyrolysis zone, where the decomposition of hemicellulose, cellulose, and lignin present in the biomass takes place. The temperature of this zone ranges between 398 and 773 K, and the products of pyrolysis include biooils and gaseous products. At temperatures up to 623 K, products such as char and tar are produced, while beyond 773 K, the tar formation gets vanished. The combustion zone occurs after the pyrolysis zone where exothermic reaction takes place, with the temperature range reaching up to 1,773 K. At this stage, the gasifying

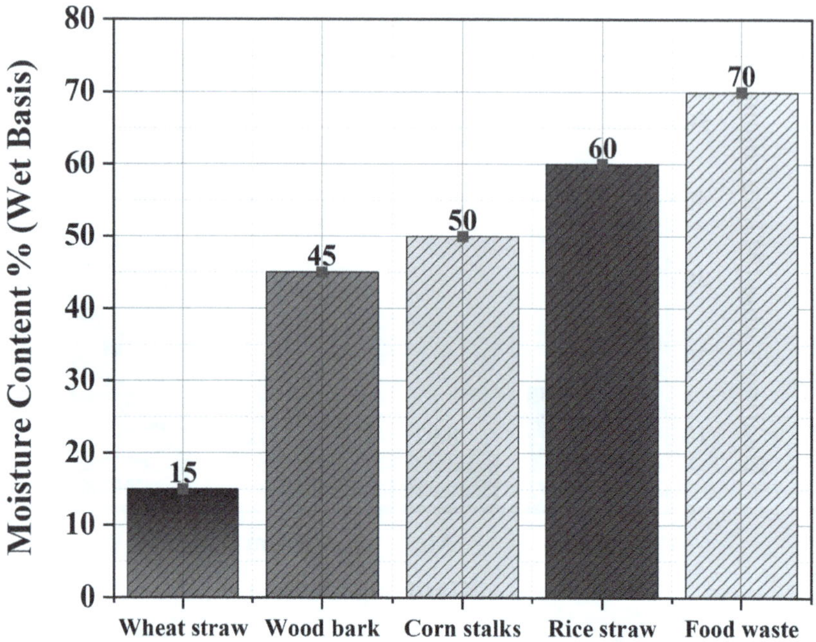


*Figure 14.4* Moisture content % (wet basis) for different feedstocks used in biomass gasification.

agent is controlled to monitor the production of CO, $CO_2$, $H_2$, and $H_2O$. After combustion, the tar produced in the fuel gas is treated by heating it up to 1,273 K in the reduction zone to free the gas from tar, which can otherwise lead to clogging the filters increasing the separation cost of plant (Anand et al., 2023).

The process of gasification of biomass is determined by several factors such as gasification temperature, gasification temperature, material of gasifier bed, and type of biomass used. Table 14.3 shows the effect of different operating parameters on the biomass gasification process.

#### *14.2.1.1.4 Thermochemical water splitting cycles*

The thermochemical water splitting method of hydrogen production requires water as raw material, which decomposes into hydrogen and oxygen through recursive chemical reactions generating intermediate products until finally converts into hydrogen and oxygen molecules. Thermal energy is utilised for generating hydrogen through this method (Safari & Dincer, 2020). Table 14.4 shows the different types of thermochemical processes present in the literature along with the advantages and challenges associated with them.

*Table 14.3* Effect of operating parameters on biomass gasification process (Mishra & Upadhyay, 2021)

| *Operating parameter* | *Effect on gasification process* |
|---|---|
| Gasification temperature | It determines the amount of syngas produced, tar formation, requirement of reactor, and economic costs for the process. Catalytic assisted gasification takes place at lower temperatures, resulting in reforming of tar produced. |
| Gasification pressure | It determines the performance and quality of product gases based on the partial pressure of gasifying mediums as well as pressure of gasification. In the case of SCWG, under super critical conditions, the dissociation of biomass occurs rapidly, resulting in faster water–gas shift reaction, which favours syngas production. |
| Material of gasifier bed | Alkali metal catalysts such as KOH, $K_2CO_3$, $Na_2CO_3$ and NaOH results in better yield of syngas and lower tar production. Ni- and Ru-based catalysts have also been found to reduce coke formation. |
| Biomass type | Biomass primarily consists of hemicellulose, lignin, and cellulose. Biomass containing higher (hemicellulose + cellulose)/lignin ratio produces higher quality syngas. Equivalence ratio, residence time, size of biomass particles, and ratio of steam/biomass govern the yield of tar formation and syngas. |

#### *14.2.1.1.5 Hybrid thermochemical cycle*

The hybrid thermochemical cycle utilises both heat and electricity to generate hydrogen. The step consuming electricity is selected to minimise electrical energy requirement. The advantages of using this method include lower maximum temperature and utilisation of waste heat from nuclear sources. The hybrid sulfur–ammonia thermochemical process is being studied recently for hydrogen production. This method uses solar irradiance (UV, visible light region) for evolution of hydrogen while (infra-red region) for oxygen production (Safari & Dincer, 2020). The reactions taking place in this process are shown in Eqs. (14.15)–(14.19):

(Chemical absorption, 25°C)

$$SO_2(g) + 2NH_3(g) + H_2O(l) \rightarrow (NH_4)_2 SO_3(aq) \tag{14.15}$$

(Electrolytic oxidation, 80°C–150°C)

$$(NH_4)_2 SO_3(aq) + H_2O(l) \rightarrow (NH_4)_2 SO_4(aq) + H_2(g) \tag{14.16}$$

(Solar thermal conversion, 400°C)

*Table 14.4* Classification of thermochemical cycles (Safari & Dincer, 2020)

| *Type of thermochemical cycle* | *Description of cycle* |
|---|---|
| Two-step cycle | In this method, metal oxide is reduced using solar radiation (endothermic process) producing oxygen. The reduced metal oxide reacts with water to produce hydrogen and regains the metal oxide (exothermic process), as shown in **Eqs. (14.6)** and **(14.7)**.<br>$MO_n \rightarrow MO_{n-\delta} + \frac{\delta}{2} O_2$ (*Endothermic reaction*) (14.6)<br>$MO_{n-\delta} + \delta H_2O \rightarrow MO_n + \delta H_2$ (*Exothermic reaction*) (14.7)<br>$CeO_2$ is considered an optimum metal oxide for two-step cycle lower costs, faster kinetics, and industrially competitive nature. Although the two-step cycle is comparatively simplistic, it requires higher working temperatures for its operation. |
| Three-step cycle | This type of cycle is a modification of the two-step cycle, where the high-temperature process is reformed into two different reactions, as shown in **Eqs. (14.8)–(14.10)**. This results in minimising the maximum temperature of the cycle.<br>$H_2SO_4 \rightarrow SO_2 + H_2O + \frac{1}{2} O_2$ (Temp. > 800°C) (14.8)<br>$I_2 + SO_2 + 2H_2O \rightarrow 2HI + H_2SO_4$ (Temp. < 120°C) (14.9)<br>$2HI \rightarrow I_2 + H_2$ (Temp. > 300°C) (14.10)<br>This method has certain limitations such as requirement of higher temperature for dissociation of $H_2SO_4$, corrosive nature of reactants, and loss of HI during their separation with $H_2SO_4$. |
| Four and higher step cycles | This type of cycles operate on lower maximum temperature compared to two-step and three-step processes. The four-step cycle takes place as shown in **Eqs. (14.11)–(14.14)**:<br>$H_2O + AB \rightarrow AH_2 + BO$ (Hydrolysis reaction) (14.11)<br>$AH_2 \rightarrow H_2 + A$ (Hydrogen evolving reaction) (14.12)<br>$BO \rightarrow 0.5O_2 + B$ (oxygen evolving reaction) (14.13)<br>$A + B \rightarrow AB$ (Reagent recycling reaction) (14.14)<br>Cu-Cl is the commonly used four-step thermochemical cycle. It offers higher hydrogen production, good process efficiency, and lower operating temperature ($T_{max}$ < 550°C). |

$$(NH_4)_2 SO_4(aq) + K_2SO_4(l) \rightarrow K_2S_2O_7(l) + 2NH_3(g) + H_2O\ (g) \quad (14.17)$$

(Solar thermal conversion, 550°C)

$$K_2S_2O_7(l) \rightarrow K_2SO_4(l) + SO_3(g) \quad (14.18)$$

(Solar thermal conversion, 850°C)

$$SO_3(g) \rightarrow SO_2(g) + \frac{1}{2}O_2(g) \tag{14.19}$$

In this process, hydrogen is produced when ammonium sulphite undergoes oxidation to ammonium sulphate under approximate ambient conditions. Potassium sulphate when reacts with ammonium sulphate in a low-temperature reactor yields potassium pyrosulphate. This potassium pyrosulphate undergoes dissociation in a low-temperature reactor to potassium sulphate and $SO_3$. $SO_3$ decomposes into $SO_2$ and oxygen in the presence of a catalyst (Safari & Dincer, 2020).

#### *14.2.1.2 Electrical methods*

These methods include water electrolysis method in which hydrogen production involves the dissociation of water molecule into hydrogen and oxygen when an electric current passes through an electrolysis cell having two submerged electrodes. The evolution of hydrogen takes place at the cathode, while oxygen evolution takes place at the anode. The primary input in the water electrolysis method is pure water, but the purity of water can vary depending on the source. For lab-scale research applications, deionised water is preferred to make precise measurements for hydrogen production. On the contrary, industrial applications prefer water based on feasibility studies of cost, availability of water, and purification methods. Water electrolysis employs different methodologies such as alkaline water electrolysis (AWE), anion exchange membrane (AEM), proton exchange membrane (PEM), and solid oxide electrolysis (SOE) (Arsad et al., 2024). The basic chemical reaction for water electrolysis takes place as shown in Eq. (14.20):

$$H_2O + \text{Energy}(\text{electrical} \,\&\, \text{heat}) \rightarrow H_2 + \frac{1}{2}O_2 \tag{14.20}$$

Theoretically, 1.23 V of cell voltage is required for the dissociation of water into hydrogen and oxygen. However, for practical applications, a cell voltage of 1.48 V is required for proper splitting to overcome the reaction kinetics and ohmic resistance of electrolyte and cell components (Horri & Ozcan, 2024). Table 14.5 shows the comparative analysis of water electrolysis methods.

##### *14.2.1.2.1 Types of water electrolysis methodologies*

###### 14.2.1.2.1.1 ALKALINE WATER ELECTROLYSIS

The AWE method is a commercially accepted hydrogen production method in which the electrolyte medium consists of alkaline potassium

*Table 14.5* Comparative analysis of water electrolysis methods (Arsad et al., 2024)

| *Parameters* | *Method of electrolysis* | | | |
|---|---|---|---|---|
| | *Alkaline water electrolysis* | *Proton exchange membrane* | *Anion exchange membrane* | *Solid oxide electrolysis* |
| Operating temperature (°C) | 30–80 | 20–80 | 30–90 | 500–900 |
| Operating pressure (bar) | 1–30 | <70 | <35 | Atmospheric pressure |
| Operating voltage (V) | 1.5–2.5 | 0.8–1.5 | 0.8–1.5 | 1.0–1.5 |
| Current density (A/cm$^2$) | 0.1–0.5 | 0.2–1.5 | 0.1–1.0 | 0.1–0.5 |
| Specific energy consumption (kWh/Nm$^3$) | 5.0–5.9 | 5.0–6.5 | 3.7–3.9 | - |
| Process efficiency (%) | 60–90 | 60–90 | 70–90 | 80–90 |
| Electrode material | Nickel-coated stainless steel (both cathode and anode) | **Cathode:** platinum-based catalyst, nickel molybdenum (Ni-Mo) **Anode:** iridium-based | **Cathode:** cobalt- and nickel-based materials **Anode:** nickel-based materials and iridium oxide | **Cathode:** lanthanum strontium manganite **Anode:** nickel-YSZ (yttria-stabilised zirconia) |
| Electrolyte type/membrane type | KOH solution (20%–30%) | Perfluorosulfonic acid polymers such as Nafion | AEMion, Evoqua AMI-SPEEK | ScSZ, GDC, LSGM |
| Seal | PTFE, EPDM | PSU, PTFE, ETFE | Silicon | Ceramic glass |
| Electrode price (USD/kW) | <50 | >500 | >500 | >500 |
| Electrolyser weight (kg) | >100 | <50 | <50 | >100 |
| Lifetime of stack (hours) | <90,000 | <20,000 | <20,000 | <5,000 |

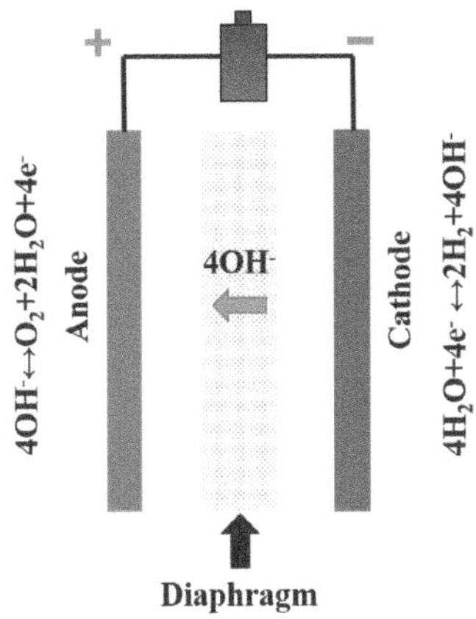


*Figure 14.5* Schematic diagram of alkaline water electrolysis.

or sodium hydroxide (20%–30%) solution with water. The typical reactions taking place in an AWE electrolyser are shown in Eqs. (14.21) and (14.22). Figure 14.5 shows a schematic diagram of AWE. This method is advantageous owing to its ability to remove the necessity of noble metals such as platinum catalysts for making electrodes. The operating temperature for AWE lies in the range of 30°C–80°C. The disadvantages of the AWE method include lower efficiency and flexibility compared to other water electrolysis technologies. Recently, non-noble metals such as nickel, nickel-molybdenum, and oxides of nickel-iron have received considerable attention for electrocatalyst synthesis. Higher stability, higher activity, and durability in hydrogen production even at lower cell voltages causing energy savings were achieved due to these synthesised novel electrocatalysts (Emam et al., 2024).

$$\text{Cathode side (reduction): } 4H_2O(l) + 4e^- \leftrightarrow 2H_2(g) + 4OH^-(aq) \quad (14.21)$$

$$\text{Anode side (oxidation): } 4OH^-(aq) \leftrightarrow O_2(g) + 2H_2O(l) + 4e^- \quad (14.22)$$

#### 14.2.1.2.1.2 PEM ELECTROLYSIS

The PEM electrolysis method uses a solid polymer electrolyte medium, which governs the various processes such as protons conduction, gas separation, and electrical insulation of electrode materials. PEM technology was developed to address the shortcomings of the conventional alkaline electrolysis method such as low current density, low pressure operation, and partial loads. Advantages of the PEM method of electrolysis includes better efficiency, higher production rate per unit area of electrodes, swift reaction

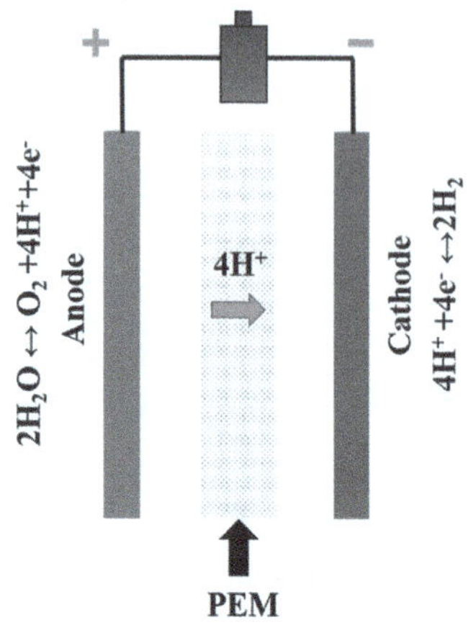


*Figure 14.6* Schematic diagram of proton exchange membrane electrolysis.

kinetics, compact design, and wider temperature range (20°C–80°C) (Goñi-Urtiaga et al., 2012). In a PEM electrolyser unit, the splitting of water takes place as shown in Eqs. (14.23) and (14.24):

$$\text{Cathode side(reduction)}:\quad 4H^+ + 4e^- \leftrightarrow 2H_2(g) \tag{14.23}$$

$$\text{Anode side(oxidation)}:\quad 2H_2O\ (l) \leftrightarrow O_2(g) + 4H^+ + 4e^- \tag{14.24}$$

PEM electrolysis utilises a polysulphonated membrane such as Nafion, which acts as a barrier between the anode and cathode to prevent the flow of electrons from the anode to the cathode. Figure 14.6 shows a schematic diagram of polymer electrolyte membrane water electrolysis. Recent innovations in PEM technology includes the development of two cell stack exhibiting superior performance. A novel technique known as reactive spray deposition technology has been developed, which incorporated platinum recombination layers compared to conventional techniques which used platinum group metal loaded membrane electrode assemblies (MEAs). This transition in design of MEA's has resulted in the improvement of the activity and durability of MEAs up to 3,000 hours in an industrial environment (Arsad et al., 2024).

#### 14.2.1.2.1.3 AEM ELECTROLYSIS

The AEM method of electrolysis works in a similar manner to PEM electrolysis for water splitting; however, the difference lies in the manner in which splitting takes place. AEM electrolysis uses a polymer-based AEM, which is responsible for the supply of $OH^-$ ions to the stack. An AEM electrolyser constitutes a polymer-based AEM, catalyst layers (CLs), gas diffusion layers (GDLs), and bipolar plates. Membrane electrode assembly (MEA) is composed of GDLs, CLs, and membranes. The complete cell of the AEM contains MEAs sandwiched between bipolar plates. When current passes through the electrodes, water flows towards the cathode side from the anode side and $OH^-$ ions flows in the opposite direction. At the cathode side, water loses electrons producing hydrogen and $OH^-$ ions, which reaches the anode side through the AEM to undergo oxidation (Liu et al., 2024). The reactions for AEM electrolysis are as shown in Eqs. (14.25) and (14.26):

$$\text{Cathode side}(\text{reduction}):\ 2H_2O\ (l) + 2e^- \leftrightarrow H_2(g) + 2OH^-,\ E_0 = -0.83\,V \tag{14.25}$$

$$\text{Anode side}(\text{oxidation}):\ 4OH^- \leftrightarrow O_2(g) + 2H_2O\ (l) + 4e^-,\ E_0 = 0.4\,V \tag{14.26}$$

Figure 14.7 shows a schematic diagram of AEM water electrolysis. Several factors such as electrolyte, pressure, and temperature conditions governs the performance of AEM electrolysis. The role of an electrolyte is to improve the electrochemically active surface area of the electrode, decrease the membrane resistance, and activate the membrane for the reaction to take place. Usually, hydroxide solutions of sodium/potassium, carbonates/bicarbonates, and demineralised water are potentially exploited as electrolytes in AEM electrolysis. Demineralised water is the most preferred electrolyte as the reaction kinetics of hydroxides and carbonates are lower. Higher pressure should be maintained to ensure optimum performance of

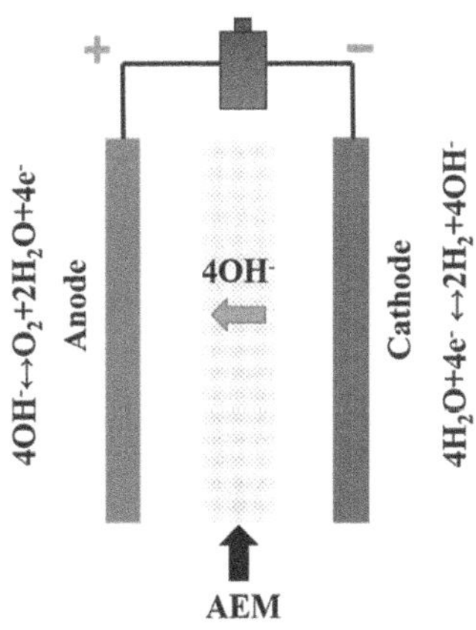


*Figure 14.7* Schematic diagram of anion exchange membrane electrolysis.

the electrolysis. Temperature improves the reaction kinetics, resulting in lower voltage, thus enhancing the performance of an electrolyser (Arsad et al., 2024).

#### 14.2.1.2.1.4 SOLID OXIDE ELECTROLYSIS

The SOE method utilises a solid oxide electrolyte medium mostly yttria-stabilised zirconia to split the water molecules into hydrogen and oxygen. Hydrogen production takes place at high pressure (up to 35 bar) and high temperature (500°C–850°C). The process is also termed high-temperature electrolysis owing to the higher temperature limits at which hydrogen production takes place. The water used in the electrolysis is in the form of steam (Horri & Ozcan, 2024). The typical SOE reaction for dissociation of water molecules is as shown in Eqs. (14.27) and (14.28):

$$\text{Cathode side (reduction)}: \ 2H_2O\ (l) + 4e^- \leftrightarrow 2H_2(g) + 2O^{2-} \tag{14.27}$$

$$\text{Anode side (oxidation)}: \ 2O^{2-} \leftrightarrow O_2(g) + 4e^- \tag{14.28}$$

Figure 14.8 shows a schematic diagram of solid oxide water electrolysis. The SOE method offers several advantages such as higher process efficiency, tolerance to feedstock (water) impurities, and versatility in hydrogen production along with the production of synthetic fuel using carbon dioxide. The limitations of using higher temperature for electrolysis includes material compatibility issues, thermal management, and long startup times as compared to other methods of electrolysis. Recent studies on SOE includes

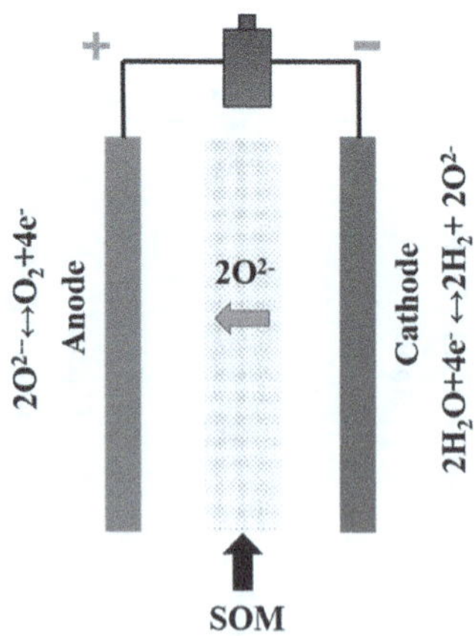


*Figure 14.8* Schematic diagram of solid oxide electrolysis.

development of electrolytes, membrane technology, and electrode materials to enhance the catalytic activity and mass transfer between electrode materials. Despite providing avenues for curbing $CO_2$ emissions by employing co-electrolysis approach and integration of renewable technology for power generation, SOE method is deprived of certain maturity to be implemented for practical applications. This is due to low conversion rates of $CO_2$ and deterioration of SOEC stacks (Arsad et al., 2024).

#### *14.2.1.3 Biochemical methods*

These methods employ biological means to produce hydrogen at ambient pressure and temperature conditions with minimal energy consumption. Hydrogen produced by using biochemical methods is termed as biohydrogen, which utilises biomass to produce hydrogen from microorganisms. The microorganisms that produce biohydrogen either in the presence or absence of sunlight. A wide variety of biomass resources are available for biohydrogen production such as organic-based starch waste, industrial waste, lignocellulosic materials such as wood, and household waste. Under anaerobic conditions, carbohydrate-rich biomass yields hydrogen in the absence (dark fermentation method) or presence (photofermentation method) of sunlight. The production of hydrogen through the dark fermentation method using biomass occurs in two steps: firstly, the high-molecular weight organic substances are converted into water soluble substances using enzymatic hydrolysis, secondly the conversion of hydrolysed substances into volatile fatty acids, hydrogen and carbon dioxide. Photofermentation method of hydrogen production has the same operating principle as dark fermentation with an exception of sunlight availability for photosynthetic bacteria to synthesise hydrogen. From an economic viewpoint, the dark fermentation method is advantageous over photofermentation as it is independent of sunlight for hydrogen production (Azwar et al., 2014). The reaction governing hydrogen production from organic substances through the dark fermentation method is shown in Eq. (14.29):

$$C_{12}H_{22}O_{11} + 9H_2O \rightarrow 4CH_3COO^- + 8H^+ + 4HCO_3^- + 8H_2 \qquad (14.29)$$

Several factors play a vital role in improving the yield of hydrogen from the dark fermentation method such as pH, organic substances, nutrients feed rate, temperature, solid retention time, and partial pressure of hydrogen. Out of all the factors mentioned, pH has a significant impact on hydrogen yield as it governs the activity of hydrogenase enzyme responsible for dark fermentation method. The hydrogen conversion efficiency of 60%–70% could be achieved by maintaining the pH range between 5.5 and 8.0. A variety of reactors are available based on the type of raw materials such as batch reactor (for glycerol, food waste, sucrose, etc.), fluidised-bed bioreactor (for sucrose substrate), continuously stirred tank reactor (for starch, glucose, sweet sorghum extract etc. substrates), anaerobic sequencing batch reactor (for food waste, liquid swine manure, cassava starch, etc. substrates), and fed batch reactors (for starch, glucose, etc., substrates) (Zhao et al., 2024).

#### *14.2.1.4 Photonic methods*

Solar energy-based hydrogen production methods are termed as photonic methods. These methods have promising prospective owing to the high theoretical efficiency and lower cost of production than other methods. A variety of hydrogen production methods are available using solar energy such as photochemical water splitting, biophotolysis, and PV electrolysis (Gopinath & Marimuthu, 2022).

##### *14.2.1.4.1 Photochemical water splitting*

The photochemical water splitting method involves the dissociation of water molecules into hydrogen and oxygen by the action of photocatalysts. This process occurs on the surface of semiconductor materials in three steps: Firstly, photons with an energy level higher than the bandgap of semiconductor material gets absorbed, leading to the formation of electrons ($e^-$) and hole pairs ($h^+$) in the semiconductor material. Secondly, the hole and electron entities generated in the semiconductor material become separated and start migration. Lastly, these charge carriers combine with various compounds such as water. The photocatalyst being used for photochemical splitting of water should occupy a highly negative position in the conduction band than the reduction potential of water to produce hydrogen. Similarly, it should occupy the higher positive position in the valence band compared to the oxidation potential of water for oxygen production (Frowijn & van Sark, 2021). The typical reaction of photochemical water splitting is shown in Eqs. (14.30) and (14.31):

$$\text{Cathodeside (reduction)}: \ 2H^+ + 2e^- \rightarrow H_2(g) \tag{14.30}$$

$$\text{Anodeside(oxidation)}: \ 2p^+ + H_2O \rightarrow 0.5O_2(g) + 2H^+ \tag{14.31}$$

The photochemical process of hydrogen offers advantages such as minimum emissions, availability of a wide variety of raw materials, easier storage, and transportation. The limitations include lower efficiency and higher solar energy requirements (Ismail & Bahnemann, 2014).

#### *14.2.1.4.2 Biophotolysis*

The biophotolysis process involves the splitting of water molecules into constituent molecules, that is, hydrogen and oxygen, in the presence of sunlight and specific microorganisms. Biophotolysis can be either direct type or indirect type. When water is dissociated into hydrogen and oxygen molecules in the presence of direct sunlight, it is termed direct biophotolysis, while if splitting takes place through organisms such as microalgae and cyanobacteria, it is termed indirect biophotolysis (Frowijn & van Sark, 2021).

##### 14.2.1.4.2.1 DIRECT BIOPHOTOLYSIS

Direct biophotolysis methods involves microalgal species such as cyanobacteria and green algae, which convert water into hydrogen and oxygen in the presence of carbon dioxide and sunlight. In this method, there are no intermediate stages of operation. Photosystems (PSI and PSII) used for this method absorb a wide range of wavelengths from solar radiations to split water molecules. Hydrogen production from direct biophotolysis occurs using hydrogenase and nitrogenase enzymes (Gopinath & Marimuthu, 2022).

##### 14.2.1.4.2.2 INDIRECT BIOPHOTOLYSIS

Indirect biophotolysis involves species such as cyanobacteria, which utilises water and carbon dioxide from soil and atmosphere. The succeeding step involves the release of oxygen to form carbohydrate molecules. In the subsequent stages, carbohydrate is broken into carbon dioxide, hydrogen, and other constituents (Gopinath & Marimuthu, 2022). The reaction takes place as shown in Eqs. (14.32), (14.33) and (14.34):

$$6H_2O + 6CO_2(g) + \text{light} \rightarrow C_6H_{12}O_6 + 6O_2 \tag{14.32}$$

$$C_6H_{12}O_6 + 2H_2O \rightarrow 4H_2 + 2CH_3COOH + 2CO_2 \tag{14.33}$$

$$2CH_3COOH + 4H_2O + \text{light} \rightarrow 8H_2 + 4CO_2 \tag{14.34}$$

#### *14.2.1.4.3 PV electrolysis*

In this method, solar energy received by photovoltaic cells gets converted into electrical energy. This electrical energy is used as a source of energy for the dissociation of water molecules to produce hydrogen and oxygen. The produced hydrogen can be used as a fuel source for numerous applications such as a power source for homes, businesses, and a transportation fuel

(Gopinath & Marimuthu, 2022). PV electrolysis employs two types of electrodes for producing hydrogen: organic and inorganic electrodes. Hydrogen production can take place at several cell voltages (1.23–12 V). The highest production of 16.4 mL/hour hydrogen occurs at 12 V for an inorganic steel electrode. PV cells have an efficiency of 20% while the electrolyser has 80% efficiency. The overall efficiency of PV electrolysis method stands at about 16%. PV electrolysis is an energy-intensive and expensive process for hydrogen production owing to high costs of PV solar panels. However, this approach of hydrogen production could be used in the textile dyeing industry, where hydrogen can be produced from wastewater, leading to efficient treatment of wastewater to generate hydrogen. The purity of hydrogen produced by PV electrolysis is very high (up to 99.99 volume percent) (Awad et al., 2024).

## 14.3 COMPARATIVE ASSESSMENT OF HYDROGEN PRODUCTION METHODS

From the available literature, it is concluded that hydrogen energy can be derived from a wide variety of feedstocks and multiple technological pathways. To implement hydrogen energy from a commercial viewpoint, it is pertinent to understand the feasibility in terms of cost, efficiency, environmental impact, and maturity level of any given technology. Technologies such as coal gasification and natural gas reforming are commercially available technologies owing to low cost of production, ease of operation, and availability of raw materials. Although these technologies (coal gasification and natural gas reforming) are widely used, they face the limitations of extreme emissions and postprocessing of output products. The water electrolysis method, on the other hand, offers clean operation with minimal emissions and high purity of hydrogen; however, it encounters the challenges of high operation costs. The optimisation of production costs, process efficiency, emissions, and technology maturity is therefore important to understand which technology can be selected to meet the requirement of hydrogen energy (Acar & Dincer, 2014, 2022).

### 14.3.1 Environmental impact comparison

The environmental impact encompasses the emissions produced as a result of a selected method of production. It mainly takes into account the effect on natural resources such as soil, water, and air. Several parameters such as global warming potential (GWP) and acidification potential (AP) and have been developed to assess the sustainability of a method based on life cycle analysis procedures following the ISO standards published by the Centre of Environmental Science of Leiden University (Acar & Dincer, 2014, 2022).

#### 14.3.1.1 Global warming potential

Greenhouse gases act as a blanket over the earth's atmosphere by absorbing the energy and preventing it from escaping into space. The term GWP was developed to compare the impact of global warming caused by different gases. GWP gives a measure of the amount of energy absorbed by 1 tonne of a gas over a period of time with respect to 1 tonne of carbon dioxide (Acar & Dincer, 2014, 2022). The expression for GWP used for the comparison of hydrogen production methods is described in Eq. (14.35).

$$\text{Global warming potential (GWP)} = \frac{\text{kg} - \text{CO}_2 \text{ equivalent}}{kg - \text{H}_2 \text{ produced}} \tag{14.35}$$

From Table 14.6, it can be understood that GWP values are the highest for coal gasification and fossil fuel reforming techniques of hydrogen production. The integration of carbon capture technologies has resulted in 86.7% and 88.8% reduction in GWP. Other technologies with the least GWP value were photonic-based and biochemical technologies.

*Table 14.6* Comparative assessment of selected hydrogen production methods (Acar & Dincer, 2014, 2022)

| | *Environmental impact* | | *Performance criteria* | | |
|---|---|---|---|---|---|
| *Method/technology* | *GWP(kg-$CO_2$ eq/kg $H_2$)* | *AP (kg-$SO_x$ eq/kg $H_2$)* | *Energy efficiency* | *Exergy efficiency* | *Production cost (USD/kg $H_2$)* |
| Coal gasification | 15.8 | 30.4 | 63 | 46 | 0.92 |
| Coal gasification (with carbon capture) | 2.1 | 30.4 | 61 | 44 | 0.96 |
| Fossil fuel reforming | 9 | 15 | 72.5 | 46 | 0.75 |
| Fossil fuel reforming (with carbon capture) | 1 | 15 | 65.8 | 42 | 2.66 |
| Biomass gasification | 5 | 35 | 42.5 | 39.2 | 1.81 |
| Thermochemical water splitting | 1 | 2 | 42 | 30 | 2.02 |
| Hybrid thermochemical cycle | 0.68 | 3.44 | 53 | 48 | 2.68 |
| Water electrolysis | 0.5 | 4 | 76 | 36 | 2.76 |
| Dark fermentation | 0.5 | 1 | 0.94 | 0.87 | 2.45 |
| Photochemical water splitting | 0.5 | 1 | 6.25 | 1.33 | 6.36 |
| Photolysis | 0.5 | 1 | 3.3 | 2.93 | 2.55 |
| PV electrolysis | 3 | 8 | 12.4 | 7 | 5.7 |

#### 14.3.1.2 Acidification potential

Acidification potential takes into account the various harmful compounds that contribute towards acid rain including gases such as sulphur dioxide ($SO_2$), nitrogen oxides ($NO_x$), nitrogen monoxide (NO), nitrogen dioxide ($N_2O$), and other harmful substances. It is usually measured in terms of $SO_2$- equivalent (Acar & Dincer, 2014, 2022). The expression for GWP used for the comparison of hydrogen production methods is described in Eq. (14.36).

$$\text{Acidification potential}\left(\text{AP}\right)=\frac{\text{kg}-\text{SO}_\text{x}\ \text{equivalent}}{\text{kg}-\text{H}_2\ \text{produced}} \tag{14.36}$$

Table 14.6 illustrates the AP values for selected hydrogen production methods. The highest AP was observed for biomass gasification, followed by coal gasification and fossil fuel reforming techniques of hydrogen production. Photonic methods recorded the least AP values.

### 14.3.2 Energy and exergy efficiency comparison

Production systems are primarily dependent on energy sources for their working. Sustainability of production systems however is decided by their efficiency to utilise the energy in the most optimised manner, that is, they should produce better output with lesser energy consumption (Acar & Dincer, 2014, 2022). The expression for energy efficiency is described as in Eq. (14.37):

$$\eta_{\text{energy}}=\frac{\dot{m}LHV_{H_2}}{\dot{E}_{\text{input}}} \tag{14.37}$$

Another term used for assessing the performance of any production method is exergy efficiency. It measures the effectiveness of the system with respect to its performance under reversible conditions. Exergy efficiency of a system is defined as the ratio between output exergy of produced hydrogen and energy supplied for production (Acar & Dincer, 2014, 2022). The expression for energy efficiency is described as in Eq. (14.38).

$$\eta_{\text{exergy}}=\frac{\dot{m}ex_{H_2}^{\text{ch}}}{\dot{E}_{\text{input}}} \tag{14.38}$$

Energy and exergy values represent the usefulness of a given hydrogen production technology. Data from Table 14.6 show that the water electrolysis method has superior performance under efficiency criteria, while the dark fermentation method has the most undesirable energy efficiency. The hybrid thermochemical cycle is the most effective method for producing hydrogen based on exergy efficiency criteria, followed by coal gasification and fossil fuel reforming method. The most undesirable method under exergy criteria is dark fermentation.

### 14.3.3 Production cost comparison

Financial consideration serves as one of the most important criteria for assessment of feasibility of any given technology. The cost of any production method is dependent on several factors including advancement in technology, availability of raw materials, and existing infrastructure. The cost of hydrogen production methods is calculated based on the cost incurred in the production of 1 kg of hydrogen from a given technology (unit: USD/kg-$H_2$) (Acar & Dincer, 2014, 2022). From Table 14.6, it can be inferred that photochemical water splitting has the highest production cost due to the semiconductor material responsible for energy supply to dissociate water into individual molecules of $O_2$ and $H_2$. The least cost was recorded for coal gasification and fossil fuel reforming methods.

## 14.4 CURRENT SCENARIO AND FUTURE EVOLUTION OF HYDROGEN PRODUCTION

Based on the "Global Hydrogen Review 2023" report published in 2023 by the International Energy Agency, the global hydrogen utilisation has reached about 95 Mt by 2022.

The majority of the produced hydrogen is used in the production of ammonia, fertilisers, and petrochemical refineries. Currently, the majority of hydrogen production worldwide is taking place through the water electrolysis method. The electrolysis of water primarily derives energy from three categories: grid connected electricity, grid along with renewable electricity, and dedicated renewable electricity. A total of 274 hydrogen production facilities are reported based on the IEA, which are subdivided into operational (141) and under construction category (133). These production facilities producing hydrogen are using different electrolysis technologies: PEM, alkaline electrolysis, SOE, and other electrolysis methods. Renewable energy-based electrolysis has a massive share in hydrogen production worldwide (about 76.3%). Solar PV, onshore wind, off shore wind, hydropower, and other renewable electricity-based technologies are used for hydrogen production using water electrolysis (International Energy Agency, 2023). Solar energy is the second largest contributor to renewable water electrolysis (31.6%), followed by onshore wind energy (16.3%) and hydropower (6.2%), respectively, as shown in Figure 14.9.

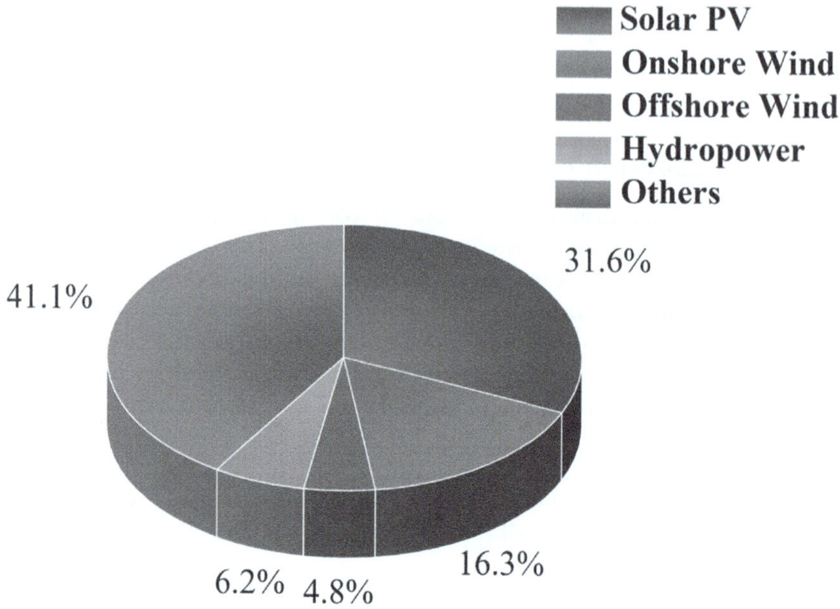


*Figure 14.9* Distribution of various renewable energy sources for water electrolysis hydrogen generation.

## REFERENCES


Acar, C., & Dincer, I. (2014). Comparative assessment of hydrogen production methods from renewable and non-renewable sources. *International Journal of Hydrogen Energy*, 39(1), 1–12. https://doi.org/10.1016/j.ijhydene.2013.10.060

Acar, C., & Dincer, I. (2022). Selection criteria and ranking for sustainable hydrogen production options. *International Journal of Hydrogen Energy*, 47(95), 40118–40137. https://doi.org/10.1016/j.ijhydene.2022.07.137

Afzanizam Samiran, N., Nazri Mohd Jaafar, M., & Tung Chong, C. (2014). Review of biomass gasification technology to produce syngas. *American-Eurasian Journal of Sustainable Agriculture*, 8(7). https://www.researchgate.net/publication/278849675

Anand, A., Kachhap, A., & Gautam, S. (2023). Synergistic effect of coal and biomass gasification and organo-inorganic elemental impact on gasification performance and product gas. *Energy*, 282. https://doi.org/10.1016/j.energy.2023.128662

Arsad, S. R., Arsad, A. Z., Ker, P. J., Hannan, M. A., Tang, S. G. H., Goh, S. M., & Mahlia, T. M. I. (2024). Recent advancement in water electrolysis for hydrogen production: A comprehensive bibliometric analysis and technology updates. *International Journal of Hydrogen Energy*, 60, 780–801. https://doi.org/10.1016/j.ijhydene.2024.02.184

Awad, M., Said, A., Saad, M. H., Farouk, A., Mahmoud, M. M., Alshammari, M. S., Alghaythi, M. L., Abdel Aleem, S. H. E., Abdelaziz, A. Y., & Omar, A. I. (2024). A review of water electrolysis for green hydrogen generation considering PV/wind/hybrid/hydropower/geothermal/tidal and wave/biogas energy systems, economic analysis, and its application. *Alexandria Engineering Journal*, 87, 213–239. https://doi.org/10.1016/j.aej.2023.12.032

Azwar, M. Y., Hussain, M. A., & Abdul-Wahab, A. K. (2014). Development of bio-hydrogen production by photobiological, fermentation and electrochemical processes: A review. *Renewable and Sustainable Energy Reviews*, 31, 158–173. https://doi.org/10.1016/j.rser.2013.11.022

Chai, Y. F., Luo, G. P., An, S. L., Peng, J., & Wang, Y. C. (2019). Influence of unburned pulverized coal on gasification reaction of coke in blast furnace. *High Temperature Materials and Processes*, 38, 733–738. https://doi.org/10.1515/htmp-2019-0016

Coppola, A., Solimene, R., Bareschino, P., & Salatino, P. (2015). Mathematical modeling of a two-stage fuel reactor for chemical looping combustion with oxygen uncoupling of solid fuels. *Applied Energy*, 157, 449–461. https://doi.org/10.1016/j.apenergy.2015.04.052

Cormos, C. C. (2023). Green hydrogen production from decarbonized biomass gasification: An integrated techno-economic and environmental analysis. *Energy*, 270. https://doi.org/10.1016/j.energy.2023.126926

Dincer, I., & Acar, C. (2014). Review and evaluation of hydrogen production methods for better sustainability. *International Journal of Hydrogen Energy*, 40(34), 11094–11111. https://doi.org/10.1016/j.ijhydene.2014.12.035

Emam, A. S., Hamdan, M. O., Abu-Nabah, B. A., & Elnajjar, E. (2024). A review on recent trends, challenges, and innovations in alkaline water electrolysis. *International Journal of Hydrogen Energy*, 64, 599–625. Elsevier Ltd. https://doi.org/10.1016/j.ijhydene.2024.03.238

Frowijn, L. S. F., & van Sark, W. G. J. H. M. (2021). Analysis of photon-driven solar-to-hydrogen production methods in the Netherlands. *Sustainable Energy Technologies and Assessments*, 48. https://doi.org/10.1016/j.seta.2021.101631

Goñi-Urtiaga, A., Presvytes, D., & Scott, K. (2012). Solid acids as electrolyte materials for proton exchange membrane (PEM) electrolysis: Review. *International Journal of Hydrogen Energy*, 37(4), 3358–3372. https://doi.org/10.1016/j.ijhydene.2011.09.152

Gopinath, M., & Marimuthu, R. (2022). A review on solar energy-based indirect water-splitting methods for hydrogen generation. *International Journal of Hydrogen Energy*, 47(89), 37742–37759. https://doi.org/10.1016/j.ijhydene.2022.08.297

Habibi, R., & Mehrpooya, M. (2021). A novel integrated Ca-Cu cycle with coal/biomass gasification unit for clean hydrogen production. *Energy Conversion and Management*, 228. https://doi.org/10.1016/j.enconman.2020.113682

Horri, B. A., & Ozcan, H. (2024). Green hydrogen production by water electrolysis: Current status and challenges. *Current Opinion in Green and Sustainable Chemistry*, 100932. https://doi.org/10.1016/j.cogsc.2024.100932

International Energy Agency (2023). Global hydrogen review 2023. www.iea.org

Ismail, A. A., & Bahnemann, D. W. (2014). Photochemical splitting of water for hydrogen production by photocatalysis: A review. *Solar Energy Materials and Solar Cells*, 128, 85–101. https://doi.org/10.1016/j.solmat.2014.04.037

Liu, C., Geng, Z., Wang, X., Liu, W., Wang, Y., Xia, Q., Li, W., Jin, L., & Zhang, C. (2024). Development of advanced anion exchange membrane from the view of the performance of water electrolysis cell. *Journal of Energy Chemistry*, 90, 348–369. https://doi.org/10.1016/j.jechem.2023.11.026

Megia, P. J., Vizcaino, A. J., Calles, J. A., & Carrero, A. (2021). Hydrogen production technologies: From fossil fuels toward renewable sources: A mini review. *Energy and Fuels*, 35(20), 16403–16415. https://doi.org/10.1021/acs.energyfuels.1c02501

Midilli, A., Kucuk, H., Topal, M. E., Akbulut, U., & Dincer, I. (2021). A comprehensive review on hydrogen production from coal gasification: Challenges and opportunities. *International Journal of Hydrogen Energy*, 46(50), 25385–25412. https://doi.org/10.1016/j.ijhydene.2021.05.088

Mishra, S., & Upadhyay, R. K. (2021). Review on biomass gasification: Gasifiers, gasifying mediums, and operational parameters. *Materials Science for Energy Technologies*, 4, 329–340. https://doi.org/10.1016/j.mset.2021.08.009

Mission Innovation (n.d.). *Guidelines for Hydrogen Valley Innovation Cluster*. Revised document for "Call for proposals on Hydrogen Valley Platform in India."

Safari, F., & Dincer, I. (2020). A review and comparative evaluation of thermochemical water splitting cycles for hydrogen production. *Energy Conversion and Management*, 205. https://doi.org/10.1016/j.enconman.2019.112182

Teoh, Y. H., How, H. G., Le, T. D., Nguyen, H. T., Loo, D. L., Rashid, T., & Sher, F. (2023). A review on production and implementation of hydrogen as a green fuel in internal combustion engines. *Fuel*, 333. https://doi.org/10.1016/j.fuel.2022.126525

Yoo, J., Park, S., Song, J. H., Yoo, S., & Song, I. K. (2017). Hydrogen production by steam reforming of natural gas over butyric acid-assisted nickel/alumina catalyst. *International Journal of Hydrogen Energy*, 42(47), 28377–28385. https://doi.org/10.1016/j.ijhydene.2017.09.148

Younas, M., Shafique, S., Hafeez, A., Javed, F., & Rehman, F. (2022). An overview of hydrogen production: Current status, potential, and challenges. *Fuel*, 316. https://doi.org/10.1016/j.fuel.2022.123317

Zhao, Z. T., Ding, J., Wang, B. Y., Bao, M. Y., Liu, B. F., Pang, J. W., Ren, N. Q., & Yang, S. S. (2024). Advances in the biomass valorization in dark fermentation systems: A sustainable approach for biohydrogen production. *Chemical Engineering Journal*, 481. https://doi.org/10.1016/j.cej.2023.148444

Chapter 15

# Solar-integrated hydrogen production and utilization in India

*Ashwani Kumar*

## 15.1 INTRODUCTION

The concept of clean energy refers to energy sources that have minimal environmental impact during their production, use, and disposal. Unlike traditional fossil fuels (such as coal, oil, and natural gas), which release greenhouse gases and pollutants, clean energy sources are either renewable or have a minimal carbon footprint. Clean energy sources like solar, wind, geothermal, and hydropower do not emit greenhouse gases during electricity generation, mitigating climate change. By avoiding the burning of fossil fuels, clean energy sources reduce air pollution, leading to improved public health and reduced respiratory illnesses [1]. Unlike finite fossil fuels, clean energy sources are either constantly replenished (renewable) or have a significantly lower environmental impact (nuclear). Increased reliance on domestic clean energy sources reduces dependence on foreign oil supplies and price fluctuations [2].

Sustainable development is defined as meeting the needs of the present without compromising the ability of future generations to meet their own needs [3]. Clean energy plays a pivotal role in achieving sustainable development. Transitioning to clean energy sources is crucial for mitigating climate change and its associated environmental and economic consequences. Clean energy sources offer a long-term, reliable energy supply that is not subject to geopolitical instability or resource depletion. Investments in clean energy technologies create new jobs and stimulate economic development in related sectors [4]. Additionally, reduced air pollution from clean energy sources leads to improved public health and reduced healthcare costs.

Within clean energies, hydrogen ($H_2$) is a versatile energy carrier with a high energy density. However, current hydrogen production methods often rely on fossil fuels, which release greenhouse gases. Solar-integrated hydrogen production offers a clean alternative by utilizing solar energy to split water ($H_2O$) into hydrogen and oxygen through a process called electrolysis [5].

DOI: 10.1201/9781003521341-15

## 15.2 SOLAR-INTEGRATED HYDROGEN PRODUCTION METHODS

India has a tremendous potential for solar-integrated hydrogen production due to its abundant solar energy. India boasts a high solar insolation rate, meaning it receives a significant amount of solar radiation throughout the year. Estimates suggest a potential of 1800 TW (terawatt) for solar power generation in the country [6]. Recognizing this potential, the Indian government launched the National Green Hydrogen Mission in early 2023.

This mission aims to make India a global leader in green hydrogen production using renewable energy. The production target of 5 million metric tons per annum of green hydrogen by 2030 [7]. Additionally, the installation of 125 GW of renewable energy capacity, primarily solar and wind, is planned to support green hydrogen production by 2030 [7]. Figure 15.1 shows the different methods of solar-integrated hydrogen production, with comparative analysis in their advantages, disadvantages, challenges, and applications (Table 15.1). While the potential is significant, there are challenges that

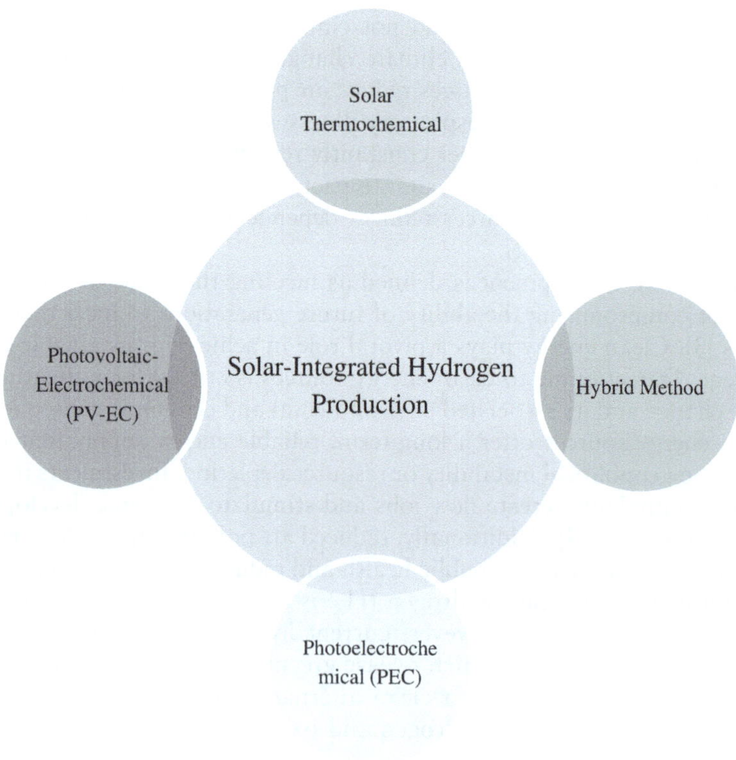


*Figure 15.1* Different solar-integrated hydrogen production methods.

*Table 15.1* Comparison of different solar-integrated hydrogen production methods [5,8–13]

| *Method* | *Working principle* | *Advantage* | *Disadvantage* | *Efficiency (solar-to-hydrogen)* | *Challenges* | *Applications (potential)* | *References* |
|---|---|---|---|---|---|---|---|
| PV-EC | Uses solar panels (PV) to generate electricity that powers an electrolyzer to split water into $H_2$ and $O_2$. | Modular and scalable. Established technology for PV and electrolysis. | Requires two separate components (PV and electrolyzer). Efficiency limited by both PV and electrolysis. | 10%–20% (current research) | Optimizing system integration. Cost reduction of electrolyzers. | Large-scale hydrogen production facilities. Grid-connected hydrogen generation. | [5,8] |
| PEC | Photoelectrodes directly convert solar energy into electricity for water splitting in a single unit. | Simpler system compared to PV-EC. Potential for high theoretical efficiency. | Achieving high STH conversion efficiency. Limited material stability and durability of photoelectrodes. | 1%–10% (current research) | Material development for efficient and stable photoelectrodes. Scalability and cost reduction. | Distributed hydrogen production for buildings. Portable hydrogen generation for remote applications. | [9,10] |
| Solar thermochemical | Concentrated solar thermal energy drives thermochemical cycles to split water into $H_2$. | High potential for theoretical efficiency. Can utilize intermittent solar energy with heat storage. | Complex system with multiple reaction steps. Requires high temperatures for efficient operation. | 10%–30% (estimated) | Developing efficient and cost-effective thermochemical cycles. High-temperature material challenges. | Large-scale industrial hydrogen production. Co-generation of hydrogen and electricity. | [11,12] |
| Hybrid methods | Combines elements of PV, PEC, or thermochemical with other technologies. | Can leverage advantages of different methods. Potential for improved efficiency and functionality. | Increased system complexity compared to single methods. Requires optimization of integration between technologies. | Varies depending on the specific hybrid approach | Overcoming limitations of individual methods. Tailoring hydrogen production for specific applications. | | [13] |

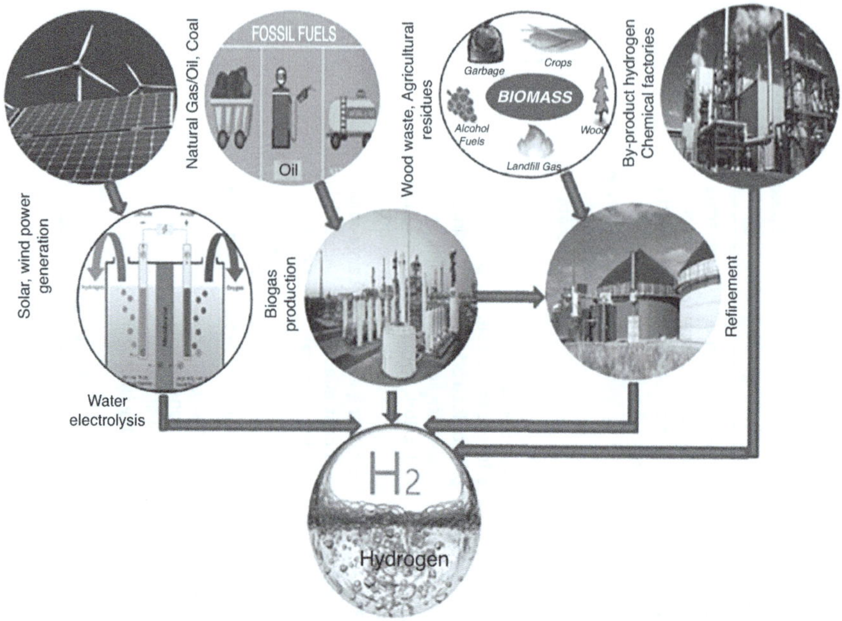


*Figure 15.2* Renewable energy-integrated hydrogen production [14].

need to be addressed. The efficiency of electrolyzers (the devices that produce hydrogen from water using electricity) needs improvement to reduce production costs. Moreover, developing efficient and cost-effective methods for storing and transporting hydrogen is crucial. India's solar-integrated hydrogen production potential is promising (Figure 15.2). Government initiatives and ongoing research and development efforts are expected to propel India toward a significant role in the global green hydrogen market.

### 15.2.1 Photovoltaic–electrochemical methods

Hydrogen ($H_2$) is a clean and versatile energy carrier with the potential to revolutionize various sectors. It can be used in transportation as a fuel for clean-running vehicles, power fuel cells for efficient electricity generation, and find applications in diverse industrial processes. However, traditional methods of hydrogen production often rely on fossil fuels like natural gas, leading to the emission of greenhouse gases and contributing to climate change. Solar-integrated hydrogen production offers a promising solution to this challenge. By harnessing the power of the sun, a clean and abundant renewable resource, we can produce hydrogen in a sustainable manner. The photovoltaic–electrochemical (PV-EC) method is one such approach that leverages this potential. It combines the established technology of PV cells, which convert sunlight directly into electricity, with electrocatalysis, a process that uses electrical energy to drive chemical reactions. In the context of PV-EC systems, this chemical reaction involves splitting water molecules ($H_2O$) into hydrogen gas ($H_2$) and oxygen gas ($O_2$). The PV-EC system itself comprises of three key components [15–17]:

**PV Cell:** This component acts as the heart of the system, capturing sunlight and converting it into an electrical current through the PV effect. This electricity provides the driving force for hydrogen production.

**Electrode Assembly:** This assembly consists of two electrodes, an anode and a cathode, immersed in an electrolyte solution. The specific materials chosen for the anode and cathode are crucial, as they need to efficiently catalyze the water-splitting reaction at their respective surfaces.

**Electrolyte:** This solution facilitates the flow of ions between the anode and cathode, completing the electrical circuit within the system. The chosen electrolyte needs to be compatible with both electrodes and possess the necessary conductivity properties for efficient operation.

During operation, sunlight strikes the PV cell, generating an electric current. This current flows through the electrical circuit and reaches the electrodes immersed in the electrolyte. At the cathode, water reduction occurs, producing hydrogen gas ($H_2$), while simultaneously, oxygen gas ($O_2$) evolves at the anode. PV-EC systems directly convert solar energy into hydrogen, eliminating the need for intermediate energy conversion steps. By utilizing sunlight, a clean and abundant renewable resource, PV-EC systems can be designed in various sizes, from small-scale units for individual applications to large-scale facilities for industrial hydrogen production. The modular design allows for flexibility in system configuration and facilitates easier maintenance.

Despite its advantages, the overall efficiency of the PV-EC system is crucial for cost-effectiveness. Improving the efficiency of both PV cells and electrocatalysts remains a critical area of research. Developing cost-effective, durable, and efficient electrode materials with high solar-to-hydrogen (STH) conversion efficiency is necessary. The chosen electrolyte needs to be stable under operating conditions and compatible with both electrodes. Optimizing the integration of PV cells, electrocatalysis, and system components for efficient operation is essential.

Research efforts are ongoing to address the challenges and improve PV-EC technology. Materials with improved light absorption properties and efficient charge transfer capabilities are being explored. Researchers are developing novel electrocatalysts with high activity and selectivity for the hydrogen evolution reaction. New electrolytes with improved conductivity, stability, and compatibility with electrode materials are under investigation. Additionally, researchers are focusing on optimizing system design and integration to enhance overall efficiency and reduce costs. While challenges remain, ongoing research and development efforts are paving the way for advancements in materials, efficiency, and system integration. As PV-EC technology matures, it has the potential to become a key player in the transition toward a clean hydrogen economy.

### 15.2.2 Photoelectrochemical process

The photoelectrochemical (PEC) process is a promising technology for generating clean hydrogen fuel directly from sunlight. A PEC cell utilizes a light-absorbing semiconductor photoanode, an electrolyte solution, and a counter electrode to achieve this. When sunlight strikes the photoanode, it excites electrons, creating electron–hole pairs [16]. These excited electrons travel through the photoanode to be collected, while the remaining holes at the photoanode surface oxidize water molecules, generating oxygen and protons. The protons migrate through the electrolyte to the counter electrode, where they combine with collected electrons to produce hydrogen gas [17]. This single-step process offers several advantages. It directly converts solar energy to hydrogen, eliminating the need for intermediate energy conversions. It utilizes a clean and abundant renewable resource – sunlight. PEC systems are scalable for both distributed and large-scale hydrogen production and are environmentally friendly, as they produce hydrogen without greenhouse gas emissions. However, challenges remain. Developing efficient and stable photoanode materials is crucial. Improving overall system efficiency is necessary for cost-effectiveness. Finding electrolytes compatible with photoanodes and promoting efficient ion transport is important [18,19]. Additionally, optimizing the design and integration of all PEC cell components is essential for enhanced performance and durability. Fortunately, significant research efforts are underway to address these challenges and improve PEC technology. These effort include developing novel photoanode materials with improved light absorption and stability, incorporating cocatalysts or doping to enhance water oxidation activity, optimizing electrolytes for conductivity and compatibility, and integrating PEC cells with solar concentrators for improved efficiency and cost reduction. As PEC technology matures, it has the potential to become a key player in the transition toward a clean hydrogen economy [20].

### 15.2.3 Solar thermochemical production

Solar thermochemical hydrogen production offers a compelling approach for generating clean hydrogen fuel by harnessing concentrated solar energy. This method diverges from solar PV systems that convert sunlight directly into electricity. Instead, it utilizes the intense heat from focused sunlight to drive a series of chemical reactions that split water molecules ($H_2O$) into hydrogen ($H_2$) and oxygen ($O_2$) [21]. A typical solar thermochemical system comprises several key components: a solar concentrator (mirrors or lenses focusing sunlight), a reactor (housing the water-splitting reactions), a heat exchanger (transferring heat to the reactor), and a product separation unit (isolating the produced hydrogen). Sunlight is concentrated, to generate high temperatures exceeding 700°C. This concentrated heat then drives thermochemical cycles within the reactor, involving specific chemical compounds. These cycles, often comprising multiple steps, facilitate the transfer of oxygen from water to other reactants, ultimately releasing hydrogen gas. The product separation unit isolates the hydrogen from other reaction products [22].

Solar thermochemical hydrogen production boasts several advantages. Compared to PV cells, these systems capture a broader spectrum of sunlight, potentially leading to higher STH conversion efficiencies. Additionally, unlike PV systems that rely on sunlight availability, thermochemical systems can store thermal energy, allowing for hydrogen production even during periods of low sunlight (dispatchable production). Water serves as the primary feedstock, similar to other clean hydrogen production methods like PEC and electrolysis. Lastly, the modular design allows for scalability, enabling deployment for both small- and large-scale facilities. However, challenges remain. Identifying and optimizing efficient thermochemical cycles with high reaction yields while minimizing energy requirements is a critical hurdle [23]. Developing materials that can withstand the high temperatures involved in the process while maintaining stability and promoting efficient chemical reactions is another key challenge. Optimizing the integration of various components like the solar concentrator, reactor, and heat exchanger for efficient heat transfer and overall system performance is essential. Finally, reducing the overall system cost, including capital and operational expenses, is crucial for widespread adoption.

Researchers are actively addressing these challenges and advancing solar thermochemical technology through various efforts. These efforts include exploring novel thermochemical cycles with improved efficiency, lower operating temperatures, and compatibility with various materials. Additionally, researchers are developing advanced materials with high thermal stability, catalytic properties for efficient chemical reactions, and cost-effectiveness. System optimization focuses on improving heat transfer, reaction rates, and overall system efficiency through better integration of different components. Lastly, researchers are exploring the integration of solar thermochemical systems with other renewable energy sources for a more robust and reliable hydrogen production approach [24,25].

Solar thermochemical hydrogen production holds significant promise for generating clean and sustainable hydrogen fuel. While challenges remain in terms of material development, cycle optimization, and cost reduction, ongoing research efforts are paving the way for advancements in this technology. As solar thermochemical systems mature, they have the potential to play a key role in the transition toward a clean hydrogen economy.

### 15.2.4 Hybrid method

Hydrogen, a clean-burning fuel with the potential to decarbonize various sectors, has gained significant attention in recent years. However, developing efficient and sustainable methods for hydrogen production remains crucial. Hybrid methods, which combine elements of PV, PEC, or thermochemical water splitting with other technologies, offer a promising approach to addressing the limitations of individual methods [26].

Existing hydrogen production methods often face limitations. Solar PV systems rely on sunlight and have limited conversion efficiencies. PEC processes struggle with efficiency and material stability. Thermochemical cycles require high temperatures and complex reactor designs. Hybrid methods aim to address these limitations by leveraging the strengths of different approaches. Combining technologies can potentially lead to improved overall efficiency for hydrogen production. For example, integrating solar concentrators with PEC cells can enhance light utilization and hydrogen generation. Hybrid methods can also offer additional functionalities beyond hydrogen production. For instance, integrating a desalination unit with a solar thermochemical system could simultaneously generate hydrogen and freshwater. Furthermore, hybrid systems can be designed with flexibility for different applications, from small-scale distributed hydrogen production or large-scale industrial facilities.

#### *15.2.4.1 Types of hybrid methods*

Several hybrid approaches are being explored, each with its unique advantages and challenges [27,28]:

- **PV-Electrolysis:** This combines solar PV electricity generation with conventional water electrolysis. The PV system converts sunlight into electricity, which powers the electrolyzer to split water into hydrogen and oxygen.
- **PV-PEC:** This integrates PV electricity with a PEC cell. The PV system provides electrical bias to enhance the efficiency of the PEC cell in driving water splitting.
- **Thermochemical–Electrolysis:** This combines thermochemical cycles with electrolysis. The thermochemical cycle generates high-temperature steam, which can be used to drive a high-temperature electrolyzer for efficient hydrogen production.
- **Solar–Biohybrid Systems:** These integrate biological processes with solar energy capture. Sunlight can be used to cultivate hydrogen-producing microorganisms or power processes that convert biomass into hydrogen.

**Challenges and Considerations:**

- **System Complexity:** Designing, integrating, and optimizing hybrid systems require careful consideration of component interactions and overall system performance.
- **Cost Optimization:** Hybrid systems may involve additional components, which can increasing the initial capital cost. Balancing these costs with efficiency and functionality is essential making the systems economically stable.
- **Technology Maturity:** Some hybrid approaches involve emerging technologies like advanced PEC materials or high-temperature electrolyzers. Further research and development are crucial for their cost reduction and performance improvement.

**Recent Research and Development Efforts:**
Researchers are actively exploring advancements in hybrid methods for hydrogen production [14,29,30]:

- **Developing Novel Materials:** Research is focused on materials for high-efficiency PEC cells, cost-effective electrolyzers for solar-powered systems, and materials that can withstand the harsh environments in thermochemical cycles.
- **System Integration Optimization:** Research efforts are aimed at optimizing the integration of different components within hybrid systems for improved heat transfer, mass flow, and overall system efficiency.
- **Hybrid System Modeling and Simulation:** Advanced modeling and simulation tools are used to predict system performance, optimize design parameters, and guide the development of efficient hybrid configurations.

Hybrid methods for hydrogen production offer a promising path toward a clean energy future. By combining the strengths of existing technologies and addressing the limitations of individual methods, hybrid approaches have the potential for high-efficiency and scalable hydrogen generation. Continued research and development in material advancements, system optimization, and cost reduction are crucial for the widespread adoption of hybrid methods in the clean hydrogen economy.

## 15.3 DISCUSSION AND FUTURE DIRECTIONS

Renewable integrated hydrogen production technology faces several challenges that need to be addressed for widespread adoption. The primary hurdle is cost. Currently, green hydrogen (produced from renewables) is more expensive than traditional methods that rely on fossil fuels [31–34]. This is due to factors like the high cost of electrolyzers, the equipment used for splitting water into hydrogen, and the lack of established infrastructure for hydrogen storage and transportation.

Another challenge is the intermittent nature of renewable energy sources. Solar and wind power generation fluctuate depending on weather conditions. Integrating renewable energy with hydrogen production requires efficient energy storage solutions to manage these fluctuations and ensure a steady supply of hydrogen [35–38].

Furthermore, large-scale hydrogen production raises concerns about water usage, particularly in arid regions. Optimizing water utilization through improved electrolysis techniques and exploring alternative water sources like seawater desalination are crucial steps. Despite these challenges, the future of renewable integrated hydrogen production is promising. Advancements in technology are continuously reducing the cost of electrolyzers and other components. Research on more efficient catalysts and materials for electrolysis holds promise for further cost reductions. Governments and private companies are also increasingly investing in hydrogen infrastructure development. Building a robust network for hydrogen storage and transportation will be essential for creating a viable hydrogen economy [39–43]. In conclusion, while challenges remain, renewable integrated hydrogen production has the potential to revolutionize the energy sector. By addressing cost concerns, developing efficient storage solutions, and optimizing water usage, we can pave the way for a clean and sustainable energy future fueled by hydrogen [44–50].

### 15.3.1 Future directions

The future of solar-integrated hydrogen production is promising and filled with potential. Researchers are actively exploring several key areas to advance this technology for a clean energy future. One crucial area of focus is material innovation. Scientists are developing novel photoelectrodes and catalysts that are not only efficient at converting sunlight into electricity but also cost-effective to produce [51–54]. These advanced materials are essential for minimizing energy losses during the hydrogen production process. A key metric for this technology's success is the STH conversion efficiency. By developing better materials, researchers aim to significantly improve this metric, making solar-integrated hydrogen production more viable.

Another important area of research is system integration and optimization. This involves seamlessly integrating the various processes involved in hydrogen production, such as capturing solar energy, splitting water molecules, and purifying the resulting hydrogen gas. Researchers are focusing on optimizing system design, heat management, and reaction conditions to maximize hydrogen production and ensure the overall system efficiency [55–59].

Scaling up current lab-based systems for real-world applications is another critical challenge. Researchers are addressing issues related to material stability, component durability, and ensuring efficient operation under various environmental conditions. Overcoming these scalability hurdles is essential for transitioning solar-integrated hydrogen production from the research stage to large-scale implementation. Hybrid and tandem technologies are showing promise. These approaches combine different methods of solar energy conversion or integrate photocatalysis with solar thermal processes. Exploring these possibilities may lead to significant improvements in both efficiency and cost-effectiveness for solar-integrated hydrogen production [60–63].

Integrating with advanced energy storage solutions is another exciting research direction. This involves combining solar-driven hydrogen production with storage options like compressed hydrogen or chemical carriers. This would help buffer the intermittent nature of solar energy and ensure a reliable supply of hydrogen fuel [64–66].

Finally, researchers are conducting comprehensive life cycle analyses to assess the environmental impact of solar-integrated hydrogen production. This includes minimizing water usage throughout the process, developing sustainable methods for sourcing materials, and ensuring responsible management of the technology at the end of its lifespan. By addressing these diverse research areas, scientists and engineers are working toward creating a cost-effective, scalable, and sustainable solar-integrated hydrogen production technology. This holds immense potential for a clean and renewable energy future [67–72].

## REFERENCES


1. Iberdrola (n.d.). What is clean energy? **Retrieved from** https://www.iberdrola.com/about-us/our-company/renewable-energies.
2. International Renewable Energy Agency (2022). World energy outlook. **Retrieved from** https://www.irena.org/.
3. UN Report (n.d.). Our common future. **Retrieved from** https://sdgs.un.org/future-we-want.
4. IRENA (2019). Renewable energy benefits: Employment and economic development. **Retrieved from** https://www.irena.org/.
5. Turner, J. A., Maness, P.-C., & Reece, J. S. (2008). Efficient solar to hydrogen conversion using an integrated photoelectrochemical and photovoltaic system. *Proceedings of the National Academy of Sciences*, 105(25), 9023–9027.
6. MNRE (Ministry of New and Renewable Energy) (2022, June). India solar capacity. https://mnre.gov.in/solar/installed-capacity (**Accessed** 24 May 2024).
7. Press Information Bureau (2024, May 13). India's green hydrogen revolution..
8. Hwang, S., Lee, J. H., & Jeon, T. H. (2021). Recent progress in photoelectrochemical water splitting technology: Materials, systems, and catalysts. *Chemical Society Reviews*, 50(11), 5832–5873.

9. Hisatomi, T., Kubota, K., & Domen, K. (2014). Recent advances in semiconductors for photocatalytic and photoelectrochemical water splitting. *Chemical Society Reviews*, 43(22), 7520–7538.
10. Ma, Q., Young, K. W., Zachariah, Z., & Sun, Z. (2020). Photoelectrochemical water splitting for hydrogen production: recent progress and future challenges. *Chemical Society Reviews*, 49(5), 1887–1919.
11. Gaya, U. M., Abdelkareem, M. A., & Alshehri, A. A. (2019). Solar thermochemical water splitting for hydrogen production: A review. *Renewable and Sustainable Energy Reviews*, 111, 851–866.
12. Zhang, X., Chen, Y., & Li, Y. (2023). Progress and challenges in solar thermochemical water splitting for hydrogen production. *International Journal of Energy Research*, 47(2), 1189–1215.
13. Luo, J., Ibrahim, S., Zhu, H., & Irvine, J. T. S. (2023). Hybrid solar-driven water splitting: A review. *Chemical Society Reviews*, 52(4), 1676–1723.
14. Benghanem, M., Mellit, A., Almohamadi, H., Haddad, S., Chettibi, N., Alanazi, A. M., Dasalla, D., & Alzahrani, A. (2023). Hydrogen production methods based on solar and wind energy: A review. *Energies*, 16(2), 757. https://doi.org/10.3390/en16020757.
15. Karayel, G. K., Javani, N., & Dincer, I. (2022). Green hydrogen production potential for Turkey with solar energy. *International Journal of Hydrogen Energy*, 47(45), 19354–**19364**. https://doi.org/10.1016/j.ijhydene.2021.10.240.
16. Dias, P., Leite, E., Mendes, A., & Mendes, A. (2020). Recent advances in photovoltaic-electrochemical water splitting: A review. *Renewable and Sustainable Energy Reviews*, 134, 110201.
17. Luo, J., Jiang, T., Zhang, Y., Li, M., & Sun, L. (2023). Perovskite photoelectrodes for solar-driven water splitting. *Advanced Materials*, 35(17), 2209242.
18. Singh, A., Singh, N., & Khan, M. A. (2020). A review on photoelectrochemical water splitting for hydrogen production: Current status and future prospects. *Journal of Materials Science: Materials in Electronics*, 31(12), 11359–11382.
19. McCrory, C. C., Junghans, C., Peters, J. C., & Bard, A. J. (2019). Benchmarking hydrogen evolution reaction and oxygen evolution reaction electrocatalysts for solar water splitting devices. *Journal of the American Chemical Society*, 141(23), 9177–9193.
20. Singh, M., & Khare, A. (2023). Solar-driven water splitting for hydrogen production: Progress and challenges. *International Journal of Hydrogen Energy*, 48(28), 13754–13770.
21. Abdallah, R., Mitsos, A., & Sattler, K. (2020). Solar thermochemical water splitting for hydrogen production: A review. *Renewable and Sustainable Energy Reviews*, 126, 110002.
22. Chen, X., Koukini, A., & Li, K. (2023). Recent progress in high-temperature thermochemical cycles for solar hydrogen production. *Chemical Society Reviews*, 52(1), 148–203.
23. Hayes, D., McDaniel, A. H., & Inman, D. J. (2020). Solar thermochemical water splitting for hydrogen production: A review of materials and reaction cycles. *Energy & Environmental Science*, 13(4), 1061–1091.
24. Steinfeld, A., Weimer, A., & Rönner, M. (2016). Solar thermochemical reactor design and materials research: A review. *Annual Review of Chemical and Biomolecular Engineering*, 7(1), 863–891.

25. Zhang, Z., Zhao, H., & Bao, M. (2022). Recent advances in solar thermochemical water splitting for hydrogen production: A review. *Chinese Journal of Chemical Engineering*, 40(2), 437–455.
26. Chen, X., Koukini, A., & Li, K. (2023). Recent progress in hybrid thermochemical-electrochemical cycles for solar hydrogen production. *Sustainable Energy & Fuels*, 7(1), 186–208.
27. Leung, D. Y. C., & Yang, X. (2020). A review of solar-driven water splitting by photoelectrochemical systems and photocatalysis. *Renewable and Sustainable Energy Reviews*, 125, 110001.
28. Llorente, J. F., Serrano, D. P., & García, H. (2015). Hybrid and integrated approaches for solar hydrogen production. *Energy & Environmental Science*, 8(8), 2018–2038.
29. Melton, R. G., Bricker, M. D., & Nocera, D. G. (2017). Combining photovoltaics with biology for solar-driven water splitting. *Science*, 357(6352), eaan6100.
30. Ulman, M., Lillia, S., & Wendt, S. (2021). Recent progress in solar-driven hybrid photoelectrochemical and photovoltaic-electrolysis systems for hydrogen production. *Chemical Society Reviews*, 50(20), 10242–10284.
31. Montoya, J. H., Seitz, L. C., Silverman, D. M., Hinrichs, J. D., & Nocera, D. G. (2017). Theoretical insights into the design of efficient transition metal electrocatalysts for fuel production. *Proceedings of the National Academy of Sciences*, 114(31), 8017–8022. DOI: 10.1073/pnas.1705098114.
32. Acar, C., & Dincer, I. (2019). Review of solar-based hydrogen production systems. *International Journal of Hydrogen Energy*, 44(12), 6046–6074. DOI: 10.1016/j.ijhydene.2018.12.182.
33. Kumar, A., et al. (2021). *Advanced Computational Methods in Mechanical and Materials Engineering*. CRC Press: Boca Raton, FL. https://doi.org/10.1201/9781003202233.
34. Prasad, A., et al. (2022). *Biodegradable Composites for Packaging Application*. CRC Press: Boca Raton, FL. https://doi.org/10.1201/9781003227908.
35. Gajrani, K. K., et al. (2022). *Advances in Sustainable Machining and Manufacturing Processes*. CRC Press: Boca Raton, FL. https://doi.org/10.1201/9781003284574.
36. Singh, S., et al. (2022). *Nanomanufacturing and Nanomaterials Design: Principles and Applications*. Taylor & Francis (CRC Press): Boca Raton, FL. https://doi.org/10.1201/9781003220602.
37. Sharma, V. K., et al. (2022). *Additive Manufacturing in Industry 4.0: Methods, Techniques, Modeling, and Nano Aspects*. CRC Press: Boca Raton, FL. https://doi.org/10.1201/9781003360001.
38. Kumar, A., et al. (2023). *Thermal Energy Systems: Design, Computational Techniques and Applications*. CRC Press: Boca Raton, FL. https://doi.org/10.1201/9781003395768.
39. Kumar, A., et al. (2023). *Laser Based Technologies for Sustainable Manufacturing*. CRC Press: Boca Raton, FL. https://doi.org/10.1201/9781003402398.
40. Kumar, R., et al. (2023). *Sustainable Smart Manufacturing Processes in Industry 4.0*. Taylor & Francis (CRC Press): Boca Raton, FL. https://doi.org/10.1201/9781003436072.

41. Awasthi, M.K., et al. (2024). *Computational Fluid Flow and Heat Transfer: Advances, Design, Control & Applications.* CRC Press: Boca Raton, FL. https://doi.org/10.1201/9781003465171.
42. Singh, V.P., et al. (2024). *Energy Efficient Vehicles: Technologies and Challenges.* CRC Press: Boca Raton, FL. https://doi.org/10.1201/9781003464556.
43. Shah, S., et al. (2024). *Sustainability in Smart Manufacturing: Trends, Scope, and Challenges.* CRC Press: Boca Raton, FL. https://doi.org/10.1201/9781003467496.
44. Verma, V., et al. (2024). *Highly Efficient Thermal Renewable Energy Systems: Design, Optimization and Applications.* CRC Press: Boca Raton, FL. https://doi.org/10.1201/9781003472629.
45. Kumar, A., et al. (2024). *Applications of Biotribology in Biomedical Systems.* Springer Nature. https://link.springer.com/book/9783031583261.
46. Awasthi, M. K., et al. (2024). Viscous correction to the potential flow analysis of Rayleigh–Taylor instability in a Rivlin–Ericksen viscoelastic fluid layer with heat and mass transfer. *Heat Transfer*, 1–17. https://onlinelibrary.wiley.com/doi/10.1002/htj.23076.
47. Dutt, N., et al. (2024). Thermo-hydraulic performance investigation of solar air heater duct having staggered D-shaped ribs: Numerical approach. *Heat Transfer*, 53(3), 1501–1531. https://onlinelibrary.wiley.com/doi/10.1002/htj.22998.
48. Awasthi, M. K., et al. (2024). Electrohydrodynamic capillary instability of Rivlin–Ericksen viscoelastic fluid film with mass and heat transfer. *Heat Transfer*, 53(1), 115–133. DOI: 10.1002/htj.22944.
49. Verma, V., et al. (2023). Ground and solar assisted heat pump systems for space heating and cooling applications in the northern region of India: A study on energy and CO2 saving potential. *SETA*, 59, 103405. https://doi.org/10.1016/j.seta.2023.103405.
50. Dutt, N., et al. (2023). Thermo-hydraulic performance of solar air heater having discrete D-shaped ribs as artificial roughness. *ESPR*. https://doi.org/10.1007/s11356-023-28247-9.
51. Kushwaha, P. K., et al. (2023). Recent advancements in augmentation of solar water heaters using nanocomposites with PCM: Past, present, and future. *Buildings*, 13, 79. https://doi.org/10.3390/buildings13010079.
52. Meena, C. S., et al. (2022). Utilization of solar energy for water heating application to improve building energy efficiency: An experimental study. *Buildings*, 12, 2166. https://doi.org/10.3390/buildings12122166.
53. Meena, C. S., et al. (2022). Innovation in green building sector for sustainable future. *Energies*, 15, 6631. https://doi.org/10.3390/en15186631.
54. Meena, C. S., et al. (2022). Review on boiling heat transfer enhancement techniques. *Energies*, 15, 5759. https://doi.org/10.3390/en15155759.
55. Bhoi, S., et al. (2022). Performance evaluation of different coating materials in delamination for micro- milling applications on high-speed steel substrate. *Micromachines*, 13, 1277. https://doi.org/10.3390/mi13081277.
56. Reddy, P. N., et al. (2023). CFD simulation and thermal performance optimization of flow in a channel with multiple baffles. *Journal of Heat and Mass Transfer Research*, 10(2), pp. 257–268. https://doi.org/10.22075/JHMTR.2023.31108.1458.

57. Paul, A. K., et al. (2022). Review on artificial neural network and its application in the field of engineering. *Journal of Mechanical Engineering: PRAKASH*, 1(1), 53–61.
58. Verma, V., et al. (eds.) (2024). Introduction to thermal energy storage: Solar, geothermal and hydrogen energy. In *Highly Efficient Thermal Renewable Energy Systems: Design, Optimization and Applications.* CRC Press: Boca Raton, FL, Chapter 01, pp. 01–22. https://doi.org/10.1201/9781003472629-1.
59. Ahmadizadeh, M., et al. (2024). Technological advancements in sustainable and renewable solar energy systems. In Verma, V., et al. (eds.), *Highly Efficient Thermal Renewable Energy Systems: Design, Optimization and Applications.* CRC Press: Boca Raton, FL, Chapter 02, pp. 23–39. https://doi.org/10.1201/9781003472629-2.
60. Chitt, M., et al. (2024). Green hydrogen productions: methods, design and smart applications. In Verma, V., et al. (eds.), *Highly Efficient Thermal Renewable Energy Systems: Design, Optimization and Applications.* CRC Press: Boca Raton, FL, Chapter 16, pp. 261–276. https://doi.org/10.1201/9781003472629-16.
61. Khashehchi, M., et al. (2024). Solar desalination techniques: Challenges and opportunities. In Verma, V., et al. (eds.), *Highly Efficient Thermal Renewable Energy Systems: Design, Optimization and Applications.* CRC Press: Boca Raton, FL, Chapter 19, pp. 305–329. https://doi.org/10.1201/9781003472629-19.
62. Kumar, A., et al. (2024). Globalization and international issues in sustainable manufacturing. In Shah, S. et al. (eds.), *Sustainability in Smart Manufacturing: Trends, Scope, and Challenges.* CRC Press: Boca Raton, FL, Chapter 01, pp. 01–18. https://doi.org/10.1201/9781003467496-1.
63. Singh, V. P., et al. (eds.) (2024). Introduction to energy-efficient vehicles for sustainable transportation: Transitions and challenges. In *Energy Efficient Vehicles: Technologies and Challenges.* CRC Press: Boca Raton, FL. https://doi.org/10.1201/9781003464556-1.
64. Upadhyay, R. K., et al. (2024). Sustainable transportation policy, planning, and implementation. In Singh, V. P., et al. (eds.), *Energy Efficient Vehicles Technologies and Challenges.* CRC Press: Boca Raton, FL, Chapter 05, pp. 74–95. https://doi.org/10.1201/9781003464556-5.
65. Singh, V. P., & Kumar, A. (2024). Techno-economic and future aspects of the HEV-EV-FCV: Decarbonization, digitalization and sustainability. In Singh, V. P., et al. (eds.), *Energy Efficient Vehicles Technologies and Challenges.* CRC Press: Boca Raton, FL, Chapter 11, pp. 212–231. https://doi.org/10.1201/9781003464556-11.
66. Pathak, A., et al. (2023). Application of laser technology in the mechanical and machine manufacturing industry. In Kumar, A., et al. (eds.), *Laser-based Technologies for Sustainable Manufacturing.* . CRC Press: Boca Raton, FL, Chapter 06, pp. 107–155. DOI: 10.1201/9781003402398-6.
67. Kundu, A., et al. (2023). Introduction to thermal energy resources and its smart application. In Kumar, A., Singh, V. P., Meena, C. S., & Dutt, N. (eds.), *Thermal Energy Systems Design, Computational Techniques, and Applications.* CRC Press: Boca Raton, FL, Chapter 1, pp. 1–15. https://dx.doi.org/10.1201/9781003395768-1.

68. Dewangan, A. K., et al. (2023). Thermal energy storage: Opportunities, challenges and future scope. In Kumar, A., Singh, V. P., Meena, C. S., & Dutt, N. (eds.), *Thermal Energy Systems Design, Computational Techniques, and Applications*. CRC Press: Boca Raton, FL, Chapter 2, **pp.** 17–28. https://dx.doi.org/10.1201/9781003395768-2.
69. Kundu, A., et al. (2023). Modelling and simulation of thermal energy system for design optimization. In Kumar, A., Singh, V. P., Meena, C. S., & Dutt, N. (eds.), *Thermal Energy Systems Design, Computational Techniques, and Applications*. CRC Press: Boca Raton, FL, Chapter 7, **pp.** 103–140. https://dx.doi.org/10.1201/9781003395768-7.
70. Singh, S., et al. (eds.) (2022). Challenges and opportunities in nanomanufacturing. In *Nanomanufacturing and Nanomaterials Design: Principles and Applications*. CRC Press, Taylor & Francis: Boca Raton, FL. DOI: 10.1201/9781003220602-2.
71. Singh, V. P., et al. (2022). Nanomanufacturing and design of high-performance piezoelectric nanogenerator for energy harvesting. In Singh, S., et al. (eds.), *Nanomanufacturing and Nanomaterials Design: Principles and Applications*. CRC Press, Taylor & Francis: Boca Raton, FL. DOI: 10.1201/9781003220602-15.
72. Kumar, A., Rana, S., Gori, Y., & Sharma, N. K. (2021). Thermal contact conductance prediction using FEM based computational techniques. In Kumar, A., Gori, Y., Dutt, N., Singla, Y. K., & Maurya, A. (eds.), *Advanced Computational Methods in Mechanical and Materials Engineering*. CRC Press: Boca Raton, FL, pp. 183–220. DOI: 10.1201/9781003202233-13.

Chapter 16

# Production and utilization of hydrogen as a future fuel for electric vehicles

## A review

*Ved Prakash, Vishal Saraswat, Ashish Dewangan, Ashok Kumar Yadav, and Ashwani Kumar*

## 16.1 INTRODUCTION

The challenges facing contemporary energy systems are numerous, encompassing issues such as security of supply and climate change. There is a recognized need to transition toward a sustainable energy system and society, with initial steps evident in political declarations and demonstration projects worldwide. One vision that has gained traction in harmonizing economic growth and environmental concerns is the concept of the hydrogen economy [1,2]. Hydrogen is hailed as a supremely clean energy carrier, potentially generated from renewable sources. It holds promise as a renewable fuel for both transportation and domestic applications, with ongoing exploration for its use in combustion engines and fuel cell electric vehicles (FCEVs). Consequently, a significant increase in hydrogen demand is anticipated in both the short and the long term. Despite being the most abundant element on Earth, gaseous hydrogen ($H_2$) does not occur naturally in the environment and must be manufactured [3]. However, producing hydrogen, like any other synthetic fuel, requires more energy compared to extracting oil or using other fossil fuels. Currently, approximately 95% of hydrogen production is derived from fossil fuel conversion, primarily through natural gas reforming. Given this context, the primary objective of the current study is to explore alternative technological methods for hydrogen production [4].

As the world pivots toward sustainable transportation, hydrogen emerges as a compelling contender for fuel EVs. Its clean combustion, high energy density, and adaptability make hydrogen a frontrunner in reducing greenhouse gas emissions and fossil fuel dependency [5]. This review delves into the diverse production methods and utilization prospects of hydrogen as a promising fuel for future EVs, highlighting its benefits, challenges, and potential trajectories.

DOI: 10.1201/9781003521341-16

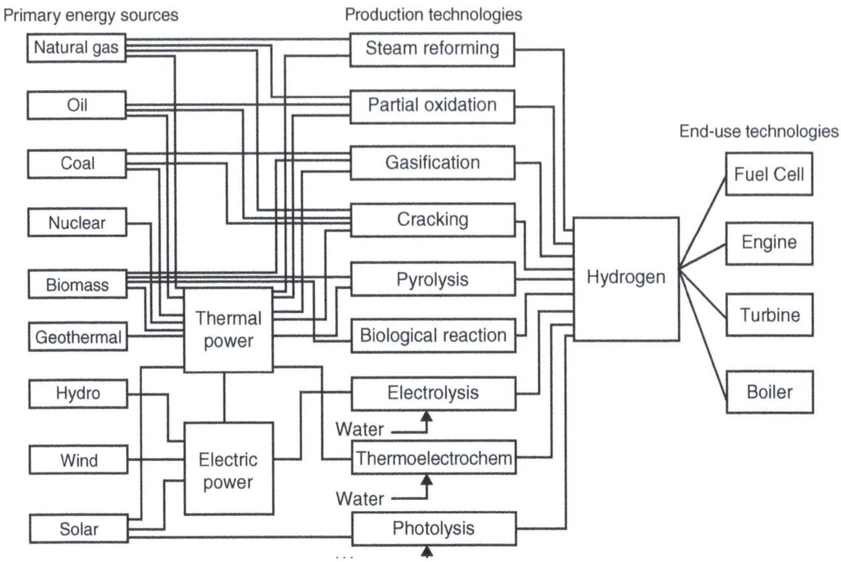


*Figure 16.1* Method of hydrogen production [6].

### 16.1.1 Methods for hydrogen production

Hydrogen can be generated through various methods, including electrochemical, thermochemical, photochemical, photocatalytic, and photoelectrochemical processes. The different methods of hydrogen production are shown in Figure 16.1. These alternative methods of hydrogen production primarily rely on different energy sources. The conventional method involves steam reforming of methane for hydrogen production [7]. However, renewable and nuclear systems offer alternatives by producing hydrogen from water through thermal or electrolytic processes. Another promising method for hydrogen recovery is bio-hydrogen production through anaerobic fermentation using renewable organic waste sources [6]. In this process, anaerobic bacteria utilize organic substances as the sole source of electrons and energy, converting them into hydrogen.

## 16.2 STEAM METHANE REFORMING (SMR)

Steam reforming of natural gas stands out as one of the most economically feasible methods for hydrogen production, currently constituting approximately half of the global hydrogen output. The costs associated with hydrogen derived from SMR vary based on factors such as feedstock expenses, production scale, and other pertinent variables. Steam reforming is a technique utilized for hydrogen production from various feedstocks

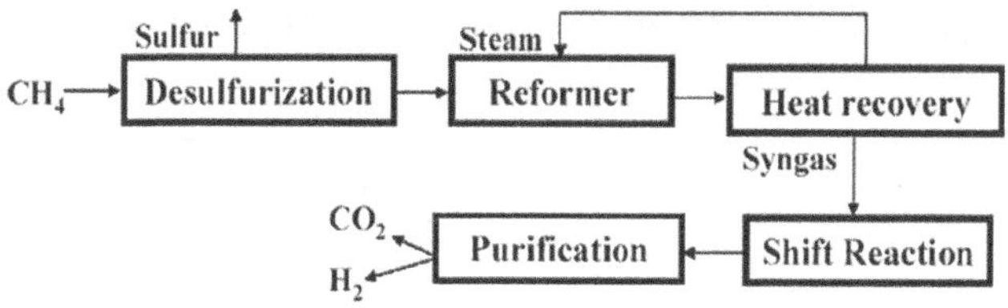


*Figure 16.2* Steam methane reforming [8].

such as higher hydrocarbons, acetone, ethanol, methanol, ethane, and methane, employing different catalysts [6]. Among these, methane is a favorable feedstock due to its superior by-product formation. This method facilitates multi-stage hydrogen production. The SMR method is illustrated in Figure 16.2.

SMR, a well-established technology, is used for hydrogen generation from natural gas by introducing steam, with its first industrial application dating back 1930. The process involves catalytic reforming or syngas production, gas purification or methanation, and water gas shift reactions [7]. Operating within a higher temperature range of 700°C–900°C, it is an endothermic process. The key chemical reactions involved in SMR are as follows [9]:

**Reformer equation:**

$$CH_4 + H_2O \rightarrow 3H_2 + CO \ \Delta H = 206.2 \text{ kJ/mol at } 298 \text{ K}$$

$$CH_4 + H_2O \rightarrow 3H_2 + CO \ \Delta H = 206.2 \text{ kJ/mol at } 298 \text{ K}$$

**Water gas shift reaction:**

$$CO + H_2O \rightarrow H_2 + CO \ \Delta H = -41.1 \text{ kJ/mol at } 298 \text{ K}$$

$$CO + H_2O \rightarrow H_2 + CO_2 \ \Delta H = -41.1 \text{ kJ/mol at } 298 \text{ K}$$

**Methanation reaction:**

$$CH_4 + 2H_2O \rightarrow 4H_2 + CO_2 \ \Delta H = 164.9 \text{ kJ/mol at } 298 \text{ K}$$

$$CH_4 + 2H_2O \rightarrow 4H_2 + CO_2 \ \Delta H = 164.9 \text{ kJ/mol at } 298 \text{ K}$$

In the steam reforming reaction, the water gas shift and reverse methanation reactions occur over a catalyst at temperatures exceeding 700°C.

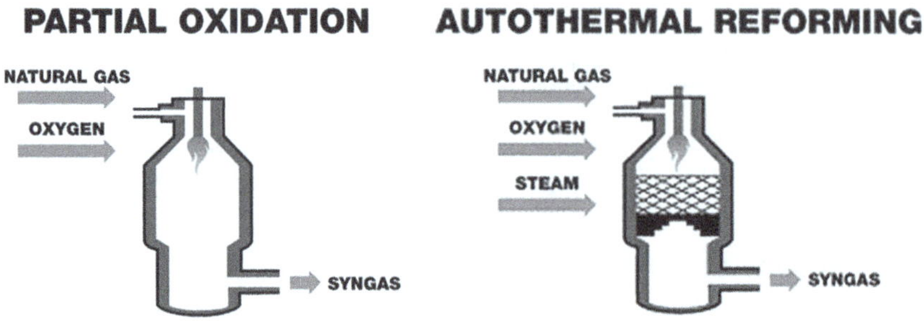


*Figure 16.3* Partial oxidation and auto thermal reforming [10].

## 16.3 PARTIAL OXIDATION AND AUTOTHERMAL REFORMING

Partial oxidation (POX) of hydrocarbons is the exothermic reaction with oxygen and steam. The amounts of oxygen and water vapor are controlled so that the reaction proceeds without the need for external energy. An example reaction for this process is as follows:

$$2C_8H_{18} + 2H_2O + 9O_2 \rightarrow 12CO + 4CO_2 + 20H_2$$

Autothermal reforming (ATR) combines both steam reforming and POX. The total reaction is exothermic, and so it releases heat. The outlet temperature from the reactor ranges from 950°C to 1,100, and the gas pressure can be as high as 100 bar [6]. ATR has the advantages of not needing external burners and being simpler and less expensive than SMR. However, ATR suffers from somewhat lower efficiency levels of 65%–75%. Several studies have been explored POX of ethanol. In a study, hydrogen production by POX of ethanol on $Pt/CeO_2$ catalysts at three different temperatures was investigated. The results showed that the $Pt/CeO_2$ catalyst performs the POX of ethanol with good stability and activity at 300°C [11]. POX and ATR are shown in Figure 16.3.

## 16.4 THERMOCHEMICAL HYBRID PROCESS

A thermochemical hybrid process (shown in Figure 16.4) integrates both thermochemical and electrolytic reactions for water splitting. This hybrid approach utilizes low-temperature reactions powered by electricity. Primary energy sources for this process could include advanced nuclear reactors or central solar receivers, potentially making large-scale hydrogen production feasible and significant for a hydrogen economy [11]. The hybrid sulfur cycle,

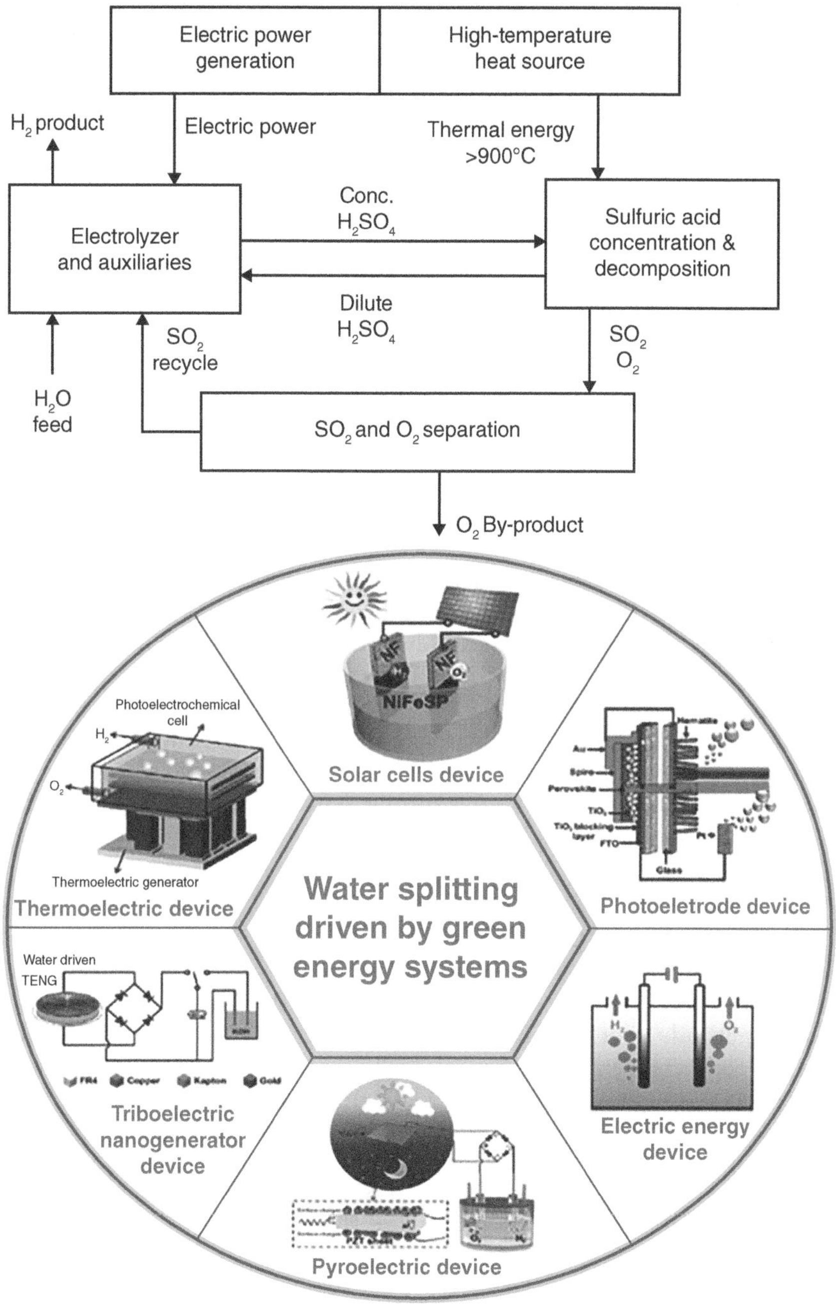


*Figure 16.4* Thermochemical hybrid process [12].

also referred to as the Westinghouse Cycle or the Sipra Mark 11 Cycle, shares the high-temperature endothermic reaction characteristic of the S-I cycle. This cycle is closed through the electrochemical oxidation of sulfur dioxide to sulfuric acid. The cycle is represented by the following reaction equations:

$$H_2SO_4 \rightarrow SO_2 + H_2O + 21O_2\,(\text{at}\,850°C,\ \text{minimum})$$

$$SO_2 + 2H_2O \rightarrow H_2SO_4 + H_2\,(\text{at}\,80°C,\ \text{electrolysis})$$

$$H_2O \rightarrow H_2 + 1/2O_2$$

A block flow schematic for the hybrid sulfur process is depicted. It comprises three main sections: The first section involves electrolysis and associated equipment; followed the second section focuses on sulfuric acid concentration and decomposition operations, which yield water and gaseous mixtures of $SO_2$ and oxygen [9].

The third section focuses on separating a clean $O_2$ product stream from the $SO_2/O_2$ mixture resulting from sulfuric acid decomposition and dissolving the remaining $SO_2$ in the anolyte. the overall net thermal efficiency of the hybrid sulfur cycle is 48.8%. This efficiency illustrates the energy efficiency of the Westinghouse sulfur process as a function of process temperature, estimated based on the thermal and electrical energy utilization from a gas turbine modular helium reactor. It is assumed that electrical energy is utilized through a He gas turbine power conversion system with a net efficiency of 42% [13].

## 16.5 COAL GASIFICATION

Hydrogen production through coal gasification POX processes is another viable method, which is shown in Figure 16.5. Although not as developed as SMR, these systems are relatively established. Syngas or artificial water gas ($CO+H_2$) can be reformed into hydrogen. As the coal rank decreases, the concentrations of hydrogen and oxygen in coal increase [4]. Steam can then be further shifted to hydrogen by establishing conditions to drive the reaction. Additional hydrogen production involve the following reactions:

$$\text{Coal}\,(\text{carbon source}) + H_2O \rightarrow CO + H_2 + \text{impurities}$$

$$CO + H_2O \rightarrow H_2 + CO_2$$

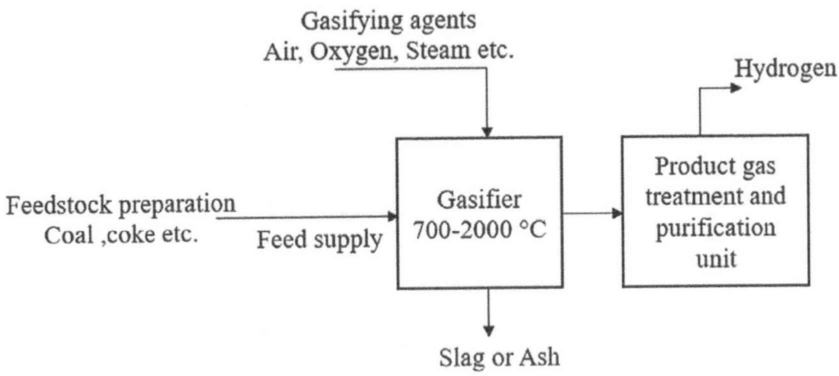


*Figure 16.5* Coal gasification process [14].

Hydrocarbon fuel reacts with oxygen in a less than stoichiometric ratio, resulting in a mixture of CO and hydrogen at temperatures ranging from 1,200°C to 1,350°C. Hydrogen can be produced from coal gasification at current delivered costs of approximately 2–2.5 $U.S. per kilogram, with potential future reductions to about 1.5 $U.S. per kilogram. The advantages of coal gasification for hydrogen production include its ability to be implemented using current technology and its relatively low cost for hydrogen production. However, disadvantages include the higher $CO_2$ production compared to other technologies, similar environmental concerns as coal-based electricity generation, centralized production, and challenges associated with the purification and separation of hydrogen at high temperatures [11].

## 16.6 STEAM IRON PROCESS

The steam iron process, dating back to the 20th century, is an older technique for hydrogen ($H_2$) generation, primarily used for balloons and airships. The steam iron process is shown in Figure 16.6. Commonly referred to as a conventional method, it involves the producing hydrogen from iron oxide at temperatures ranging from approximately 540°C to 900°C. It is a cyclic process where water ($H_2O$) is split using oil and coal. This method aims to produce reducing gas through coal gasification, yielding a mixture of $H_2$ and CO. This gas then reacts with iron oxides such as magnetite ($Fe_3O_4$), hematite ($Fe_2O_3$), and wustite (FeO) to form reduced iron oxide [6]. The regenerated iron is utilized in the steam iron reactor, where it undergoes oxidation through steam, resulting in the generation of abundant $H_2$ gas and $Fe_3O_4$ at temperatures ranging from 1,088–1,143 K. This chemical reaction is represented as follows:

$$3FeO + H_2O \rightarrow Fe_3O_4 + H_2$$

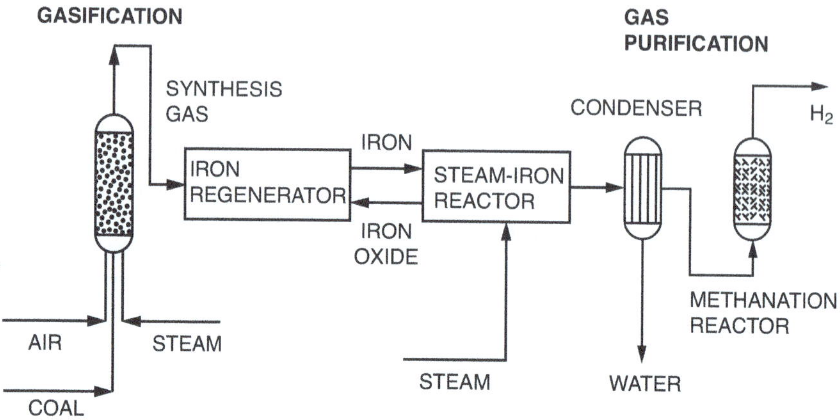


*Figure 16.6* Steam iron process.

The steam iron reactor comprises small traces of $N_2$ and $CO_2$, along with 61% steam and 37% $H_2$. After condensation, the steam is processed to achieve a $H_2$ concentration of approximately 96%, with 1.6% $CO_2$ and 2.5% $N_2$.

## 16.7 BIOLOGICAL TECHNIQUES OF $H_2$ PRODUCTION

Hydrogen production can be achieved through various biological processes, offering an environmentally friendly alternative that requires less energy and operates at atmospheric pressure and temperature. Agricultural waste serves as a promising candidate for bio-fuel production through anaerobic fermentation processes. Biomass, including waste materials like cellulose, sucrose, and starch, is increasingly gaining attention for hydrogen production, with glucose being the primary source historically [15].

## 16.8 DARK FERMENTATION

Dark fermentation (Figure 16.7) involves anaerobic bacteria and green algae under dark conditions. This biochemical process can occur with or without the presence of oxygen, resulting in the production of $CO_2$, acetone, alcohols, and $H_2$ through microbial transformation of organic materials (feedstock). Glucose, a carbohydrate, serves as an example of organic carbon [16]. The following equation demonstrates that one mole of glucose can produce four moles of hydrogen when acetic acid ($CH_3COOH$) is the fermentation by-product, and two moles of hydrogen when butyric acid ($CH_3CH_2CH_2COOH$) is the fermentation by-product.

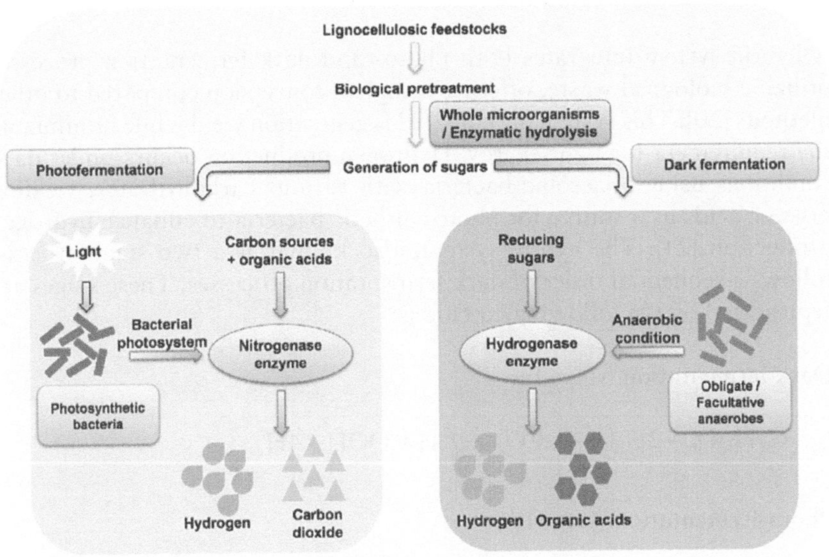


*Figure 16.7* Fermentation process [17].

$$C_6H_{12}O_6 + 2H_2O \rightarrow 4H_2 + 2CH_3COOH + 2CO_2$$

$$C_6H_{12}O_6 \rightarrow 2H_2 + CH_3C_2H_4COOH + 2CO_2$$

Hydrogen formation in dark fermentation depends on key parameters such as hydraulic retention time, pH value, and gaseous partial pressure. Maintaining a pH value between 5 and 6 is essential for optimal $H_2$ production [18].

## 16.9 PHOTO-FERMENTATION

Photo-fermentation involves biochemical processes utilizing organic acids and solar energy under nitrogen-deficient conditions. Photosynthetic bacteria can convert organic acids like butyric and acetic acids into $CO_2$ and $H_2$ in the presence of nitrogenase. For instance, photosynthetic non-sulfur (PNS) bacteria are capable of converting volatile fatty acids (VFAs) into $CO_2$ and $H_2$ under oxygenic conditions. Carbon sources such as sucrose, glucose, and succinate can also be utilized by PNS bacteria for hydrogen production, instead of VFAs [19]. When acetic acid is present in the fermentation medium, one mole of acetic acid can generate four moles of $H_2$, as shown in equation [8].

$$C_6H_{12}O_6 + 2H_2O \rightarrow 2CO_2 + 2CH_3COOH + 4H_2$$

## 16.10 HYBRID SYSTEM

A hybrid system integrates both photo- and dark fermentation processes utilizing biological waste, offering complete conversion compared to other methods [20]. This system maximizes $H_2$ generation yield while minimizing the requirement for light energy. Hydrogen production occurs under dark conditions using anaerobic bacteria with various carbohydrates, yielding organic acids as a source for photosynthetic bacteria to enhance hydrogen production [21]. The hybrid system, also known as a two-stage process, follows a sequential order of dark fermentation processes. These stages are represented by the following equations:

**Dark fermentation (stage I):**

$$C_6H_{12}O_6 + 2H_2O \rightarrow 2CO_2 + 2CH_3COOH + 4H_2$$

**Photo-fermentation (stage II):**

$$2CH_3COOH + 4H_2 \rightarrow 4CO_2 + 8H_2$$

**The overall reaction combining both stages is depicted in the following equation:**

$$C_6H_{12}O_6 + 6H_2O \rightarrow 12H_2 + 6CO_2$$

## 16.11 BIO-PHOTOLYSIS

Bio-photolysis (see Figure 16.8) harnesses the ability of plants, mainly algae, to dissociate water during photosynthesis, resulting in hydrogen and oxygen production in the presence of sunlight [22]. This biological process involves the decomposition of water using photons and biological catalysts, known as photolysis of water. Bio-photolysis is categorized into direct and indirect methods [23].

### 16.11.1 Direct bio-photolysis

Direct bio-photolysis utilizes microalgae (green algae with cyanobacteria) under solar energy. Photosynthetic reactions transform water into hydrogen, as represented by the general reaction:

$$C_6H_{12}O_6 + 6H_2O \rightarrow 12H_2 + 6CO_2$$

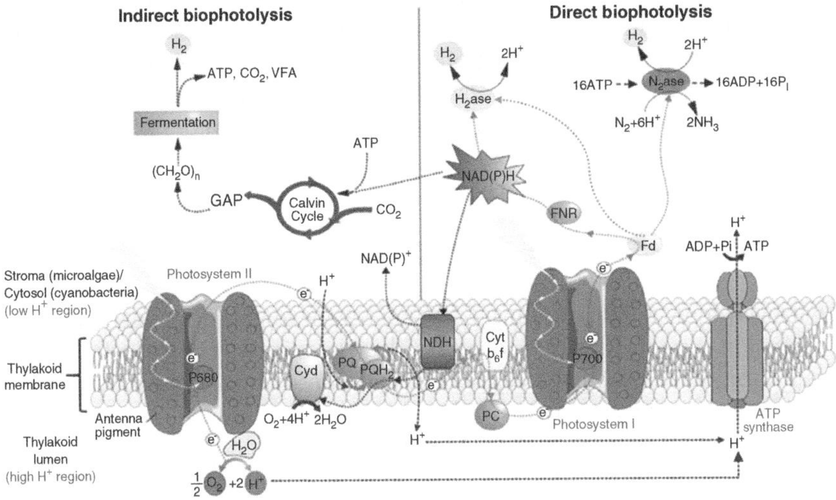


*Figure 16.8* Bio-photolysis [24].

Green algae dissociate water ($H_2O$) molecules into hydrogen ions and oxygen ions during photosynthesis. *Anabaena cylindrica* is a notable example of cyanobacteria for direct bio-photolysis [25].

## 16.11.2 Indirect bio-photolysis

Indirect bio-photolysis occurs in two steps. In the first step, photosynthesis generates glucose and $O_2$, while in the second step, $CO_2$ and $H_2$ are produced from glucose, water, and solar energy, as represented by following the general reactions:

$$12H_2O + 6CO_2 + \text{Solar energy} \rightarrow 6O_2 + C_6H_{12}O_6$$

$$C_6H_{12}O_6 + 12H_2O + \text{Solar energy} \rightarrow 6CO_2 + 12H_2$$

Both hydrogenase and nitrogenase enzymes contribute to hydrogen formation, with green algae exhibiting a higher production rate of hydrogenase enzymes [26].

Thus the Hydrogen can be sourced through various methods, including electrolysis, steam reforming of natural gas, biomass gasification, and solar-driven processes. Electrolysis, particularly powered by renewable sources like solar and wind energy, stands out for its emissions-free hydrogen production. While steam reforming is prevalent due to its economic viability, it emits carbon unless integrated with carbon capture and storage [27]. Biomass gasification presents a renewable route, utilizing organic materials for hydrogen production, potentially mitigating fossil fuel reliance. Solar-driven processes, although in nascent stages, offer sustainable hydrogen generation through photoelectrochemical and photobiological pathways [28].

## 16.12 UTILIZATION OF HYDROGEN IN EVs

Hydrogen finds application in FCEVs), where it undergoes electrochemical conversion into electricity, propelling electric motors. FCEVs offer advantages over battery EVs (BEVs), such as rapid refueling and extended driving ranges, addressing battery-related limitations like range anxiety and prolonged charging times [29]. Moreover, FCEVs emit only water vapor and heat, augmenting air quality and diminishing the carbon footprint of transportation.

### 16.12.1 Fuel Cell Electric Vehicles (FCEVs)

FCEVs represent a notable application of hydrogen in transportation. These vehicles harness hydrogen fuel cells, where hydrogen combines with oxygen from the air to generate electricity, powering the vehicle's electric motor. FCEVs offer distinct advantages including extended driving ranges and swift refuel times, setting them apart from BEVs [30,31]. Moreover, FCEVs produce only water vapor and heat as emissions, underscoring their environmental friendliness.

### 16.12.2 Commercial vehicles

Hydrogen fuel cell technology is gaining traction in commercial transportation sectors, encompassing buses, trucks, and even trains. These vehicles leverage hydrogen's high energy density and rapid refueling capabilities, making them suitable for demanding applications in long-haul transportation and heavy-duty sectors [32].

### 16.12.3 Hydrogen-powered aircraft

While still in experimental stages, hydrogen fuel cells hold promise for powering aircraft. Initial prototypes and concepts suggest that hydrogen-powered aircraft could reduce emissions and noise pollution in aviation, hinting at a greener future for air travel [30]. Hydrogen-powered aircraft is shown in Figure 16.9.

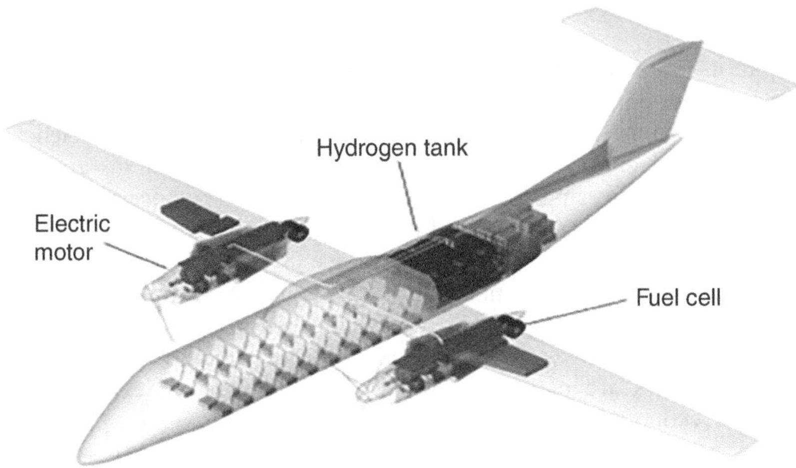


*Figure 16.9* Hydrogen-powered aircraft [33].

### 16.12.4 Hydrogen-powered boats

Maritime vessels, including boats and ferries, can benefit from hydrogen fuel cells. Hydrogen offers zero-emission capabilities and extended range, providing sustainable solutions for marine transportation [34,35].

### 16.12.5 Hydrogen-powered two-wheelers

Although less prevalent, hydrogen fuel cell technology is being explored for motorcycles and bicycles. These vehicles leverage the lightweight nature of hydrogen fuel cells, offering zero-emission transportation solutions ideal for urban environments [36,37].

### 16.12.6 Hydrogen-powered internal combustion engines

In addition to fuel cells, hydrogen can optimize the efficiency and emissions of internal combustion engines (ICEs). By blending hydrogen with traditional fuels or using it as a standalone fuel in modified ICEs, combustion efficiency can be improved, consequently reducing carbon emissions [10,14,17,24,31,33,35,37–54].

These applications underscore the versatility of hydrogen as a fuel for diverse vehicle types, offering potential avenues for emissions reduction and sustainable transportation. However, widespread adoption faces challenges such as infrastructure development, cost considerations, and scalability. Continued research, development, and investment are essential to unlocking hydrogen's full potential in revolutionizing the transportation landscape [39,55–88].

## 16.13 CHALLENGES AND FUTURE PROSPECTS

Despite its promise, the widespread hydrogen adoption faces challenges. Infrastructure development necessitates substantial investments in production, distribution, and refueling stations. Enhancing energy efficiency in hydrogen production, notably electrolysis, is imperative for sustainability. Additionally, addressing hydrogen storage and safety concerns, given its flammability, demands specialized handling and storage systems to ensure public acceptance and regulatory compliance [89].

Technological advancements, bolstered by supportive policies and environmental consciousness, are driving hydrogen-powered transportation forward. Collaborative endeavors between governments, industries, and research entities are pivotal in overcoming barriers and expediting the hydrogen economy transition. Technological innovation in production, storage, and utilization, alongside infrastructure enhancements and cost competitiveness, are instrumental in unlocking hydrogen's full potential for EVs, fostering a sustainable and low-carbon transportation paradigm [90]. Hydrogen emerges as a compelling, sustainable fuel for EVs, poised to reshape the transportation landscape and combat climate change. However, realizing these potential mandates concerted efforts to address technological, economic, and regulatory challenges. With ongoing advancements and collaborative initiatives, hydrogen-powered EVs could emerge as a linchpin solution in creating a greener and resilient transportation ecosystem in the foreseeable future [12].

## 16.14 CONCLUSIONS

Hydrogen production is at the forefront of global efforts to transition toward a sustainable energy future, offering a promising solution to address climate change, energy security, and air pollution. Previous studies have explored various methods of hydrogen production, each presenting unique advantages and challenges. Significant advancements have been made in electrolysis, SMR, renewable hydrogen production, biomass gasification, photoelectrochemical processes, thermochemical cycles, biological pathways, and nuclear-driven approaches.

In conclusion, achieving sustainable hydrogen production requires a multidimensional approach, integrating technological innovation, supportive policies, and collaborative efforts across sectors. The researchers mentioned in the literature review have played instrumental roles in advancing the field, pushing the boundaries of knowledge and technology. Their contributions underscore the importance of interdisciplinary collaboration and knowledge exchange in driving progress toward sustainable energy solutions.

Moving forward, it is essential to prioritize investments in hydrogen infrastructure, research, and development to overcome remaining challenges such as cost competitiveness, efficiency optimization, and environmental sustainability. Policy frameworks and regulatory incentives can provide the necessary support to accelerate the deployment of hydrogen technologies on a global scale. International cooperation and partnerships will also play a critical role in facilitating technology transfer and standardization, enabling widespread adoption of hydrogen as a clean energy carrier.

The integration of hydrogen fuel cell technology into EVs offers a promising avenue toward achieving sustainable transportation and energy systems. As outlined in the literature review, the development of hydrogen production methods has seen significant advancements, propelled by the innovative efforts of researchers globally. These breakthroughs have opened up diverse applications, including the utilization of hydrogen fuel cells in EVs.

Hydrogen fuel cell vehicles present numerous advantages, including emission-free operation, extended driving ranges, and quick refueling times, making them highly adaptable for various transportation requirements. Moreover, the versatility of hydrogen fuel cells extends to powering heavy-duty trucks, buses, and marine vessels, thereby playing a vital role in decarbonizing both freight and passenger transportation sectors. Additionally, hydrogen fuel cells can serve as backup power sources and off-grid energy solutions, enhancing energy reliability and facilitating the integration of renewable energy sources.

In the realm of EVs, hydrogen fuel cell technology complements BEVs by offering prolonged driving ranges and faster refueling capabilities, mitigating some of the limitations associated with battery range and charging infrastructure. Furthermore, the utilization of renewable hydrogen as a fuel source for hydrogen fuel cells contributes to further reductions in carbon emissions and environmental impact.

Overall, the adoption of hydrogen fuel cell technology in EVs signifies a significant step toward sustainable mobility and energy transition. With ongoing research and development efforts, coupled with collaborative initiatives across sectors, hydrogen-powered EVs have the potential to play a pivotal role in realizing a cleaner, greener future for transportation and energy systems.

## REFERENCES


1. Ishaq, H., Dincer, I., Crawford, C., A review on hydrogen production and utilization: Challenges and opportunities, *International Journal of Hydrogen Energy*, 2022, 47(62), 26238–26264.
2. Acar, C., Dincer, I., Comparative assessment of hydrogen production methods from renewable and non-renewable sources, *International Journal of Hydrogen Energy*, 2014 39(1), 1–12.
3. Akhlaghi, N., Najafpour-Darzi, G., A comprehensive review on biological hydrogen production, *International Journal of Hydrogen Energy*, 45(43), 22492–22512, 2020.
4. Çelik, D., Yıldız, M., Investigation of hydrogen production methods in accordance with green chemistry principles, *International Journal of Hydrogen Energy*, 2017, 42(36), 23395–23401.

5. Holladay, J. D., Hu, J., King, D. L., Wang, Y., An overview of hydrogen production technologies, *Catalysis Today*, 2009, 139(4), 244–260.
6. Wang, M., Wang, G., Sun, Z., Zhang, Y., Xu, D., Review of renewable energy-based hydrogen production processes for sustainable energy innovation, *Global Energy Interconnection*, 2019, 2(5), 436–443.
7. Dincer, I., Acar, C., Innovation in hydrogen production, *International Journal of Hydrogen Energy*, 2017, 42(22), 14843–14864.
8. Khan, O., Alsaduni, I., Equbal, A., Parvez, M., Yadav, A. K., Performance and emission analysis of biodiesel blends enriched with biohydrogen and biogas in internal combustion engines, *Process Safety and Environmental Protection*, 2024, 183, 1013–1037.
9. Dawood, F., Anda, M., Shafiullah, G. M., Hydrogen production for energy: An overview, *International Journal of Hydrogen Energy*, 2020, 45(7), 3847–3869.
10. Ahmed, A., Yadav, A. K., Singh, A., Enhancement of biogas yield from dual organic waste using hybrid statistical approach and its effects on ternary fuel blend (biodiesel/n-butanol/diesel), *Powered Diesel Engine Environmental Progress & Sustainable Energy*, 2023 DOI: 10.1002/ep.14163.
11. Xia, A., Cheng, J., Murphy, J. D., Innovation in biological production and upgrading of methane and hydrogen for use as gaseous transport biofuel, *Biotechnology Advances*, 2016, 34(5), 451–472.
12. Equbal, A., Equbal, A., Ahmad, R., Kumar, M., Khan, O., Yadav, A. K., RSM based statistical approach to enhance the compressive performance of ABS P400 parts fabricated via FDM technology, *Journal of Surface Review and Letters*, DOI: 10.1142/S0218625X25500210.
13. Nikolaidis, P., Poullikkas, A., A comparative overview of hydrogen production processes, *Renewable and Sustainable Energy Reviews*, 2017, 67, 597–611.
14. Singh, D., Singh, S., Yadav, A. K., Khan, O., Dewangan, A., Alkahtani, M. Q., Islam, S., From theory to practice: A sustainable solution to water scarcity by using a hybrid solar distiller with a heat exchanger and aluminum oxide nanoparticles, *ACS Omega*, 2023, 37(8), 33543–33553.
15. Schlapbach, L., Züttel, A., Hydrogen-storage materials for mobile applications, *Nature*, 2001, 414(6861), 353–358.
16. Ahmed, A., Yadav, A. K., Parametric analysis of wastewater electrolysis for green hydrogen production: A combined RSM, genetic algorithm, and particle swarm optimization approach, *International Journal of Hydrogen Energy*, 59, 2024, 51–62.
17. Ahmed, A., Yadav, A. K., Singh, A., Optimization of biogas yield from anaerobic co-digestion of dual waste for environmental sustainability: ANN, RSM, and GA, approach, *International Journal of Oil, Gas and Coal Technology*, 33(1),75, 2023.
18. Ahmad, A., Yadav, A. K., Hasan, S., Enhanced production of methane enriched biogas through intensified co-digestion process and its effective utilization in a biodiesel/biohydrogen fueled engine with duel injection strategies: ML-RSM based an efficient optimization approach, *International Journal of Hydrogen Energy*, 2024, 65, 671–686.

19. Ahmed, A., Yadav, A. K., Singh, A., Singh, D. K., A machine learning–genetic algorithm based models for optimization of intensification process through microwave reactor: A new approach for rapid and sustainable synthesis of biodiesel from novel Hiptage benghalensis seed oil, *Fuel*, 2024, 374,132470.
20. Khan, O., Khan, M. P., Yadav, A. K., A comparative analysis of various Nanocomposites for the remediation of heavy metals from biomass liquid Digestate using multi-criteria decision methods, *Biomass and Bioenergy*, 2024, 186, 107281.
21. Ahmed, A., Yadav, A. K., Singh, A., Singh, D. K., A machine learning-response surface optimization approach to enhance the performance of diesel engine using novel blends of aloe vera biodiesel with MWCNT nanoparticles and hydrogen, *Process Safety and Environmental Protection*, 2024, 186, 738–755, DOI: 10.1016/j.psep.2024.04.013.
22. Ahmad, A., Yadav, A. K., Singh, A., Singh, D. K., Ağbulut, Ü., A hybrid RSM-GA-PSO approach on optimization of process intensification of linseed biodiesel synthesis using an ultrasonic reactor: Enhancing biodiesel properties and engine characteristics with ternary fuel blends, *Energy*, 2024, 288, 129077.
23. Khan, O., Parvez, M., Yahya, Z., Alhodaib, A., Yadav, A. K., Hoang, A. T., Ağbulut, Ü., Waste-to-energy power plants: Multi-objective analysis and optimization of landfill heat and methane gas by recirculation of leachate, *Process Safety and Environmental Protection*, 2024, 186, 957–968.
24. Das, D., Yadav, A. K., Ahmed, A., Numerical simulation of the combustion chamber and performance analysis of microalgae-based biodiesel/diesel-powered CI engine: An Axisymmetric flow approach, *Journal of Process Mechanical Engineering: Part E*, 2023, https://doi.org/10.1177/09544089231218121.
25. Ahmed, A., Yadav, A. K., Hasan, S., Enhancing efficiency and environmental impacts of a nano-enhanced oleander biodiesel-biohydrogen dual fuel engine equipped with EGR, through operational parameter optimization using RSM-MOGA technique, *International Journal of Hydrogen Energy*, 2024, 78, 1157–1172.
26. Khanam, S., Khan, O., Ahmad, S., Sherwani, A. F., Khan, Z. A., Yadav, A. K., Ağbulut, Ü., A Taguchi-based hybrid multi-criteria decision-making approach for optimization of performance characteristics of diesel engine fuelled with blends of biodiesel-diesel and cerium oxide nano-additive, *Journal of Thermal Analysis and Calorimetry*, 2024, 149(8), 3657–3676.
27. Ahmed, A., Yadav, A. K., Hasan, S., Biogas as a sustainable and viable alternative fuel for diesel engines: A comprehensive review of production, purification, economic analysis, and performance evaluation, *Journal of Process Mechanical Engineering, Part E*, https://doi.org/10.1177/09544089241255930.
28. Singh, B., Yadav, A. K., Saha, A., Fabrication and characterization of AA6082-T6 produced by friction stir additive manufacturing, *Surface Review and Letters*, DOI: 10.1142/S0218625X25500222.
29. Dewangan, A., Mallick, A., Yadav, A. K., Islam, S., Saleel, C. A., Shaik, S., Ağbulut, Ü., Production of oxy-hydrogen gas and the impact of its usability on CI engine combustion, performance, and emission behaviors, *Energy*, 2023, 278, 127937.

30. Khan, O., Parvez, M., Kumari, P., Yadav, A. K., Akram, W., Ahmad, S., ... & Idrisi, M. J., Modelling of compression ignition engine by soft computing techniques (ANFIS-NSGA-II and RSM) to enhance the performance characteristics for leachate blends with nano-additives, *Scientific Reports*, 2023, 13(1), 15429.
31. Ahmed, A., Yadav, A. K., Singh, A., Optimization of cavitation-assisted biodiesel production and fuel properties from neochloris oleoabundans microalgae oil using genetic algorithm and response surface methodology, *Journal of Process Mechanical Engineering: Part E*, 2023, DOI: 10.1177/09544089231159832.
32. Dewangan, A., Ahmad, A., Yadav, A. K., Innovative research on waste tire recycling for sustainable biofuel production: Assessment of its usability on multi-cylinder diesel engine employing constant injection of oxyhydrogen and biogas through a premixing device, *International Journal of Hydrogen Energy*, 2024, 84, 155–163.
33. Dewangan, A., Mallick, A., Yadav, A.K., Ahmad, A., Alqahtani, D., Islam, S., Combined effect of operating parameters and nano particles on performance of a diesel engine: Response surface methodology coupled genetic algorithm Approach, *ACS Omega*, 2023, 8(27), 24586–24600.
34. Ahmed, A., Yadav, A. K., Singh, A., Process optimization of spirulina microalgae biodiesel synthesis using RSM coupled GA technique: A performance study of a biogas powered dual fuel engine, *International Journal of Environmental Science and Technology,* 2023, 21, 169–188.
35. Ahmed, A., Yadav, A. K., Singh, A., Application of machine learning and genetic algorithm for prediction and optimization of biodiesel yield from waste cooking oil, *Korean Journal of Chemical Engineering*, 2023, 40, 2941–2956.
36. Ahmad, A., Yadav, A. K., Singh, A., Singh, D. K., Multi-response optimization of a microalgae-spirulina-fueled VCR diesel engine: A comprehensive RSM-GA approach. Environment, Development and Sustainability, 2023, 1–29.**
37. Yadav, M., Karimi, M. N., Yadav, A. K., Effect of graphene oxide nanoparticles dispersed biodiesel on combustion, performance, and emission characteristics of a diesel engine, *Journal of Process Mechanical Engineering: Part E*, 2023, https://doi.org/10.1177/09544089231209139.
38. Khan, O., Ali, V., Parvez, M., Alhodaib, A., Yahya, Z., Yadav, A. K., Ağbulut, Ü., Exploring the performance of biodiesel-hydrogen blends with diverse nanoparticles in diesel engine: A hybrid machine learning K-means clustering approach with weighted performance metrics, *International Journal of Hydrogen Energy*, 2024, 78, 547–563.
39. Dwivedi, S. P., Yadav, A. K., Saxena, A., Dwivedi, V. K., Tribo-mechanical, physical and thermal behaviour of Al/Si3N4 composite with and without the addition of Cu, Ni and Cr entropy elements, *Proceedings of the Institution of Mechanical Engineers, Part E: Journal of Process Mechanical Engineering*, 2023, 09544089231189663. https://doi.org/10.1142/S0218625X24500744.
40. Ahmed, A., Yadav, A. K., Singh, A., An environmental impact assessment and optimization study of biodiesel production from microalgae, *International Journal of Global Warming*, 2023, 31(3), 294–313.

41. Ahmed, A., Yadav, A. K., Singh, A., Biodiesel yield optimization from a third-generation feedstock (microalgae spirulina) using a hybrid statistical approach, *International Journal of Ambient Energy*, 2023, 44(1), 1202–1213.
42. Ahmed, A., Yadav, A. K., Singh, A., Modelling and optimisation of VCR diesel engine parameters fuelled with microalgae methyl-ester: A RSM-coupled Taguchi approach, *International Journal of Ambient Energy*, 2023, 44(1), 2498–2506.
43. Ahmed, A., Yadav, A. K., Singh, A., Biodiesel production from Mahua oil: Characterization, optimization, and modeling with a hybrid statistical approach, *International Journal of Ambient Energy*, 2023, 44(1), 2600–2609.
44. Khan, T. A., Khan, T. A., Yadav, A. K., Khan, M. E., Experimental studies on utilization of neochloris oleoabundans microalgae biodiesel as an alternative fuel for diesel engine, *International Journal of Ambient Energy*, 2023, 44(1), 115–123.
45. Singh, D., Yadav, A. K., Kumar, A., Samsher, Energy matrices and life cycle conversion analysis of N-identical hybrid double slope solar distiller unit using Al2O3 nanoparticle, *Journal of Water and Environmental Nanotechnology*, 2023, 8(3), 267–284.
46. Khan, T. A., Yadav, A. K., Khan, T. A., A hydrodynamic cavitation-assisted system for optimization of biodiesel production from green microalgae oil using a genetic algorithm and response surface methodology approach, *Environmental Science and Pollution Research*, 2022, 29, 49465–49477.
47. Dewangan, A., Yadav, A. K., Mallick, A., Singh, S., Comparative study of Manilkara Zapota and Karanja based biodiesel properties and its effect on multi-cylinder diesel engine characteristics, *Energy Sources, Part A*, 2022, 44(2), 5143–5153.
48. Yadav, A. K., Khan, T. A., Khan, T. A., Kumar, S., Methyl ester of Gmelina arborea oil as a substitute for petroleum diesel: An experimental study on its performance and emissions in a diesel engine, *Energy Sources, Part A*, 2021, 43(11), 1307–1314.
49. Khan, O., Yadav, A. K., Khan, M. E., Parvez, M., Characterization of bio-ethanol obtained from Eichhornia Crassipes plant; its emission and performance analysis on CI engine, *Energy Sources, Part A*, 2021, 43(14), 1793–1803.
50. Khan, I. A., Singh, S. K., Pal, A., Yadav, A. K., Efficient production of biodiesel from Cannabis Sativa oil using intensified transestericationn (hydrodynamic cavitation) method, *Energy Sources, Part A*, 2020, 42(20), 2461–2470.
51. Dewangan, A., Yadav, A. K., Mallick, A., Tyagi, A., Optimisation of biodiesel production and engine performance from Simarouba oil in compression ignition engine, *International Journal of Oil, Gas and Coal Technology*, 2020, 25(3), 357–376.
52. Khan, M. E., Khan, T. A., Yadav, A. K., Pal, A., A comprehensive review on performance, emission and combustion characteristics of diesel engines fuelled with algae oil and methyl ester, *WEENTECH Proceedings in Energy*, 2020, 6(2), 160–168.

53. Yadav, A. K., Khan, M. E., Pal, A., Custard apple seed (Annona Squamosa) oil as a promising feedstock for biodiesel production using advanced techniques and its experimental investigations on an agricultural diesel engine, *International Journal of Oil, Gas and Coal Technology*, 2019, 20(4), 473–492.
54. Yadav, A. K., Pal, A., Ghosh, U., Gupta, S. K., Comparative study of biodiesel production methods from yellow oleander oil and its performance analysis on an agricultural diesel engine, *International Journal of Ambient Energy*, 2019, 40(2), 152–157.
55. Khan, I. A., Singh, S. K., Yadav, A. K., Sharma, D., Enhancement in the Performance of a diesel engine fueled with Pongamia methyl ester and n-butanol as oxygenated additive, *International Journal of Ambient Energy*, 2019, 40(8), 842–846.
56. Yadav, A. K., Khan, O., Khan, M. E., Utilization of high FFA landfill waste (leachates) as a feedstock for sustainable biodiesel production: Its characterization and engine performance evaluation, *Environmental Science and Pollution Research*, 2018, 25(32), 32312–32320.
57. Yadav, A. K., Dewangan, A., Mallick, A., Effect of n-butanol and diethyl ether additives on performance and emission characteristics of a diesel engine fuelled with diesel-Pongamia biodiesel blends, *Journal of Energy Engineering*, 2018, 144(6), 04018062.
58. Vinay, S. B., Yadav, A. K., Optimization of performance and emission characteristics of CI engine fuelled with Mahua oil methyl ester-diesel blend using response surface methodology, *International Journal of Ambient Energy*, 2020, 44(6), 674–685.
59. Yadav, A. K., Dewangan, A., Mallick, A., Synthesis and stability study of biodiesel from Kachnar seed oil, *Journal of Energy Engineering*, 2018, 144(5), 04018053.
60. Yadav, A. K., Pal, A., Dubey, A. M., Experimental studies on utilization of Prunus Armeniaca L. (wild apricot) biodiesel as an alternative fuel for CI engine, *Waste and Biomass Valorization*, 2018, 9(10), 1961–1969.
61. Yadav, A. K., Khan, M. E., Pal, A., Ultrasonic assisted production of biodiesel from Karabi oil using heterogeneous catalyst, *Biofuels*, 2018, 9(1), 101–112.
62. Yadav, A. K., Vinay, Singh, B., Optimization of biodiesel production from Annona Squamosa seeds oil using response surface methodology and its characterization, *Energy Sources, Part A*, 2018, 40(9), 1051–1059.
63. Dewangan, A., Yadav, A. K., Mallick, A., Current scenario of biodiesel development in India: Prospects and challenges, *Energy Sources, Part A*, 2018, 40(20), 2494–2501.
64. Yadav, A. K., Khan, M. E., Pal, A., Performance and emission characteristics of a stationary diesel engine fuelled by Schleichera Oleosa oil methyl ester (some) produced through hydrodynamic cavitation Process, *Egyptian Journal of Petroleum*, 2018, 27(1), 89–93.
65. Yadav, A. K., Khan, M. E., Pal, A., Performance, emission and combustion characteristics of an Indica diesel engine operated with Yellow Oleander (Thevetia Peruviana) oil biodiesel produced through hydrodynamic cavitation method, *International Journal of Ambient Energy*, 2018, 39(4), 365–371.

66. Dewangan, A., Mallick, A., Yadav, A. K., Effect of metal oxide nanoparticles and engine parameters on the performance of a diesel engine: A review, *Materials Today: Proceedings*, 2020, 21, 1722–27.
67. Yadav, A. K., Khan, M. E., Pal, A., Dubey, A. M., Experimental investigations of performance and emissions characteristics of Kusum (Schleichera Oleosa) biodiesel in a multi-cylinder transportation diesel engine, *Waste Biomass Valor*, 2017, 8(4), 1331.
68. Yadav, A. K., Khan, M. E., Pal, A., Biodiesel production from Oleander (Thevetia Peruviana) oil and its performance testing on a diesel engine, *Korean Journal of Chemical Engineering*, 2017, 34(2), 340–345.
69. Dubey, A. M., Kumar, A., Yadav, A. K., Wear behaviour of friction stir weld joint of cast Al (4–10%) Cu alloy welded at different operating parameters, *Journal of Materials Processing Technology*, 2017, 240, 87–97.
70. Khan, O., Khan, M. E., Yadav, A. K., Sharma, D., The ultrasonic assisted optimization of biodiesel production from Eucalyptus oil, *Energy Sources, Part A*, 2017, 39(13), 1323–1331.
71. Dewangan, A., Yadav, A. K., Wax deposition during production of Waxy Crude oil and its remediation, *Petroleum Science and Technology*, 2017, 35(18), 1831–1838.
72. Yadav, A. K., Khan, M. E., Pal, A., Kaner biodiesel production through hybrid reactor and its performance testing on a CI engine at different compression ratios. *Egyptian Journal of Petroleum*, 2017, 26(2), 525–532.
73. Yadav, A. K., Khan, M. E., Pal, A., Dubey, A. M., Performance and emission characteristics of a transportation diesel engine operated with non-edible vegetable oils biodiesel, *Case Studies in Thermal Engineering*, 2016, 8, 236–244.
74. Yadav, A. K., Khan, M. E., Pal, A., Biodiesel production from Nerium Oleander (Thevetia Peruviana) oil through conventional and ultrasonic irradiation methods, *Energy Sources, Part A*, 2016, 38(23), 3447–3452.
75. Yadav, A. K., Khan, M. E., Pal, A., A comparative study on ultrasonic cavitation and mechanical stirring method towards efficient production of biodiesel from non-edible oils and performance testing on a C.I. engine, *International Journal of Environment and Waste Management*, 2016, 18(4), 349–367.
76. Dewangan, A., Mallick, A., Yadav, A. K., Kumar, R., Combustion-generated pollutions and strategy for its control in CI engines: A review, *Materials Today: Proceedings*, 2020, 21, 1728–1733.
77. Yadav, A. K., Khan, M. E., Pal, A., Sharma, D., Optimization of biodiesel production from bitter groundnut oil using Taguchi method and its performance and emissions characteristics on a 4-cylinder Tata Indica engine, *International Journal of Sustainable Agricultural Management and Informatics*, 2015, 1, 285–300.
78. Khan, F., Khan, O., Parvez, M., Ahmad, S., Yahya, Z., Alhodaib, A., Yadav, A. K., Ağbulut, Ü., K-means clustering optimization of various quantum dots and nanoparticles-added biofuels for engine performance, emission, vibration, and noise characteristics, *Thermal Science and Engineering Progress*, 2024, 50, 102815. https://doi.org/10.1016/j.tsep.2024.102815.

79. Khan, O., Parvez, M.., Yadav, A. K., et al., Investigation of the thermal properties of lauric acid melting as a phase change material using priority based K-means clustering method, *Energy Storage,* 2024, DOI: 10.1002/est2.70026.
80. Sharma, G., Dewangan, A. K., Yadav, A. K., Ahmed, A., Current status of research on hydrogen generation, storage and transportation technologies: A state-of-the-art review towards sustainable energy, *Process Safety and Environmental Protection,* 2024, 91, 1445–1460.
81. Singh, S., Singh, B., Kumar, S., Yadav, A. K., Temperature-dependent dynamic hysteresis scaling of ferroelectric hysteresis parameters of lead free [(Ba0.825+xCa0.175-x) (Ti1 - x Snx)O3] ceramics, *Ferroelectrics*, 2019, 551(1), 133–142.
82. Kumar, N., Yadav, A. K., Dewangan, A., Kumar, M., CFD based investigation of thermophoresis effect on microparticles in micro channel. In S. Yadav, P. K. Jain, P. K. Kankar, Y. Shrivastava (eds.), *Advances in Mechanical and Energy Technology*, Lecture Notes in Mechanical Engineering. Springer, 2022, pp. 285–296, https://doi.org/10.1007/978-981-19-1618-2_31.
83. Ahmad, A., Singh, A., Singh, D. K., Yadav, A. K., An experimental investigation of the effect of diethyl ether as an additive on the performance of a single cylinder diesel engine. In A. Singh, D. P. Mishra, G. Bhat (eds.), *Recent Trends in Thermal and Fluid Sciences*, Lecture Notes in Mechanical Engineering. Springer, 2022, pp. 69–75, DOI: 10.1007/978-981-19-3498-8_7.
84. Singh, S., Yadav, A. K., Kumar, N., Choudhury, U. K., Kumar, M., Investigation on lead free ferroelectric $[(Ba_{0.825}+xCa_{0.175-x})(Ti_{1-x}Sn_x)O_3]$ ceramics for energy storage density and thermal energy harvesting capacity, In A. Singh, D. P. Mishra, G. Bhat (eds.), *Recent Trends in Thermal and Fluid Sciences*, Lecture Notes in Mechanical Engineering. Springer, pp. 497–504, DOI: 10.1007/978-981-19-8517-1_38.
85. Gupta, G., Agarwal, K., Yadav, A., Yadav, A. K., Sinha, D. K., Design and fabrication of PLA-printed wearable exoskeleton with 7 DOF for upper limb physiotherapy training and rehabilitation. In R. Sharma, R. Kannojiya, N. Garg, S. S. Gautam (eds.), *Advances in Engineering Design*, Lecture Notes in Mechanical Engineering. Springer, 2023. https://doi.org/10.1007/978-981-99-3033-3_5.
86. Gupta, G., Yadav, A. K., Sinha, D. K., Studying current safety systems for accident prevention and wellbeing of powered two-wheeler community: Prevalence of safety components. In R. Sharma, R. Kannojiya, N. Garg, S. S. Gautam (eds.), *Advances in Engineering Design*, Lecture Notes in Mechanical Engineering. Springer, 2023. https://doi.org/10.1007/978-981-99-3033-3_64.
87. Dewangan, A., Singh, B., Srivastava, A., Srivastava, A., Yadav, A. K., A broad review of biodiesel feedstocks with competency to replace diesel. In S. Yadav, P. K. Jain, P. K. Kankar, Y. Shrivastava (eds.), *Advances in Mechanical and Energy Technology*, Lecture Notes in Mechanical Engineering. Springer, 2022, pp. 323–331, https://doi.org/10.1007/978-981-19-1618-2_28.
88. Dubey, A. M., Ram, R., Yadav, A. K., Ion cell performance using single particle representation of battery electrode, *Journal of Electroanalytical Chemistry*, 2016, 768, 11–17.

89. Alhodaib, A., Yahya, Z., Khan, O., Equbal, A., Equbal, M. S., Parvez, M., Idrisi, M. J., Sustainable coatings for green solar photovoltaic cells: Performance and environmental impact of recyclable biomass digestate polymers, *Scientific Reports*, 2024, 14(1), 11221.
90. Singh, A. A., Singh, R. K., Yadav, A. K., Saha, A., Dewangan, A., & Singh, B., Experimental investigations on the electrochemical machining of D3 die steel material and multi-objective parameter optimization, *Surface Review and Letters*, 2024, 31 (9), 2450074. https://doi.org/10.1142/S0218625X24500744.

Chapter 17

# Comparative evaluation of carbon emissions from various vehicles

## Integrating AHP techniques for ranking

*Farzan Ashraf, Mohammad Shajar, M. Emran Khan, Azhar Equbal, Osama Khan, Ashok Kumar Yadav, and Mohd Parvez*

### 17.1 INTRODUCTION

Electric vehicles (EVs) present concrete advantages over vehicles powered by traditional fossil fuels. First, EVs contribute to a significant reduction in greenhouse gas emissions. Studies have shown that EVs produce lower emissions over their lifetime than conventional vehicles, even when factoring in electricity generation from fossil fuels [1,2]. For example, a study by the Union of Concerned Scientists found that EVs produce less than half the emissions of comparable gasoline-powered vehicles. Secondl, EVs are inherently more energy-efficient. According to the U.S. Department of Energy, EVs convert about 59%–62% of the electrical energy from the grid to power at the wheels, while conventional gasoline vehicles convert only about 17%–21% of the energy stored in gasoline [3,4]. This higher efficiency translates to lower energy consumption per mile driven, resulting in cost saving on fuel for drivers. Additionally, EVs offer quieter operation and smoother acceleration due to the absence of an internal combustion engine (ICE) [5,6].

The expansion of charging infrastructure has also addressed concerns about range limitations, with fast-charging stations allowing for rapid recharging on the go. Moreover, EVs contribute to energy security by reducing dependence on imported oil, as electricity can be generated domestically from a variety of sources, including renewable energy [7,8]. With ongoing advancements in battery technology, such as increased energy density and faster charging capabilities, the appeal of EVs continues to grow, making them a compelling choice for environmentally conscious consumers [9,10]. The global surge in EV sales, surpassing 7.8 million in 2021 with a staggering 109% increase from the previous year, signals a rapid shift in the automotive industry toward electric propulsion. Countries like Norway are

 DOI: 10.1201/9781003521341-17

setting ambitious targets to ban ICE vehicle (ICEV) sales by 2025, highlighting the accelerating momentum toward electrification that is poised to transform the transportation sector [11,12].

EVs are increasingly recognized as a superior and sustainable alternative to traditional fossil fuel-powered vehicles, offering myriad environmental, economic, and technological benefits. By harnessing electricity, often sourced from renewable sources, EVs play a vital role in mitigating greenhouse gas emissions and reducing environmental impacts associated with conventional fuels [13,14]. Moreover, their higher energy efficiency compared to ICEs translates to reduced energy consumption per mile driven. In addition to environmental advantages and reduced reliance on fossil fuels, widespread adoption of EVs promises advancements in smart grid technology, job creation in renewable energy sectors, and enhanced energy security [15,16]. As the demand for sustainable transportation solutions grows, EVs are emerging as frontrunners, ushering in an era of cleaner, greener, and more efficient mobility.

The availability of various charging options, including home charging stations, public infrastructure, and rapid charging networks, offers users with convenient refueling solutions [17,18]. Furthermore, powering EVs with renewable energy sources like solar or wind power provides a sustainable alternative to finite fossil fuels, contributing to environmental preservation. Continued advancements in battery technology, leading to increased energy density and longer driving ranges, further solidify the position of battery-operated vehicles in the future of transportation [19,20]. However, despite the numerous advantages, EVs face challenges, notably the limited range of many models on a single charge, often leading to "range anxiety." Balancing affordability with battery costs remains a persistent hurdle, and the availability and convenience of charging infrastructure continue to present barriers to widespread adoption [21,22]. Environmental concerns surrounding the manufacturing and disposal of lithium-ion batteries, including resource consumption and environmental damage, persist. Initial costs of EVs are typically higher than those of conventional combustion engine vehicles, necessitating ongoing improvements in battery technology, infrastructure, and manufacturing processes to fully unlock their potential for promoting a cleaner, more sustainable future [23,24].

Efficient energy management in vehicles is facilitated by mechatronic systems, which integrate sensors, actuators, and control algorithms to optimize power allocation, maximize energy efficiency, and minimize carbon emissions [25,26]. These systems enable regenerative braking, which converts kinetic energy into electrical energy to recharge batteries, thus reducing energy wastage and lowering carbon emissions. Mechatronics also play a crucial role in optimizing motor performance and battery usage to enhance drivetrain efficiency in EVs [27,28]. Additionally, they contribute to vehicle weight reduction through the integration of lightweight materials and advanced structural designs, without compromising safety or performance, ultimately reducing energy consumption during operation [29,30]. Smart thermal management solutions regulate temperature within the vehicle's battery pack and power electronics, improving battery performance and longevity while minimizing energy consumption and carbon emissions throughout the vehicle's lifespan [30,31].

Farzaneh et al. [32] explored how to reduce greenhouse gas emissions from transportation by transitioning from ICEVs to EVs. A life cycle analysis comparing ICEV (Ford Transit) and EV (Ford E-Transit) models over a 150,000 km lifespan revealed significantly lower carbon emissions per kilometer for EVs. The findings suggest that using renewable energy sources for electricity generation further reduces EVs' carbon footprint, with longer vehicle lifespans favoring EV efficiency. Saenz-Esteruelas et al. [33] studied the electrification of last-mile distribution with electric tricycles in urban areas like downtown Portland, OR, significantly reducing greenhouse gas emissions and congestion, this promoting smart cities. An analysis of a tricycle logistics company (B-Line) revealed substantial emissions reduction compared to traditional diesel-powered vans, making tricycle systems twice as efficient. Onat et al. [34] explored how EVs, plug-in hybrid EVs, and hybrid EVs (HEVs) are often favored for their lower greenhouse gas emissions and energy consumption. However, choosing among them is complex due to factors like electricity sources and driving patterns. This study compared these options across 50 states, finding that EVs were the least carbon-intensive in 24 states, while HEVs were the most energy-efficient in 45 states [35].

## 17.2 MATERIALS

Battery electric vehicles (BEVs) represent a transformative shift in automotive propulsion technology, relying solely on electric power stored in onboard batteries to propel the vehicle. The heart of a BEV lies in its electric motor, which converts electrical energy from the battery into mechanical energy to drive the vehicle. Unlike traditional ICEs, BEV engines produce zero tailpipe emissions, offering significant environmental benefits by reducing greenhouse gas emissions and air pollution [36–50]. The simplicity of the electric motor design, characterized by fewer moving parts compared to ICEs, results in reduced maintenance requirements and enhanced reliability. Moreover, BEV engines provide instant torque delivery, enabling smooth and silent acceleration, which contributes to a refined driving experience. As advancements in battery technology continue to increase energy density and reduce costs, BEVs are poised to play a pivotal role in the transition toward sustainable transportation systems, offering a cleaner and more efficient alternative to conventional gasoline-powered vehicles [51–60].

Hybrid engine technology has gained significant attention in recent years as a promising solution to reduce carbon emissions in the transportation sector. By integrating both conventional ICEs with electric propulsion systems, hybrid vehicles offer improved fuel efficiency and lower greenhouse gas emissions compared to traditional gasoline- or diesel-powered vehicles [61–69]. The hybrid engine operates by seamlessly switching between power sources based on driving conditions, such as acceleration, cruising, or braking, to optimize energy usage and minimize environmental impact. Studies have shown that hybrid engines can achieve substantial reductions in fuel consumption and $CO_2$ emissions, particularly in urban driving scenarios where frequent starts and stops occur [70–90]. Moreover, advancements in hybrid technology, such as regenerative braking and intelligent energy management systems, continue to enhance the performance and environmental benefits of hybrid vehicles, positioning them as a key transitional technology toward a more sustainable transportation future.

Diesel engines have been a focal point of environmental scrutiny due to their significant contribution to air pollution and greenhouse gas emissions. Studies examining the environmental impact of diesel engines have highlighted their higher emissions of nitrogen oxides ($NO_x$) and particulate matter (PM) than gasoline engines. Despite advancements in diesel engine technology aimed at reducing emissions, challenges persist in achieving substantial improvements in air quality and mitigating climate change effects. The life cycle assessment of diesel engines encompasses various stages, including extraction, refining, combustion, and disposal, revealing their complex environmental footprint. Strategies to address diesel engine emissions include adopting cleaner fuel alternatives, such as biodiesel and synthetic fuels, along with stringent emission regulations and the implementation of emission control technologies like selective catalytic reduction and diesel particulate filters. However, the effectiveness and feasibility of these measures warrant ongoing research and policy interventions to promote sustainable transportation systems and reduce the environmental burden associated with diesel engines.

Petrol engines, also known as gasoline engines, have been the cornerstone of automotive propulsion for over a century [91–95]. These ICEs operate by igniting a mixture of air and fuel within a combustion chamber to generate mechanical power. Despite their widespread use, petrol engines are associated with significant carbon emissions and environmental impacts. Research efforts have focused on understanding and mitigating these emissions through advancements in engine design, fuel efficiency, and emission control technologies. Comparative studies have examined the carbon footprint of petrol engines in relation to alternative propulsion systems, such as diesel engines, HEVs, and fully EVs. While petrol engines continue to dominate the automotive market due to their affordability and convenience, ongoing research aims to reduce their environmental footprint and transition toward more sustainable transportation options [96–102].

Biodiesel engines have garnered considerable attention as a sustainable alternative to traditional petroleum-based engines due to their potential to reduce greenhouse gas emissions and dependence on fossil fuels. Research

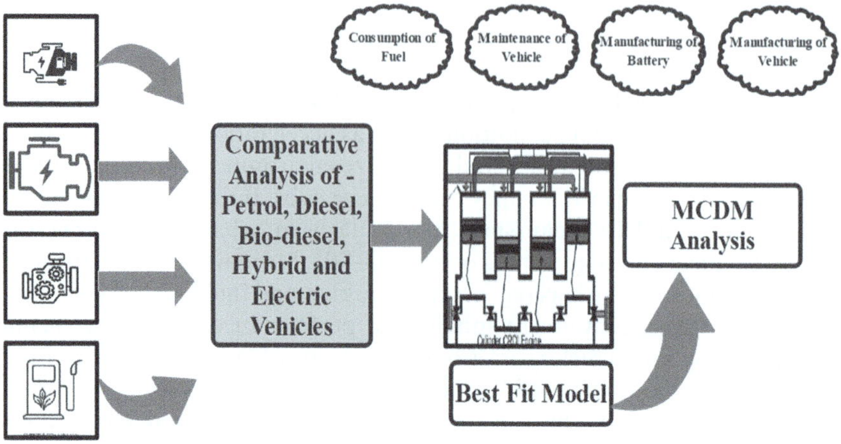


*Figure 17.1 Experimental setup.*

on biodiesel engines encompasses various aspects, including fuel production, engine performance, and environmental impact assessment. Studies have investigated the combustion characteristics of biodiesel blends, examining parameters such as ignition delay, combustion stability, and emissions profiles. Additionally, advancements in engine design and technology have aimed to optimize the compatibility of biodiesel fuels with existing engine systems, addressing challenges such as cold start issues and injector coking. Environmental assessments have demonstrated the potential benefits of biodiesel engines in terms of reducing carbon dioxide and PM emissions compared to conventional diesel engines, thereby contributing to efforts aimed at mitigating climate change and improving air quality. Ongoing research continues to explore ways to enhance the efficiency, reliability, and sustainability of biodiesel engines, facilitating their integration into transportation systems and promoting the transition toward cleaner energy sources (Figure 17.1).

## 17.3 METHODOLOGY

### 17.3.1 AHP methodology

The analytic hierarchy process (AHP) technique can enhance the Cold Metal Transfer (CMT) welding technique by providing a structured approach to identify and prioritize the most important factors that influence the quality of the weld. The AHP enables the identification of key input parameters such as welding speed, wire feed rate, and heat input, that affect the quality of the weld, and it estimates their relative importance. By understanding the relative importance of these parameters, the welding process can be optimized to achieve the desired outcome. For example, the AHP can help identify which parameter should be given the most attention to achieve the desired result. Suppose the AHP indicates that the wire feed rate is the most critical input parameter affecting the quality of the weld, the welding process can be optimized by adjusting the wire feed rate to achieve the desired outcome. Overall, the AHP serves a valuable decision-making tool for engineers and welding experts, enabling them to improve the quality of the CMT welding process by identifying the most critical input parameters and optimizing the welding process to achieve the desired outcome. To apply the AHP methodology in this study, the following steps can be followed:

- **Define the Hierarchy**: Create a hierarchy of the criteria and sub-criteria involved in the study. For example, the criteria could include welding speed, wire feed rate, and heat input, and the sub-criteria could include tensile strength, hardness, impact toughness, radiation, convection, and conduction.
- **Pairwise Comparison**: Evaluate the importance of each criterion and sub-criterion by performing pairwise comparisons using a scale from 1 to 9, where 1 indicates the two factors are equally important and 9 indicates one factor is extremely important compared to the other. Perform these comparisons for all the factors and create a pairwise comparison matrix for each level of the hierarchy.
- **Calculate the Weights**: Use the pairwise comparison matrices to calculate the weights of each criterion and sub-criterion using the eigenvector method. The eigenvector method involves calculating the dominant eigenvector of each matrix, which represents the relative importance of the criteria or sub-criteria.
- **Check Consistency**: Check the consistency of the pairwise comparison matrices using the consistency index (CI) and consistency ratio (CR) metrics. The CI measures the consistency of the pairwise comparisons, while the CR measures the consistency relative to random judgments. If the CR is >0.1, the pairwise comparisons may need revision to improve consistency.
- **Finalize Weights**: Once the weights are calculated and consistency ensured, use the AHP results to estimate the relative importance of the inputs in the study.

Overall, the AHP methodology provides a systematic and structured approach for estimating the weights of the inputs considered in the study, improving the accuracy and consistency of the results.

## 17.4 RESULTS AND DISCUSSION

### 17.4.1 Impact of operating parameters on vehicles

Operating parameters have a profound impact on the performance, efficiency, and environmental footprint of vehicles. First, factors such as vehicle speed, acceleration, and driving patterns significantly influence fuel

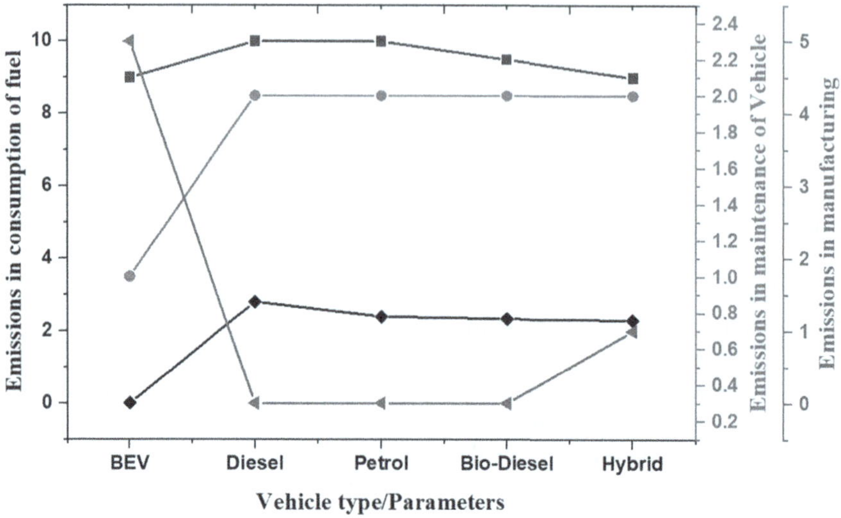


*Figure 17.2* Impact of operating parameters on vehicles.

consumption and emissions. High speeds and aggressive driving behaviors typically result in increased fuel consumption and higher emissions of pollutants such as carbon dioxide ($CO_2$), nitrogen oxides ($NO_x$), and PM. Conversely, driving at steady speeds and practicing smooth acceleration and deceleration can improve fuel efficiency and reduce emissions, lowering environmental impact and operational costs. Additionally, operating conditions such as temperature, humidity, and altitude can affect engine performance and emissions output. For instance, cold starts tend to increase emissions due to incomplete combustion and higher frictional losses, while extreme temperatures or high altitudes can negatively impact engine efficiency and fuel consumption (Figure 17.2).

Secondly, vehicle load and payload play a crucial role in determining fuel consumption and emissions levels. Heavier loads require more energy for propulsion, leading to increased fuel consumption and higher emissions of greenhouse gases and pollutants. Furthermore, vehicle design and aerodynamics also influence fuel efficiency and emissions performance. Aerodynamically optimized vehicles experience less air resistance at higher speeds, resulting in improved fuel economy and reduced emissions. Advances in engine technologies such as turbocharging, direct injection, and variable valve timing can enhance combustion efficiency and power output, leading to lower fuel consumption and cleaner emissions. Lastly, maintenance and driving behavior are essential factors in optimizing vehicle performance and reducing environmental impact. Regular maintenance and tune-ups ensure that engines operate at peak efficiency, minimizing fuel consumption and emissions output. Proper tire inflation, wheel alignment, and vehicle servicing also contribute to improved fuel economy and reduced emissions. Additionally, eco-driving techniques such as anticipating traffic flow, avoiding unnecessary idling, and maintaining steady speeds can further enhance fuel efficiency and lower emissions. By understanding and optimizing operating parameters, vehicle owners, manufacturers, and policymakers can collectively mitigate environmental impact, reduce greenhouse gas emissions, and promote sustainable transportation solutions for a cleaner and greener future.

### 17.4.2 Comparison of all vehicles

Comparing vehicles in terms of battery production emissions and overall environmental impact requires a nuanced assessment of various factors, as shown in Figure 17.3. First, EVs typically have lower emissions related with battery production than ICE vehicles, primarily due to the absence of direct emissions during operation [103–110]. However, the environmental impact of EV batteries depends on factors such as battery chemistry, manufacturing processes, and raw material extraction. Lithium-ion batteries, which are commonly used in EVs, require significant amounts of energy and resources for production, including mining and refining of lithium, cobalt, and nickel. These activities can result in environmental degradation and carbon emissions. Nonetheless, advancements in battery technology and recycling initiatives are improving the sustainability of EV batteries, reducing their overall environmental footprint over their life cycle [111–115].

On the other hand, ICEVs rely on fossil fuels for propulsion, leading to direct emissions of greenhouse gases and pollutants during operation. While ICEVs do not have the same upfront emissions associated with battery production, their life cycle emissions are significantly higher compared to EVs, particularly in regions where electricity generation is less carbon-intensive. Moreover, the extraction, refining, and transportation of fossil fuels contribute to environmental degradation and air pollution, further exacerbating the environmental impact of ICEVs. However, advancements in engine efficiency and emissions control technologies have led to reductions in tailpipe emissions for modern ICEVs, albeit still higher than their electric counterparts.

Hybrid vehicles represent a middle ground between EVs and ICEVs, with lower emissions than traditional ICEVs while avoiding some of the environmental concerns associated with battery production. By combining ICEs with electric propulsion systems, hybrid vehicles can achieve improved fuel efficiency and reduced emissions during operation. However, hybrid vehicles still rely on fossil fuels for a portion of their energy needs, resulting in ongoing emissions and environmental impact. Nevertheless, hybridization represents a transitional technology that can help bridge the gap toward a cleaner and more sustainable transportation future, offering a balance between the range and convenience of ICEVs and the environmental benefits of EVs.

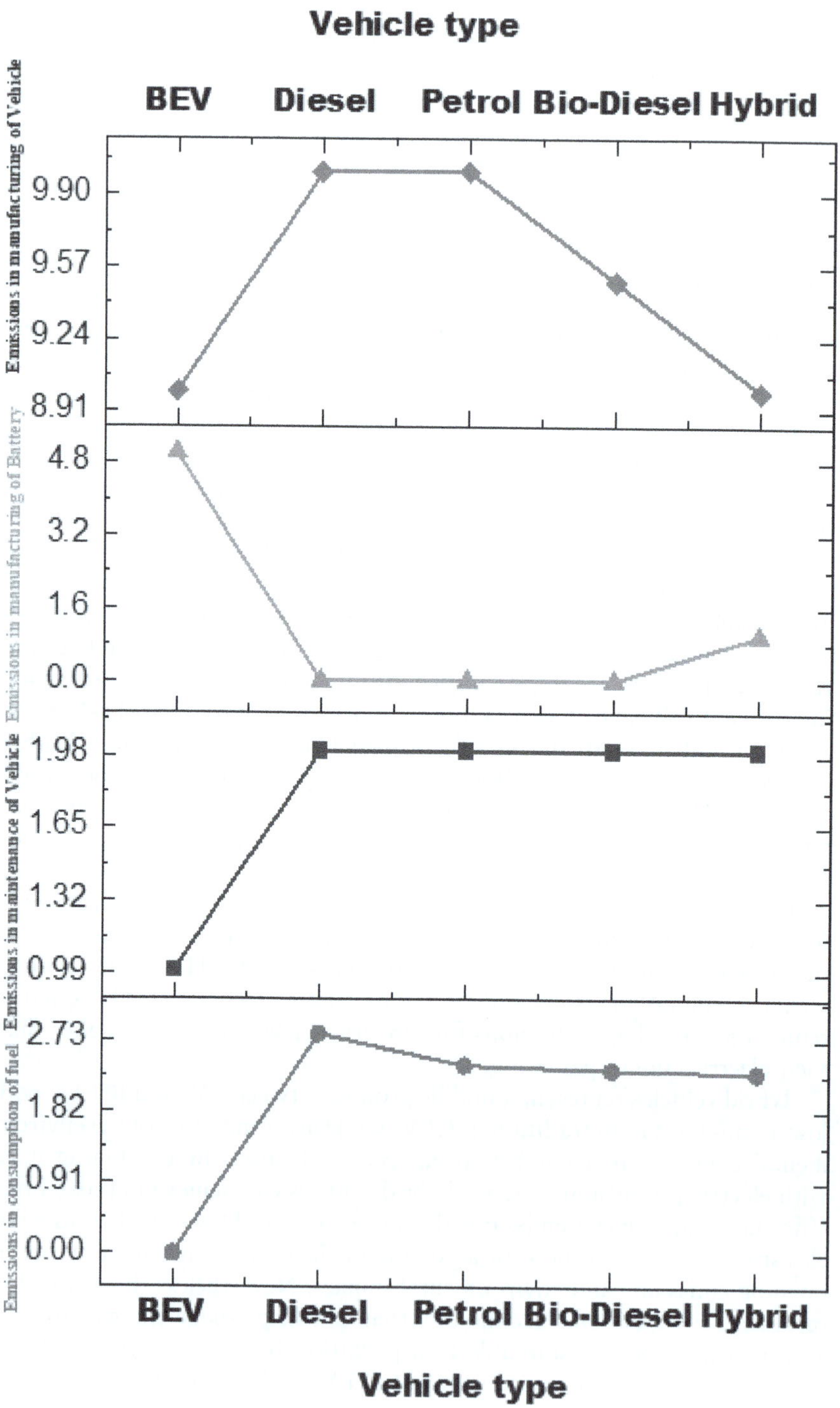


*Figure 17.3* Comparison of all vehicles.

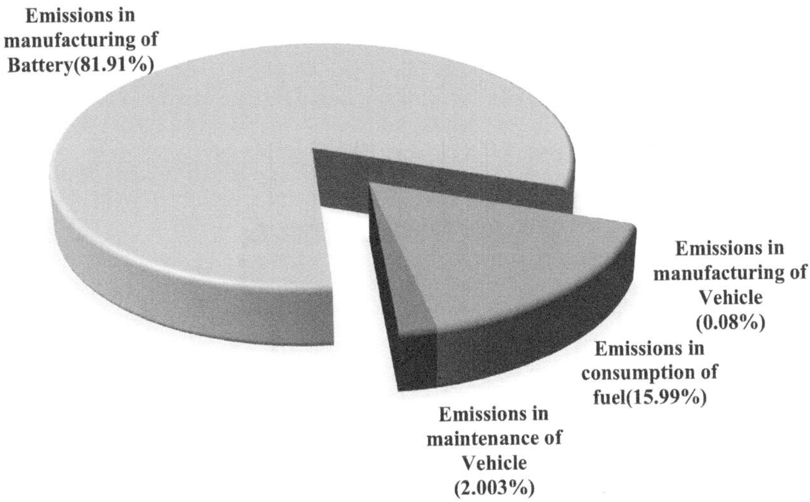


*Figure 17.4* Weightage analysis.

### 17.4.3 Weightage analysis

The application of the AHP method to normalized data revealed that hybrid vehicles demonstrate the most favorable environmental performance, characterized by the lowest emissions. Following hybrids, biodiesel vehicles demonstrated lower emissions, followed by petrol vehicles, with diesel vehicles ranking higher in emissions, as depicted in Figure 17.4. Surprisingly, BEVs were found to have the highest emissions among the considered vehicle types, despite their lack of emissions during fuel consumption. This is primarily attributed to the emissions incurred during the manufacturing of batteries. The findings emphasize the significance of considering not only direct emissions during vehicle operation but also emissions incurred throughout the vehicle life cycle. Additionally, it underscores the potential environmental advantages of hybrid and biodiesel vehicles over conventional petrol and diesel options, as well as the necessity for further enhancements in the manufacturing processes of BEVs to mitigate their environmental impact.

Hybrid vehicles achieved the highest score, followed by biodiesel and petrol vehicles, as shown in Figure 17.5. This ranking underscores the superior environmental performance of hybrid vehicles compared to biodiesel and petrol counterparts. Physically, hybrid vehicles combine the benefits of both ICEs and electric propulsion systems, resulting in lower emissions and improved fuel efficiency. The hybrid architecture allows for optimized power delivery and reduced fuel consumption during operation, contributing to lower emissions of pollutants such as carbon dioxide ($CO_2$),

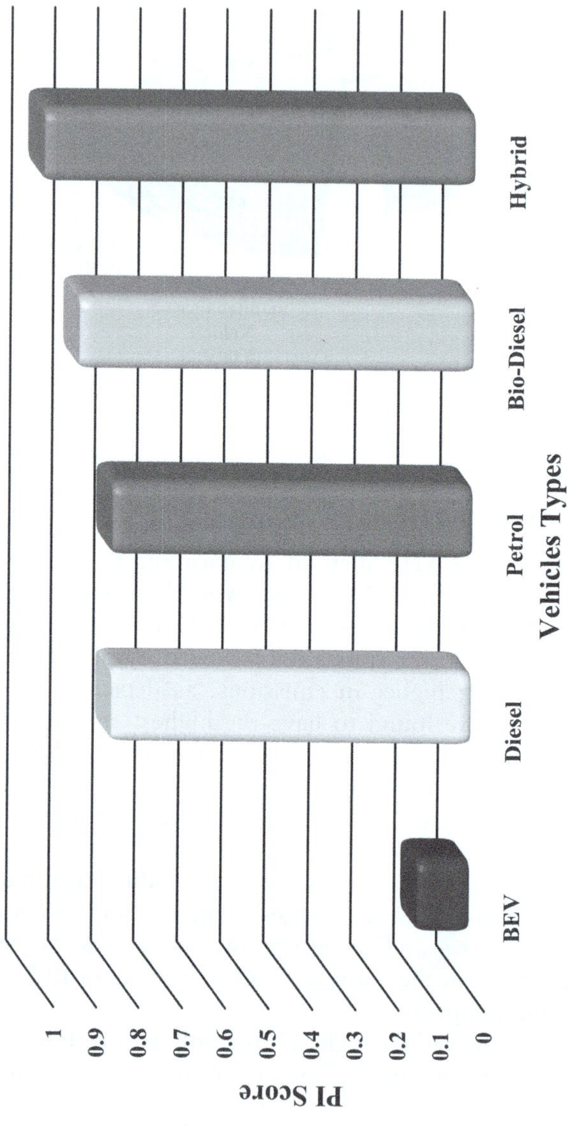


*Figure 17.5* Prioritization analysis.

unburned hydrocarbons, and carbon monoxide (CO). Although biodiesel vehicles utilize renewable fuel sources, they may still face challenges in achieving emissions reductions comparable to hybrids due to factors such as fuel composition and combustion efficiency. Similarly, petrol vehicles, although widely used, typically exhibit higher emissions than hybrids due to their reliance on fossil fuels. Thus, the higher score attained by hybrid vehicles signifies their potential to offer cleaner and more sustainable transportation solutions than biodiesel and petrol alternatives.

## 17.5 CONCLUSIONS

In today's world, it is paramount to compare vehicles based on their carbon footprint due to pressing environmental concerns such as global warming and urban pollution. With climate change increasingly impacting ecosystems and human health, reducing greenhouse gas emissions has become a global imperative. By assessing the carbon footprint of different vehicles, we can make informed decisions to mitigate their environmental impact. This comparison is crucial as transportation is a significant contributor to carbon emissions, accounting for a large portion of total emissions in many regions. Understanding the environmental impact of various vehicle technologies, from traditional ICEVs to hybrid and electric alternatives, allows policymakers, consumers, and industry leaders to prioritize low-carbon options and steer toward a more sustainable transportation future. Moreover, with advancements in technology and shifts toward renewable energy sources, accurately evaluating the carbon footprint of vehicles enables us to identify opportunities for further emissions reductions and promote cleaner mobility solutions. Therefore, comprehensive comparisons based on carbon footprint serve as a vital tool in addressing climate change and fostering sustainable development in the transportation sector.

- In the AHP method, criteria weights are assigned based on relative importance, with "emissions in manufacturing of the vehicle" likely receiving the least weight due to its consistent impact compared to parameters with greater variability.
- Hybrid vehicles emerged as the best option with the least carbon emissions due to their balanced performance across fuel consumption, maintenance, and manufacturing emissions, making them a favorable choice for reducing environmental impact.
- The AHP method is preferred for weightage calculation due to its ability to systematically prioritize criteria based on their relative importance, providing a more structured and consistent approach to decision-making.
- Hybrid vehicles were found to have the lowest carbon footprint among all vehicle types considered.

One limitation of this study is that it may not fully account for real-world variations and factors such as driving conditions, user behavior, and technological advancements over time, which could impact the accuracy of the results. The future scope of this study could involve integrating dynamic factors like advancements in vehicle technology and changes in energy sources to provide a more comprehensive and up-to-date analysis of environmental impacts.

## REFERENCES

1. Khan, O., Parvez, M., Seraj, M., Yahya, Z., Devarajan, Y., & Nagappan, B. (2024). Optimising building heat load prediction using advanced control strategies and Artificial Intelligence for HVAC system. *Thermal Science and Engineering Progress*, 49, 102484.
2. Shaad, M., Gupta, P., Parvez, M., Khan, O., & Kumari, P. (2024). Recent trends and future potential of sustainable energy efficient materials for commercial buildings. In Kulkarni, N. V. & Kharissov, B. I. (eds.), *Handbook of Emerging Materials for Sustainable Energy* (pp. 725–739). Elsevier: Amsterdam.
3. Seraj, M., Parvez, M., Khan, O., & Yahya, Z. (2024). Optimizing smart building energy management systems through industry 4.0: A response surface methodology approach. *Green Technologies and Sustainability*, 2(2), 100079.
4. Khan, O., Alsaduni, I., Equbal, A., Parvez, M., & Yadav, A. K. (2024). Performance and emission analysis of biodiesel blends enriched with biohydrogen and biogas in internal combustion engines. *Process Safety and Environmental Protection*, 183(2), 1013–1037.
5. Ahamad, T., Parvez, M., Lal, S., Khan, O., & Idrisi, M. J. (2023). 4-E analysis and multiple objective optimizations of a novel solar-powered cogeneration energy system for the simultaneous production of electrical power and heating. *Scientific Reports*, 13(1), 22246.
6. Khan, O., Mufazzal, S., Sherwani, A. F., Khan, Z. A., Parvez, M., & Idrisi, M. J. (2023). Experimental investigation and multi-performance optimization of the leachate recirculation based sustainable landfills using Taguchi approach and an integrated MCDM method. *Scientific Reports*, 13(1), 19102.
7. Khan, O., Parvez, M., Kumari, P., et al., (2023). Modelling of compression ignition engine by soft computing techniques (ANFIS-NSGA-II and RSM) to enhance the performance characteristics for leachate blends with nano-additives. *Scientific Reports*, 13, 15429. https://doi.org/10.1038/s41598-023-42353-1
8. Seraj, M., Parvez, M., Ahmad, S., & Khan, O., (2023). Sustainable energy transition and decision-making for enhancing the performance of building equipment in diverse climatic conditions. *Green Technologies and Sustainability*, 1(3), 100043.
9. Khan, F., Karimi, M. N., & Khan, O., (2023). Exploring the scalability and commercial viability of biosynthesized nanoparticles for cooling panels with the help of artificial intelligence and solar energy systems. *Green Technologies and Sustainability*, 1(3), 100036.

10. Akram, W., Parvez, M., Rafi, K. M., & Khan, O., (2022). First and second law analyses of an organic rankine cycle to recover the waste heat from the solar-operated combined cycle power plants with grid. In *International Conference on Nanotechnology: Opportunities and Challenges* (pp. 369–376). Springer Nature Singapore: Singapore.
11. Parvez, M., Khan, O., & Howari, H., (2023). Investigation of (3E) energy, exergy, and environmental analysis of syngas production for combined power and cooling plant by testing six different biomasses. *Biomass Conversion and Biorefinery*, 1–15.
12. Singh, D., Singh, S., Yadav, A. K., Khan, O., Dewangan, A., Alkahtani, M. Q., & Islam, S., (2023). From theory to practice: A sustainable solution to water scarcity by using a hybrid solar distiller with a heat exchanger and aluminum oxide nanoparticles. *ACS Omega*, 8(37), 33543–33553.
13. Khan, S. & Tomar, S. (2023). Analysis of factors influencing entrepreneurial intention among Indian women in computing industry- An NVIVO approach. *Journal of Industrial Integration and Management: Innovation and Entrepreneurship*, (In press).
14. Khan, O., Khan, M. Z., Bhatt, B. K., Alam, M. T., & Tripathi, M., (2023). Multi-objective optimization of diesel engine performance, vibration and emission parameters employing blends of biodiesel, hydrogen and cerium oxide nanoparticles with the aid of response surface methodology approach. *International Journal of Hydrogen Energy*, 48(56), 21513–21529.
15. Khan, F., Karimi, M. N., & Khan, O., (2023). Exploring the scalability and commercial viability of biosynthesized nanoparticles for cooling panels with the help of Artificial Intelligence and solar energy systems. *Green Technologies and Sustainability*, 1(3), 100036.
16. Bhatt, B. K., Akram, W., Khan, O., Parvez, M., & Ahmad, S., (2023). A methodology for multi-criteria assessment of chiller plants for building setup utilising waste heat and solar energy. *Sustainability Analytics and Modeling*, 3, 100022.
17. Nasim, M. A. G., Khan, O., Parvez, M., & Bhatt, B. K., (2023). Optimizing ultrasonic reactor operating variables using intelligent soft computing models for increased biodiesel production, *Green Technologies and Sustainability*, 1(3), 100033, https://doi.org/10.1016/j.grets.2023.100033.
18. Nasim, M. A. G., Parvez, M., Khan, O., Warsi, G. H., Hassaan, M., & Dewangan, A. K., (2023). Performance evaluation of diesel engine with fuels prepared from hydrogen and nanoparticle blended biodiesel by varying injection pressure. In *Thermal Energy Systems: Design, Computational Techniques, and Applications* (p. 231), CRC Press.
19. Khan, O., Islam, S. M. U., Hassan, T., Ahmad, H., Rahman, M. A. U., & Dewangan, A. K., (2023). Optimization of solar collector system based on different nanofluids. In *Thermal Energy Systems: Design, Computational Techniques, and Applications* (p. 41), CRC Press.
20. Nasim, M. A. G., Khan, O., Parvez, M., & Chouhan, P. S., (2023). Maximizing biodiesel production: An intelligent software-driven optimization of ultrasonic reactor operating variables. *Bioresource Technology Reports*, 101474, https://doi.org/10.1016/j.biteb.2023.101474.
21. Khan, O., Parvez, M., Kumari, P., Parvez, S., & Ahmad, S., (2023). The future of pharmacy: How AI is revolutionizing the industry, *Intelligent Pharmacy*, https://doi.org/10.1016/j.ipha.2023.04.008.

22. Fatima, M., Sherwani, N. U. K., & Singh, V., (2023). Mindful medicine: Exploring the benefits of mindfulness-based stress reduction for doctors using statistical techniques, *Intelligent Pharmacy,* https://doi.org/10.1016/j.ipha.2023.04.009.
23. Fatima, M., Sherwani, N. U. K., & Singh, V., (2023). Comparative analysis among doctors working in private and government hospitals in identifying and prioritizing essential stress factors during COVID-19- an AHP-TOPSIS approach, *Intelligent Pharmacy,* https://doi.org/10.1016/j.ipha.2023.04.005.
24. Howari, H, Parvez, M, Khan, O, Alhodaib, A, Mallah, A, & Yahya, Z., (2023). Multi-objective optimization for ranking waste biomass materials based on performance and emission parameters in a pyrolysis process—an AHP–TOPSIS Approach. *Sustainability*, 15(4), 3690. https://doi.org/10.3390/su15043690
25. Khan, O., Khan, M. Z., Alam, M. T., et al., (2023). Comparative study of soft computing and metaheuristic models in developing reduced exhaust emission characteristics for diesel engine fueled with various blends of biodiesel and metallic nanoadditive mixtures: An ANFIS–GA–HSA approach. *ACS Omega,* 8, 7344–7367. https://doi.org/10.1021/acsomega.2c05246.
26. Patra, B., Nema, P., Khan, M. Z., & Khan, O. (2023). Optimization of solar energy using MPPT techniques and industry 4.0 modelling, *Sustainable Operations and Computers*, 4, 22–28, https://doi.org/10.1016/j.susoc.2022.10.001
27. Ahmad, S., Parvez, M., Khan, T. A., & Khan, O., (2022). A hybrid approach using AHP–TOPSIS methods for ranking of soft computing techniques based on their attributes for prediction of solar radiation, *Environmental Challenges,* https://doi.org/10.1016/j.envc.2022.100634.
28. Ahmad, S., Parvez, M., Khan, T A., Siddiqui, S. A., & Khan, O., (2022). Performance comparison of solar powered cogeneration and trigeneration systems via energy and exergy analyses. *International Journal of Exergy,* 39(4), 395–409.
29. Islam, S. M. U., Khan, S., Ahmad, H., Rahman, M. A. U., Tomar, S., & Khan, M. Z., (2022). Assessment of challenges and problems in supply chain among retailers during COVID-19 epidemic through AHP-TOPSIS hybrid MCDM technique. *Internet of Things and Cyber-Physical Systems*, 2, 180–193, https://doi.org/10.1016/j.iotcps.2022.10.001.
30. Ahmad, S., Alam, M. T., Bilal, M., Khan, O., & Khan, M. Z., (2022). Analytical modelling of HVAC-IoT systems with the aid of UVGI and solar energy harvesting. In *Energy Harvesting* (pp. 65–80).
31. Khan, O., Khan, M. Z., Khan, E., Bhatt, B. K., Afzal, A., Ağbulut, U., & Shaik S., (2022). An enhancement in diesel engine performance, combustion, and emission attributes fueled with *Eichhornia crassipes* oil and copper oxide nanoparticles at different injection pressures. *Energy Sources, Part A: Recovery, Utilization, and Environmental Effects*, 44(3), 6501–6522.
32. Farzaneh, F., & Jung, S., (2023). Lifecycle carbon footprint comparison between internal combustion engine versus electric transit vehicle: A case study in the US. *Journal of Cleaner Production*, 390, 136111.
33. Saenz-Esteruelas, J., Figliozzi, M., Serrano, A., & Faulin, J., (2016). Electrifying last-mile deliveries: A carbon footprint comparison between internal combustion engine and electric vehicles. In Alba, E., Chicano, F., & Luque, G. (eds.), *Smart Cities. Smart-CT 2016.* Lecture Notes in Computer Science (vol. 9704). Springer: Cham. https://doi.org/10.1007/978-3-319-39595-1_8

34. Onat, N. C., Kucukvar, M., & Tatari, O., (2015). Conventional, hybrid, plug-in hybrid or electric vehicles? State-based comparative carbon and energy footprint analysis in the United States. *Applied Energy*, 150, 36–49.
35. Pipitone, E., Caltabellotta, S., & Occhipinti, L., (2021). A life cycle environmental impact comparison between traditional, hybrid, and electric vehicles in the European context. *Sustainability*, 13(19), 10992.
36. Khan, O., Khan, M. Z, Khan, M. E, Goyal, A., Bhatt, B. K, Khan, A., & Parvez, M., (2021). Experimental analysis of solar powered disinfection tunnel mist spray system for coronavirus prevention in public and remote places. *Materials Today: Proceedings*, 46(15), 6852–6858.
37. Zeeshan, M., Alam, M. R., Khan, O., Parvez, M., & Azad, A. S., (2021). Thermal load calculation of building envelope set up with the help of a drone setup. In *2021 First International Conference on Advances in Computing and Future Communication Technologies (ICACFCT)* (pp. 186–190). IEEE.
38. Parvez, M., Khalid, F., & Khan, O., (2020). Thermodynamic performance assessment of solar-based combined power and absorption refrigeration cycle. *International Journal of Exergy*, 31(3), 232–248.
39. Seraj, M., Khan, O., Khan, M. Z., Parvez, M., Bhatt, B. K., Ullah, A., & Alam, M. T., (2022). Analytical research of artificial intelligent models for machining industry under varying environmental strategies: An industry 4.0 approach. *Sustainable Operations and Computers*, 3, 176–187.
40. Parvez, M., & Khan, O. (2020). Parametric simulation of biomass integrated gasification combined cycle (BIGCC) power plant using three different biomass materials. *Biomass Conversion and Biorefinery*, 10(4), 803–812.
41. Meraj, M., Khan, M. E., Tiwari, G. N., & Khan, O. (2019). Optimization of electrical power of solar cell of photovoltaic module for a given peak power and photovoltaic module area. In *Advances in Fluid and Thermal Engineering* (pp. 417–433).
42. Khan, S., Tomar, S., Fatima, M., & Khan, M. Z. (2022). Impact of artificial intelligent and industry 4.0 based products on consumer behaviour characteristics: A meta-analysis-based review, *Sustainable Operations and Computers*, 3, 218–225.
43. Fatima, M., Sherwani, N. U. K., Khan, S., & Khan, M. Z. (2022). Assessing and predicting operation variables for doctors employing industry 4.0 in health care industry using an adaptive neuro-fuzzy inference system (ANFIS) approach, *Sustainable Operations and Computers*, 3, 286–295.
44. Khan, O., Khan, M. E., Parvez, M., Ahmed, K. A. A. R., & Ahmad . I. (2022). Extraction and experimentation of biodiesel produced from leachate oils of landfills coupled with nano-additives aluminium oxide and copper oxide on diesel engine. In *Nanomaterials for Innovative Energy Systems and Devices* (pp. 319–332).
45. Parvez, M., Khan, M. E., Khalid, F., Khan, O., & Akram, W., (2021). A novel energy and exergy assessments of solar operated combined power and absorption refrigeration cogeneration cycle. In *Electric Vehicles* (pp. 213–229).
46. Khan, O., Khan, M. E., Yadav, A. K., & Sharma, D., (2017). The ultrasonic-assisted optimization of biodiesel production from eucalyptus oil. *Energy Sources, Part A: Recovery, Utilization, and Environmental Effects*, 39(13), 1323–1331.

47. Khan, O., Yadav, A. K., Khan, M. E., & Parvez, M., (2019). Characterization of bioethanol obtained from Eichhornia Crassipes plant; its emission and performance analysis on CI engine. *Energy Sources, Part A: Recovery, Utilization, and Environmental Effects*, 43, 1–11.
48. Khan, O., Khan, M. Z., Ahmad, N., Qamer, A., Alam, M. T., & Siddiqui A. H., (2019). Performance and emission analysis on palm oil derived biodiesel coupled with aluminium oxide nanoparticles. *Materials Today: Proceedings*, 46, 6781–6786.
49. Ahmed A, & Yadav, A. K., (2024). Parametric analysis of wastewater electrolysis for green hydrogen production: A combined RSM, genetic algorithm, and particle swarm optimization approach, *International Journal of Hydrogen Energy*, 59, 51–62.
50. Ahmad, A., Yadav, A. K., & Hasan, S., (2024). Enhanced production of methane enriched biogas through intensified co-digestion process and its effective utilization in a biodiesel/biohydrogen fueled engine with duel injection strategies: ML-RSM based an efficient optimization approach. *International Journal of Hydrogen Energy*, 65, 671–686.
51. Ahmed A, Yadav, A. K, Singh A., & Singh D. K., (2024). A machine learning–genetic algorithm based models for optimization of intensification process through microwave reactor: A new approach for rapid and sustainable synthesis of biodiesel from novel *Hiptage benghalensis* seed oil, *Fuel*, 374, 132470.
52. Khan, O, Khan, M. P., & Yadav A. K. (2024). A comparative analysis of various nanocomposites for the remediation of heavy metals from biomass liquid digestate using multi-criteria decision methods, *Biomass and Bioenergy*, 186, 107281.
53. Ahmed A, Yadav, A. K, Singh A., & Singh D. K., (2024). A machine learning-response surface optimization approach to enhance the performance of diesel engine using novel blends of aloe vera biodiesel with MWCNT nanoparticles and hydrogen, *Process Safety and Environmental Protection*, 186, 738–755, https://doi.org/10.1016/j.psep.2024.04.013
54. Ahmed A., Yadav, A. K., Singh D.K., & Singh A., (2024). A comprehensive machine learning-coupled response surface methodology approach for predictive modeling and optimization of biogas potential in anaerobic co-digestion of organic waste, *Biomass and Bioenergy*, 180, 106995.
55. Ahmad, A., Yadav, A. K., Singh, A., Singh, D. K., & Ağbulut, Ü., (2024). A hybrid RSM-GA-PSO approach on optimization of process intensification of linseed biodiesel synthesis using an ultrasonic reactor: Enhancing biodiesel properties and engine characteristics with ternary fuel blends. *Energy*, 288, 129077.
56. Ahmed, A., Yadav, A. K., & Hasan S., (2024). Enhancing efficiency and environmental impacts of a nano-enhanced oleander biodiesel-biohydrogen dual fuel engine equipped with EGR, through operational parameter optimization using RSM-MOGA technique, *International Journal of Hydrogen Energy*, 78, 1157–1172.
57. Khan, O., Parvez, M., Yahya, Z., Alhodaib, A., Yadav, A. K., Hoang, A. T., & Ağbulut, Ü., (2024). Waste-to-energy power plants: Multi-objective analysis and optimization of landfill heat and methane gas by recirculation of leachate. *Process Safety and Environmental Protection*, 186, 957–968.

58. Khanam, S., Khan, O., Ahmad, S., Sherwani, A. F., Khan, Z. A., Yadav, A. K., & Ağbulut, Ü. (2024). A Taguchi-based hybrid multi-criteria decision-making approach for optimization of performance characteristics of diesel engine fuelled with blends of biodiesel-diesel and cerium oxide nano-additive. *Journal of Thermal Analysis and Calorimetry*, 149(8), 3657–3676.
59. Ahmed A., Yadav, A. K., & Singh A., (2024). Enhancing waste cooking oil biodiesel yield and characteristics through machine learning, response surface methodology, and genetic algorithm for optimal utilization in CI Engines, *International Journal of Green Energy*, 21, 1345–1365.
60. Singh, A. A., Singh, R. K., Yadav, A. K., Saha, A., Dewangan, A., & Singh, B. (2024). Experimental investigations on the electrochemical machining of D3 die steel material and multi-objective parameter optimization. *Surface Review and Letters*, 31(9), 1–12.
61. Ahmed A., Yadav, A. K., & Hasan S., (2024). Biogas as a sustainable and viable alternative fuel for diesel engines: A comprehensive review of production, purification, economic analysis, and performance evaluation, *Proceedings of the Institution of Mechanical Engineers, Part E: Journal of Process Mechanical Engineering*, https://doi.org/10.1177/09544089241255930
62. Khan, F., Karimi, M. N., Khan, O., Yadav, A. K., Alhodaib, A., Gürel, A. E., & Ağbulut, Ü. (2024). A hybrid MCDM optimization for utilization of novel set of biosynthesized nanofluids on thermal performance for solar thermal collectors. *International Journal of Thermofluids*, 22, 100686.
63. Alhodaib, A., Yahya, Z., Khan, O., Equbal, A., Equbal, M. S., Parvez, M., Yadav, A. K., & Idrisi, M. J. (2024). Sustainable coatings for green solar photovoltaic cells: performance and environmental impact of recyclable biomass digestate polymers. *Scientific Reports*, 14(1), 11221.
64. Ojha, A., Agarwal, A. P., Singh, R. K., Saha, A., Yadav, A. K., & Gupta, G., (2024). Fabrication and characterization of AA6082-T6 produced by friction stir additive manufacturing, *Surface Review and Letters*, https://doi.org/10.1142/S0218625X25500222
65. Equbal, A., Equbal, A., Ahmad, R., Kumar, M., Khan O., & Yadav A. K., (2024) RSM based statistical approach to enhance the compressive performance of ABS P400 parts fabricated via FDM technology, *Surface Review and Letters*, https://doi.org/10.1142/S0218625X25500210
66. Dewangan, A., Ahmad, A., & Yadav, A.K., (2024). Innovative research on waste tire recycling for sustainable biofuel production: Assessment of its usability on multi-cylinder diesel engine employing constant injection of oxy-hydrogen and biogas through a premixing device. *International Journal of Hydrogen Energy*, 84, 155–163.
67. Dewangan, A., Mallick, A., Yadav, A. K., Islam, S., Saleel, C. A., Shaik, S., & Ağbulut, Ü., (2023). Production of oxy-hydrogen gas and the impact of its usability on CI engine combustion, performance, and emission behaviors. *Energy*, 278, 127937.
68. Ahmad, A., Yadav, A. K., Singh, A., & Singh, D. K. (2023). Multi-response optimization of a microalgae-spirulina-fueled VCR diesel engine: A comprehensive RSM-GA approach. Environment, *Development and Sustainability*, 1–29.
69. Ahmed A., Yadav, A. K., & Singh A., (2023). Process optimization of spirulina microalgae biodiesel synthesis using RSM coupled GA technique: A performance study of a biogas powered dual fuel engine, *International Journal of Environmental Science and Technology*, 21, 169–188.

70. Khan, O., Ali, V., Parvez, M., Alhodaib, A., Yahya, Z., Yadav, A. K., & Ağbulut, Ü., (2024). Exploring the performance of biodiesel-hydrogen blends with diverse nanoparticles in diesel engine: A hybrid machine learning K-means clustering approach with weighted performance metrics. *International Journal of Hydrogen Energy*, 78, 547–563.
71. Ahmed A., Yadav, A. K., & Singh, A., (2023). Application of machine learning and genetic algorithm for prediction and optimization of biodiesel yield from waste cooking oil, *Korean Journal of Chemical Engineering*, 40, 2941–2956.
72. Ahmed A., Yadav, A. K., & Singh, A., (2023). Enhancement of biogas yield from dual organic waste using hybrid statistical approach and its effects on ternary fuel blend (biodiesel/n-butanol/diesel) powered diesel engine. *Environmental Progress & Sustainable Energy*, 42(5), e14163. https://doi.org/10.1002/ep.14163
73. Ahmed A., Yadav, A. K., & Singh A., (2023). Optimization of cavitation-assisted biodiesel production and fuel properties from *Neochloris oleoabundans* microalgae oil using genetic algorithm and response surface methodology, *Proceedings of the Institution of Mechanical Engineers, Part E: Journal of Process Mechanical Engineering*, https://doi.org/10.1177/09544089231159832
74. Yadav, M., Karimi, M. N., & Yadav, A. K., (2023). Effect of graphene oxide nanoparticles dispersed biodiesel on combustion, performance, and emission characteristics of a diesel engine, *Proceedings of the Institution of Mechanical Engineers, Part E: Journal of Process Mechanical Engineering*, https://doi.org/10.1177/09544089231209139
75. Das, D. K., Ahmad, A., Singh, A., Singh, D. K., & Yadav, A. K., (2023). Numerical simulation of the combustion chamber and performance analysis of microalgae-based biodiesel/diesel-powered CI engine: An axisymmetric flow approach, *Proceedings of the Institution of Mechanical Engineers, Part E: Journal of Process Mechanical Engineering*, https://doi.org/10.1177/09544089231218121
76. Dewangan, A., Mallick, A., Yadav, A. K., Ahmad, A., Alqahtani, D., & Islam, S., (2023). Combined effect of operating parameters and nano particles on performance of a diesel engine: Response surface methodology coupled genetic algorithm approach, *ACS Omega*, 8(27), 24586–24600.
77. Ahmed A., Yadav, A. K., & Singh A., (2023). Optimization of biogas yield from anaerobic co-digestion of dual waste for environmental sustainability: ANN, RSM, and GA, approach, *International Journal of Oil, Gas and Coal Technology*, 33(1), 75.
78. Ahmed A., Yadav, A. K., & Singh A., (2023). An environmental impact assessment and optimization study of biodiesel production from microalgae. *International Journal of Global Warming*, 31(3), 294–313.
79. Ahmed A., Yadav, A. K., & Singh A., (2023). Biodiesel yield optimization from a third-generation feedstock (microalgae spirulina) using a hybrid statistical approach, *International Journal of Ambient Energy*, 44(1), 1202–1213.
80. Ahmed, A., Yadav, A. K., & Singh, A., (2023). Modelling and optimisation of VCR diesel engine parameters fuelled with microalgae methyl-ester: A RSM-coupled Taguchi approach, *International Journal of Ambient Energy*, 44(1) 2498–2506.
81. Ahmed, A., Yadav, A. K., & Singh A., (2023). Biodiesel production from Mahua oil: Characterization, optimization, and modeling with a hybrid statistical approach, *International Journal of Ambient Energy*, 44(1), 2600–2609.

82. Khan, T. A, Khan, T. A., Yadav, A. K., & Khan, M. E., (2023). Experimental studies on utilization of *Neochloris oleoabundans* microalgae biodiesel as an alternative fuel for diesel engine, *International Journal of Ambient Energy*, 44(1), 115–123.
83. Singh D., Yadav A. K, Kumar A., & Samsher, (2023). Energy matrices and life cycle conversion analysis of N-identical hybrid double slope solar distiller unit using Al2O3 nanoparticle. *Journal of Water and Environmental Nanotechnology*, 8(3), 267–284.
84. Khan, T. A., Yadav, A. K., & Khan T. A., (2022). A hydrodynamic cavitation-assisted system for optimization of biodiesel production from green microalgae oil using a genetic algorithm and response surface methodology approach, *Environmental Science and Pollution Research*, 29, 49465–49477.
85. Dewangan, A., Yadav, A. K., Mallick, A., & Singh S., (2022). Comparative study of *Manilkara zapota* and Karanja based biodiesel properties and its effect on multi-cylinder diesel engine characteristics. *Energy Sources, Part A*, 44(2), 5143–5153.
86. Yadav A. K., Khan T. A., Khan T. A., & Kumar S., (2021). Methyl ester of *Gmelina arborea* oil as a substitute for petroleum diesel: An experimental study on its performance and emissions in a diesel engine, *Energy Sources, Part A*, 43(11), 1307–1314.
87. Khan, I. A., Singh S. K., Pal A., & Yadav A. K., (2020). Efficient production of biodiesel from *Cannabis sativa* oil using intensified transesterication (hydrodynamic cavitation) method, *Energy Sources, Part A*, 42(20), 2461–2470.
88. Dewangan, A., Yadav, A. K., Mallick, A., & Tyagi, A., (2020). Optimisation of biodiesel production and engine performance from Simarouba oil in compression ignition engine. *International Journal of Oil, Gas and Coal Technology*, 25(3), 357–376.
89. Khan, M. E., Khan T. A., Yadav, A. K., & Pal A., (2020). A comprehensive review on performance, emission and combustion characteristics of diesel engines fuelled with algae oil and methyl ester, *WEENTECH Proceedings in Energy*, 6(2), 160–168.
90. Yadav, A. K., Khan, M. E., & Pal A., (2019). Custard apple seed (*Annona squamosa*) oil as a promising feedstock for biodiesel production using advanced techniques and its experimental investigations on an agricultural diesel engine, *International Journal of Oil, Gas and Coal Technology*, 20(4), 473–492.
91. Yadav, A. K., Pal, A., Ghosh, U., & Gupta, S. K., (2019). Comparative study of biodiesel production methods from Yellow Oleander oil and its performance analysis on an agricultural diesel engine, *International Journal of Ambient Energy*, 40(2), 152–157.
92. Khan, I. A., Singh S. K., Yadav A. K., & Sharma D., (2019). Enhancement in the performance of a diesel engine fueled with Pongamia methyl ester and n-butanol as oxygenated additive, *International Journal of Ambient Energy*, 40(8), 842–846.
93. Yadav, A. K., Khan O., & Khan, M. E., (2018). Utilization of high FFA landfill waste (leachates) as a feedstock for sustainable biodiesel production: Its characterization and engine performance evaluation, *Environmental Science and Pollution Research*, 25(32), 32312–32320.

94. Yadav A. K., Dewangan A., & Mallick A., (2018). Effect of n-butanol and diethyl ether additives on performance and emission characteristics of a diesel engine fuelled with diesel-pongamia biodiesel blends. *Journal of Energy Engineering*, 144(6), 04018062.
95. Vinay, Singh B., & Yadav, A. K., (2020). Optimization of performance and emission characteristics of CI engine fuelled with Mahua oil methyl ester-diesel blend using response surface methodology, *International Journal of Ambient Energy*, 44(6), 674–685.
96. Yadav A. K., Dewangan A., & Mallick A., (2018). Synthesis and stability study of biodiesel from Kachnar seed oil, *Journal of Energy Engineering*, 144(5), 04018053.
97. Yadav, A. K., Pal A., & Dubey A. M., (2018). Experimental studies on utilization of *Prunus armeniaca* L. (Wild Apricot) biodiesel as an alternative fuel for CI engine, *Waste and Biomass Valorization*, 9(10), 1961–1969.
98. Yadav, A. K., Khan, M. E., & Pal A., (2018). Ultrasonic assisted production of biodiesel from Karabi oil using heterogeneous catalyst, *Biofuels*, 9(1), 101–112.
99. Yadav, A. K., & Vinay, Singh, B., (2018). Optimization of biodiesel production from *Annona squamosa* seeds oil using response surface methodology and its characterization. *Energy Sources, Part A.*, 40(9), 1051–1059.
100. Dewangan, A., Yadav A. K., & Mallick A., (2018). Current scenario of biodiesel development in India: Prospects and challenges, *Energy Sources, Part A.*, 40(20), 2494–2501.
101. Yadav, A. K., Khan, M. E., & Pal A., (2018). Performance and emission characteristics of a stationary diesel engine fuelled by *Schleichera oleosa* oil methyl ester (SOME) produced through hydrodynamic cavitation process, *Egyptian Journal of Petroleum*, 27(1), 89–93.
102. Yadav, A. K., Khan, M. E., & Pal A., (2018). Performance, emission and combustion characteristics of an Indica diesel engine operated with Yellow Oleander (*Thevetia peruviana*) oil biodiesel produced through hydrodynamic cavitation method, *International Journal of Ambient Energy*, 39(4), 365–371.
103. Dewangan, A., Mallick, A., & Yadav, A. K., (2020). Effect of metal oxide nanoparticles and engine parameters on the performance of a diesel engine: A review, *Materials Today: Proceedings*, 21, 1722–1727.
104. Yadav, A. K., Khan, M. E., Pal, A., & Dubey, A. M., (2017). Experimental investigations of performance and emissions characteristics of Kusum (*Schleichera oleosa*) biodiesel in a multi-cylinder transportation diesel engine, *Waste and Biomass Valorization*, 8(4), 1331.
105. Yadav, A. K., Khan, M. E., & Pal, A., (2017). Biodiesel production from oleander (*Thevetia Peruviana*) oil and its performance testing on a diesel engine, *Korean Journal of Chemical Engineering*, 34(2), 340–345.
106. Dewangan A., & Yadav A. K., (2017). Wax deposition during production of waxy crude oil and its remediation, *Petroleum Science and Technology*, 35(18), 1831–1838.
107. Yadav, A. K., Khan, M. E., & Pal A., (2017). Kaner biodiesel production through hybrid reactor and its performance testing on a CI engine at different compression ratios. *Egyptian Journal of Petroleum*, 26(2), 525–532.

108. Yadav, A. K., Khan, M. E., Pal A., & Dubey A. M., (2016). Performance and emission characteristics of a transportation diesel engine operated with non-edible vegetable oils biodiesel, *Case Studies in Thermal Engineering*, 8, 236–244.
109. Yadav, A. K., Khan, M. E., & Pal A., (2016). Biodiesel production from *Nerium oleander* (*Thevetia peruviana*) oil through conventional and ultrasonic irradiation methods, *Energy Sources, Part A*, 38(23), 3447–3452.
110. Yadav, A. K., Khan, M. E., & Pal A., (2016). A comparative study on ultrasonic cavitation and mechanical stirring method towards efficient production of biodiesel from non-edible oils and performance testing on a C.I. engine, *International Journal of Environment and Waste Management*, 18(4), 349–367.
111. Dewangan, A., Mallick, A., Yadav, A. K., & Kumar, R., (2020). Combustion-generated pollutions and strategy for its control in CI engines: A review. *Materials Today: Proceedings*, 21, 1728–1733.
112. Yadav, A. K., Khan, M. E., Pal A., & Sharma D., (2015). Optimization of biodiesel production from bitter groundnut oil using Taguchi method and its performance and emissions characteristics on a 4-cylinder Tata Indica engine. *International Journal of Sustainable Agricultural Management and Informatics*, 1, 285–300.
113. Khan, F., Khan, O., Parvez, M., Ahmad, S., Yahya, Z., Alhodaib, A., Yadav, A. K., & Ağbulut, Ü., (2024). K-means clustering optimization of various quantum dots and nanoparticles-added biofuels for engine performance, emission, vibration, and noise characteristics. *Thermal Science and Engineering Progress*, 54, 102815.
114. Khan, O., Parvez, M., Kumari, P., et al., (2024). Optimization of Thermal Performance in Lauric Acid-Based Phase Change Materials Using a Priority Clustering Approach, *Energy Storage*, 6, e70026. https://doi.org/10.1002/est2.70026.
115. Singh, S., Singh, B., Kumar, S., & Yadav, A. K., (2019). Temperature-dependent dynamic hysteresis scaling of ferroelectric hysteresis parameters of lead free [(Ba0.825 + xCa0.175-x) (Ti1-x Snx)O3] ceramics, *Ferroelectrics*, 551(1), 133–142.

Chapter 18

# Agro-residue-derived biochar for environmental management

## Conversion techniques and applications

*Anshul Tyagi, Deepika, and Anil Kumar Haritash*

## 18.1 INTRODUCTION

Agriculture and its allied sectors contribute 17.32% to India's gross domestic product and provide livelihoods to approximately 60% of the population [1]. Agrochemicals, high-yielding variety seeds, and machinery have significantly increased agriculture production. However, as agricultural production increases, so does the pollution associated with the sector [2,3]. Modern agricultural practices contribute directly to carbon emissions through fuel use in on-farm activities like ploughing [4], pumping of water for irrigation [5], and the production, transportation, and use of chemical fertilizers [6]. While chemical fertilizers enhance crop yields, poor management can have adverse impacts on the environment [7,8]. The production of agrochemicals and mismanaged agro-waste are major sources of carbon emissions within the agriculture sector [9]. Excessive use of fertilizers not only contributes to climate change but also leads to nutrient losses, causing soil pollution, surface and groundwater contamination [10,11], and eutrophication of freshwater systems and coastal zones [12,13]. Furthermore, agriculture residue management remains a major concern in India. According to the Indian Ministry of New and Renewable Energy, 500 million tons of agro-residue is generated annually in India. While 70%–75% of this waste is used for industrial processes, fodder, fuel, etc., a surplus of 140 million tons is burned, causing environmental hazards. Crop residue burning releases hazardous air pollutants such as particulate matter, ammonia, $SO_2$, $NO_x$, and $CO_2$ [14–16], leading to the loss of nutrients like nitrogen and organic carbon that would otherwise be retained in the soil [17]. Reportedly, 1.41 Mt of $CO_2$ is released per Mt of crop residue burned [18]. Given the current scenario, it is evident that intensive agriculture practices, combined with increasing food demands due to population growth, have placed tremendous pressure on agriculture, significantly increasing waste and carbon emissions from the sector. This increases underscores the need for technological interventions to manage agro-waste for sustainable production from agroecosystems.


DOI: 10.1201/9781003521341-18

In the past few years, biochar, a carbonized biomass derived from various lignocellulosic materials such as agro-waste and wood, has drawn attention as a low-cost and environment-friendly approach to managing agricultural waste. Biochar is considered a promising negative emission technology. Due to its large surface area, catalysis, and recalcitrant properties, biochar provides a range of environmental applications, including soil remediation [19], wastewater treatment [20,21], and carbon sequestration [22,23].

The application of biochar to improve environmental quality has gained significant interest in the past few years. However, most previous studies have focused on the synthesis of biochar from materials like wood, bamboo, and biosolids. The use of agro-residue for biochar synthesis is still in its infancy. This study aims to analyze and evaluate the versatile potential of agro-residue-derived biochar across different environmental matrices including air, water, and soil. The role of biochar in the context of climate change and sustainability is also discussed. This chapter aims to analyze and summarize the environmental applications of biochar based on previous research including pollution abatement, soil remediation, soil quality improvement, and carbon sequestration. To better understand the recent developments in biochar application, synthesis techniques, and characteristics of biochar derived from agriculture waste are also discussed.

## 18.2 PROCESSES FOR SYNTHESIS OF BIOCHAR FROM AGRO-WASTE

Biomass feedstock and production conditions are the most important factors regulating biochar synthesis [24,25]. The composition of feedstock significantly impacts the biochar yield. Feedstocks with a higher lignin content produce higher biochar yield than cellulose or hemicellulose. However, hemicellulose and cellulose content contribute to the formation of functional groups by thermal modification [26]. Synthesis conditions including residence time, heating rate, and temperature affect the physicochemical properties of the biochar produced. Among the different production conditions, thermochemical conversion is the most common technique for converting agro-residues into biochar [25,26]. The method includes fast and slow pyrolysis, hydrothermal and flash carbonization, gasification, and torrefaction (Table 18.1). Pyrolysis is the thermal decomposition of organic substances like crop residue, wood chips, and sewage sludge at temperatures ranging from 250°C to 900°C under oxygen-free conditions. Bio-oil and syngas are also produced along with biochar, and the concentration of the end products depends on the experimental conditions [27]. In this process, the lignocellulosic compounds in the biomass feedstock (lignin, cellulose, and hemicellulose) undergo depolymerization and fragmentation to yield solid (biochar), liquid (bio-oil), and gaseous products (syngas), along with $CO_2$, $H_2$, and carbon monoxide (CO). Generally, an increase in temperature is negatively correlated with biochar yield and positively correlated with syngas production [28].

*Table 18.1* Overview of methods of synthesis of biochar using agro-waste

| *Method* | *Heating rate* | *Temperature (°C)* | *Residence time* | *% Yield* | *References* |
|---|---|---|---|---|---|
| Slow pyrolysis | 0.1–1°C/ seconds | 300–700 | 1 hour – days | 35 | [38] |
| Fast pyrolysis | 10–200°C/ seconds | 500–1,000 | 1–2 seconds | 12 | [39] |
| Torrefaction | ≤50°C/ minutes | 290 | 20–120 minutes | 80 | [40] |
| Gasification | 50–100°C/ minutes | 600–1,200 | 30–60 minutes | 10 | [41] |
| Hydrothermal carbonization | - | 180–300 | 1–16 hours | 70 | [42] |
| Flash carbonization | >1000°C/ seconds | 300–600 | 30 minutes | 37 | [43] |

Furthermore, pyrolysis can be categorized into four types depending on residence time, pyrolysis temperature, and heating rate.

Slow pyrolysis is the process of synthesis of biochar at 300°C–700°C with a long residence time and a heating rate of 0.1–1°C/seconds. This process yields about 35%–40% biochar. The conditions help produce approximately equal proportions of solid, liquid, and gaseous products. On the other hand, fast pyrolysis employs short residence times and operates at temperatures ranging from 500°C to 1,000°C. The heating rate ranges from 10°C/seconds to 200°C/seconds. Bio-oil is the dominant product of the process, yielding 75%, 13%, and 12% of bio-oil, non-condensable gases, and biochar, respectively. Therefore, the primary focus of fast pyrolysis is the production of bio-oil, rather than solid or any gaseous product. Slow pyrolysis is economically viable for biochar production due to its lower investment costs, higher yield, and wide applicability in the agriculture sector [24,28,29].

Gasification is the partial combustion of biomass at high temperatures of 700°C–800°C in an oxygen-controlled environment. This process leads to the formation of syngas and producer gas, along with a low yield of biochar (10%) [30]. However, the biochar derived from gasification is not suitable for environmental remediation as it contains inorganic compounds and polyaromatic hydrocarbons, which are potential environmental toxicants.

Hydrothermal carbonization operates at relatively low temperatures (180°C–260°C) [31]. The process occurs under saturation vapor pressure of water, with a residence time ranging from 1 to 16 hours. As the process takes place in the presence of water, it eliminates the pre-drying of feedstock, thereby potentially reducing the energy requirements and costs of biochar production. The biochar yield from hydrothermal carbonization is 70% [32].

Torrefaction is performed at 200°C–300°C under inert conditions at atmospheric pressure, with residence times of 20–120 minutes. It is a mild pyrolysis process that produces a higher product yield (70%–80%). However, the torrefied biomass has properties of both raw material and biochar; therefore, it is not used as a stand-alone process for biochar production. Instead, it is applied as a pre-treatment to improve the characteristics of biomass feedstock such as the removal of moisture and densification of biomass [33–36]. Combining torrefaction pre-treatment with pyrolysis has been shown to produce a high yield of biochar with improved physiochemical properties [37].

## 18.3 EFFECT OF BIOMASS FEEDSTOCK AND TEMPERATURE ON BIOCHAR PHYSICOCHEMICAL PROPERTIES

The properties of biochar can fluctuate significantly depending on the elemental composition. Biochar can exhibit reducing or oxidizing properties, be hydrophilic or hydrophobic, and have a basic or acidic nature. Carbon (60%–70%) is the predominant element of biochar, followed by hydrogen and oxygen. The carbon is stacked as stable aromatic rings of alcohols, esters, and fatty acids. Other elements include oxygen, nitrogen, sulfur, sodium, potassium, magnesium, and phosphorous. The content of these elements is highly dependent on the type of biomass feedstock and synthesis conditions [28]. At higher pyrolysis temperatures, volatile compounds, organic surface functional groups, acidic constituents, and water are lost [44]. This decreases the hydrogen and oxygen content while concentrating other elements in the final product. Increased temperatures also increases the surface area due to the shrinking of the solid matrix and large pores. Therefore, feedstocks containing a higher amount of cellulose exhibit the lowest biochar mass yield, as they lack aromatic rings and contain high concentrations of functional groups like OH, which are easily volatilized during pyrolysis. The availability of nitrogen and phosphorous is highly dependent on and negatively correlated with the pyrolysis temperature [45].

The raw materials used for biochar generation can be classified as woody or non-woody. Woody biomass, derived from tree and forestry residues, has higher bulk density and calorific value, fewer voids, and lower moisture and ash content. On the other hand, non-woody biomass, derived from agricultural residues like rice husk, wheat straw, and sugarcane bagasse, exhibits low calorific value, high ash and moisture content, and low bulk density. Biochar derived from agriculture residue tends to have lower carbon and higher plant-available nutrients than wood-derived biochar [46]. Furthermore, biochar from crop residue and garden wastes has higher cation exchange capacity (CEC) and pH due to oxygenated surface and inorganic functional groups formed during the pyrolysis process. The increased CEC can be also be attributed to an increase in pH and higher ash content, which contains insoluble precipitates that act as binding sites [28].

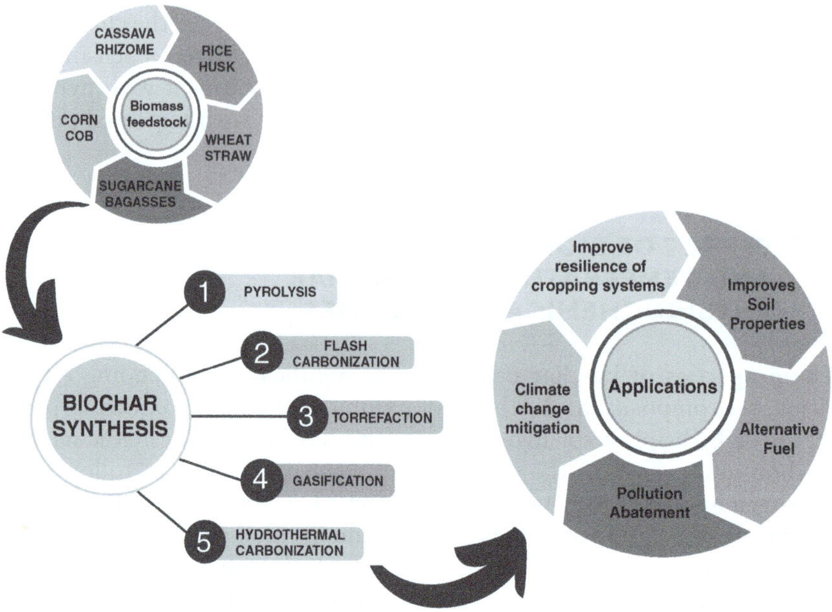


*Figure 18.1* Biomass feedstock for biochar synthesis, processes, and applications.

## 18.4 APPLICATIONS OF AGRO-RESIDUE-BASED BIOCHAR

Biochar has gained significant attention for its vast application in waste management, sustainable agriculture, and energy production (Figure 18.1). In agriculture, biochar significantly improves soil health by enhancing fertility, water retention, and microbial activity, which collectively contribute to increased crop yields. It also plays an important role in pollution control by adsorbing heavy metals and organic pollutants from soils and water bodies, reducing their bioavailability and toxicity. Conversion of agro residue to biochar not only helps in waste management but also provides as an excellent source of renewable energy. it is also utilized for carbon sequestration due to its ability to store carbon for long periods, thus mitigating climate change.

### 18.4.1 Water and wastewater treatment

The application of biochar in wastewater management has emerged as a highly efficient technique for treating a range of contaminants. In the field of wastewater treatment, biochar can act as a catalyst for the removal of emerging contaminants of concern like antibiotics [47] and dyes [47,48], filter media for the removal of suspended debris [49] and heavy metals [50], and additive media in anaerobic digestion. Numerous studies have demonstrated the potential of biochar to reduce heavy metal contamination by decreasing the bioavailability of heavy metals, thereby improving the water quality [51–53]. Table 18.2 presents the potential of agro-waste biochar to produce wastewater treatment. The intrinsic properties of biochar, such as high CEC and surface area, coupled with oxygen-rich functional groups help in removing inorganic and organic contaminants. Biochar synthesized at lower temperatures has a high surface area and micropores, which are effective for sorption of organic contaminants [41]. In addition to surface area and porosity, oxygen-containing functional groups and inorganic components in biochar catalyze the process of adsorption of contaminants like dyes onto biochar. Polar compounds, such as norflurazon and fluridone, are adsorbed on the oxygen-containing moieties by hydrogen bonding [54]. The adsorption mechanism of contaminants on biochar includes surface complexation, physical adsorption, ion exchange, and electrostatic interactions.

### 18.4.2 Soil amendment and carbon sequestration

In recent years, biochar has emerged as a soil amendment due to its ability to immobilize heavy metals and organic contaminants [58]. Soil pollution abatement using agro-waste biochar is dependent on soil type, amendment rates, and type of pollutant species. The immobilization of heavy metals and other contaminants is facilitated through decomposition, surface adsorption, hydrogen bonding, and precipitation [59–61]. Biochar helps in sustainable agricultural systems by increasing crop yield and improving soil quality. The benefits as soil amendment include long-term carbon storage, decreased biomass decomposition, and improved physicochemical and biological properties [62]. Moreover, biochar reduces soil acidity and increases the availability of micro- and macronutrients for plants. Its porous and chemically reactive surface enhances the soil nutrient and water-holding capacity, which supports the development of beneficial bacteria and fungal hyphae. Several studies have reported the positive effect of microflora on soil carbon and crop yield [63].

### 18.4.3 Climate change mitigation

Apart from its application as a soil quality enhancer, biochar has significant potential to mitigate climate change by sequestrating carbon and reducing greenhouse gas (GHG) emissions. The carbon in biochar is highly stable and resistant to decomposition. When applied to the soil, it is resistant to bacterial decomposition due to its condensed aromatic structure and forms a recalcitrant pool that keeps carbon sequestered for a long period of time. Moreover, the synthesis of biochar from agriculture residues can significantly reduce $CO_2$ and $CH_4$ emissions that would have otherwise been

*Table 18.2* Application of biochar in pollution abatement

| *Pollutant class* | *Target contaminant* | *Biomass feedstock* | *Method of production* | *Removal ability* | *References* |
|---|---|---|---|---|---|
| Heavy metals | Pb and Cd | Rice husk | Pyrolysis 550°C, 2 hours | 96.4% and 94.7% removal efficiency for Pb and Cd, respectively, from wastewater | [51] |
| | Pb | Wheat straw | Pyrolysis, 300°C and 700°C, 4 hours | 26.8%–40.1% removal efficiency of Pb from contaminated soil at 300°C and 25.1%–42% removal efficiency at 700°C | [52] |
| | Cr | Walnut shell | Pyrolysis, 450°C, 2 hours | Biochar dose of 0.7 g/L led to 93% Cr removal from wastewater | [53] |
| Dyes | Reactive red 141 | Peanut shells | Pyrolysis, 800°C, 1 hour | 85% efficiency was achieved using peanut biochar as adsorbent and 23% using raw peanut shells | [48] |
| Pesticides | Thiacloprid | Maize straw | Pyrolysis at 300°C, 500°C and 700°C, 4 hours | Maximum removal rate of 8 mg/gm | [31] |
| Inorganic ions | $PO_4^{2-}$ | Rice husk and wood | Pyrolysis | Maximum removal rate of 25–28 mg/g | [55] |
| | $NH^{4+}$ | Bamboo | Pyrolysis, 370°C | Removal rate 6.4 mM/gm | [56] |
| Antibiotic | Ciprofloxacin | Sugarcane bagasse | Carbonization 60°C, 24 hours | The absorption capacity of 16 mg/gm at equilibrium | [47] |
| | Sulfamethoxazole | Corn cob | Carbonization 60°C, 24 hours | Removal efficiency of 85% at equilibrium | [57] |

released through natural decomposition or burning of agro-waste. A global meta-analysis by Ref. [64] found that biochar amendments in agriculture soils significantly reduced $CO_2$, $N_2O$, and $CH_4$ emissions. For example, applying biochar at a rate of 4.2 t/ha/year in sugarcane fields increased soil carbon sequestration by 2.35±0.4t C/ha/year [22]. Soils in arid and semi-arid regions are increasingly affected by water scarcity, especially during the lean periods. Biochar acts as a mulch layer and enhances water retention and improves plant-available water-holding capacity. This reduces the irrigation cycle and ensures the year-round availability of water for crops [65].

### 18.4.4 Alternative fuel

Biochar has emerged as a viable and clean alternative energy source [66,67]. Overall, fifty percent of the energy from biomass is stored in biochar, making it a highly carbonaceous compound with an energy content even higher than that of high-rank coal. Additionally, the thermochemical treatment of agro-residue releases less concentrations of $SO_2$ and $NO_x$ than fossil fuels. Furthermore, the heterogeneous reaction of solid carbonaceous compounds with oxygen is slower, easier to control, and therefore safer than homogeneous oxidation. Compared to raw agro-residue, biochar has a higher ignition temperature and is combustible at a wide temperature range. These factors promote combustion efficiency and reduce the emission of toxic gases. Swapan and Gautam [68], demonstrated that biochar has higher carbon content and gross calorific value, while possessing lower volatile matter and ash content than raw biomass [69]. This makes biochar a better and cleaner fuel than raw biomass. Furthermore, biochar plays a significant role as a catalyst in biodiesel production. Chemically activated biochar used for biodiesel production results in high yields [70]. The biochar-based catalyst can be reused and recycled, contributing to the sustainable production of biodiesel (Table 18.3).

## 18.5 CONCLUSIONS AND FUTURE SCOPE

Biochar has emerged in recent decades as a promising tool, offering a plethora of environmental and socioeconomic benefits. Its properties and applicability significantly depend on synthesis conditions like temperature, biomass feedstock, heating rate, and residence time. Biochar has found extensive application as an absorbent in wastewater treatment, as a catalyst, and as a biofilter for the removal of gaseous pollutants, and to improve soil quality and crop productivity, as well as sequester carbon in soils. Due to its high calorific value, it can be used as a clean and cost-effective source of fuel. Its application in nutrient-deficient soil enhances crop yield by promoting plant nutrient uptake as it contains nutrients essential for plant growth.

*Table 18.3* Applications of biochar in environment management

| *S. no.* | *Environment matrix/component* | *Application* | *References* |
|---|---|---|---|
| 1. | Water | Wastewater treatment | [20,51,53] |
| | | Removal of contaminants of emerging concern | [47,57] |
| | | Nutrient removal | [55,56] |
| | | Water retention in soil | [62] |
| 2. | Soil | Soil conditioner/fertilizer | [71] |
| | | Immobilization of heavy metals | [58] |
| | | Improving physicochemical properties | [72] |
| | | Adsorption of toxic contaminants | [73] |
| 3. | Air | Use as biofilter | [74] |
| | | Reduce oxidative damage on plants | [75] |
| 4. | Vegetation | Crop yield and productive enhancement | [63] |
| | | Increase leaf photosynthesis rate | [76] |

Additionally, it increases the water retention capacity of the soil, thereby decreasing the number of irrigation cycles and retaining water during lean periods. As a soil amendment, biochar also reduces GHG emissions by decreasing the use of chemical fertilizers and directly adding carbon to the soil. Biochar derived from agro-residues promotes regenerative agriculture and provides a sustainable option for waste utilization.

## REFERENCES


1. GOI (2021) Economic Survey of India 2020-2021, **2** (9), 319–345. Accessed from: https://www.indiabudget.gov.in/budget2021-22/economicsurvey/doc/echapter_vol2.pdf.
2. Liu, Y., Wang, P., Gojenko, B., Yu, J., Wei, L., Luo, D. and Xiao, T. (2021) A review of water pollution arising from agriculture and mining activities in Central Asia: Facts, causes and effects. *Environmental Pollution* **291**, 118209.
3. Harizanova-Bartos, H. and Stoyanova, Z. (2018) Impact of agriculture on air pollution. *CBU International Conference Proceedings* 6, 1071–1076.
4. Willems, A. B., Augustenborg, C. A., Hepp, S., Lanigan, G., Hochstrasser, T., Kammann, C. and Müller, C. (2011) Carbon dioxide emissions from spring ploughing of grassland in Ireland. *Agric Ecosyst Environ* **144**(1), 347–351.
5. Mishra, V., Asoka, A., Vatta, K. and Lall, U. (2018) Groundwater depletion and associated $CO_2$ emissions in India. *Earths Future* **6**(12), 1672–1681.
6. Menegat, S., Ledo, A. and Tirado, R. (2022) Greenhouse gas emissions from global production and use of nitrogen synthetic fertilisers in agriculture. *Scientific Reports* **12**(1), 14490.
7. Li, D. P. and Wu, Z. J. (2008) Impact of chemical fertilizers application on soil ecological environment. *Ying Yong Sheng Tai Xue Bao* **19**(5), 1158–1165.

8. Chai, R., Ye, X., Ma, C., Wang, Q., Tu, R., Zhang, L. and Gao, H. (2019) Greenhouse gas emissions from synthetic nitrogen manufacture and fertilization for main upland crops in China. *Carbon Balance Manag* **14**(1), 20.
9. Kitamura, R., Sugiyama, C., Yasuda, K., Nagatake, A., Yuan, Y., Du, J., Yamaki, N., Taira, K., Kawai, M. and Hatano, R. (2021) Effects of three types of organic fertilizers on greenhouse gas emissions in a grassland on andosol in Southern Hokkaido, Japan. *Frontiers in Sustainable Food Systems* **5**, 649613.
10. Bijay-Singh and Craswell, E. (2021) Fertilizers and nitrate pollution of surface and ground water: an increasingly pervasive global problem. *SN Applied Sciences* 3(4), 518.
11. Deepika and Haritash, A. K. (2023) Phytoremediation potential of ornamental plants for heavy metal removal from contaminated soil: A critical review. *Horticulture, Environment, and Biotechnology* 64, 709–734.
12. Yang, X., Wu, X., Hao, H. and He, Z. (2008) Mechanisms and assessment of water eutrophication. *Journal of Zhejiang University Science B* **9**(3), 197–209.
13. Shan, V., Singh, S. K. and Haritash, A. K. (2020) Water crisis in the Asian countries: Status and future trends. In Kumar, M., Munoz-Arriola, F., Furumai, H. and Chamind, T. (eds.), *Resilience, Response, and Risk in Water Systems* (pp. 173–194). Singapore: Springer Singapore.
14. Lan, R., Eastham, S. D., Liu, T., Norford, L. K. and Barrett, S. R. H. (2022) Air quality impacts of crop residue burning in India and mitigation alternatives. *Nature Communications* **13**(1), 6537.
15. Bhuvaneshwari, S., Hettiarachchi, H. and Meegoda, J. (2019) Crop residue burning in India: Policy challenges and potential solutions. *International Journal of Environmental Research and Public Health* **16**(5), 832.
16. Raza, M. H., Abid, M., Faisal, M., Yan, T., Akhtar, S. and Adnan, K. M. M. (2022) Environmental and health impacts of crop residue burning: Scope of sustainable crop residue management practices. *International Journal of Environmental Research and Public Health* **19**(8), 4753.
17. Ministry of Agriculture (GOI) (2014) National Policy for Management of Crop Residues. Department of Agriculture & Cooperation. Government of India. Accessed from: https://agriwelfare.gov.in/Documents/NPMCR_1.pdf
18. Chandra, M. S., Naresh, R. K., Rajput, P., Chandana, P., Kumar Singh, P. and Shivashankar, K. (2020) Carbon footprint as prominent indicator of agricultural sustainability in diverse agro-climatic zones of Northern India: A critical review. *International Journal of Current Microbiology and Applied Sciences* **9**(11), 3077–3102.
19. El-Naggar, A., Ahmed, N., Mosa, A., Niazi, N. K., Yousaf, B., Sharma, A., Sarkar, B., Cai, Y. and Chang, S. X. (2021) Nickel in soil and water: Sources, biogeochemistry, and remediation using biochar. *Journal of Hazardous Materials* **419**, 126421.
20. Enaime, G., Baçaoui, A., Yaacoubi, A. and Lübken, M. (2020) Biochar for wastewater treatment: Conversion technologies and applications. *Applied Sciences* **10**(10), 3492.
21. Wang, X., Guo, Z., Hu, Z. and Zhang, J. (2020) Recent advances in biochar application for water and wastewater treatment: A review. *PeerJ* **8**, e9164.

22. Lefebvre, D., Williams, A., Meersmans, J., Kirk, G. J. D., Sohi, S., Goglio, P. and Smith, P. (2020) Modelling the potential for soil carbon sequestration using biochar from sugarcane residues in Brazil. *Scientific Reports* **10**(1), 19479.
23. Wijitkosum, S. and Jiwnok, P. (2019) Elemental composition of biochar obtained from agricultural waste for soil amendment and carbon sequestration. *Applied Sciences* **9**(19), 3980.
24. Sahoo, S. S., Vijay, V. K., Chandra, R. and Kumar, H. (2021) Production and characterization of biochar produced from slow pyrolysis of pigeon pea stalk and bamboo. *Cleaner Engineering and Technology* **3**, 100101.
25. Tomczyk, A., Sokołowska, Z. and Boguta, P. (2020) Biochar physicochemical properties: Pyrolysis temperature and feedstock kind effects. *Reviews in Environmental Science and Bio/Technology* **19**(1), 191–215.
26. Mukherjee, A., Patra, B. R., Podder, J. and Dalai, A. K. (2022) Synthesis of biochar from lignocellulosic biomass for diverse industrial applications and energy harvesting: Effects of pyrolysis conditions on the physicochemical properties of biochar. *Frontiers in Materials* 9, 870184.
27. Oni, B. A., Oziegbe, O. and Olawole, O. O. (2019) Significance of biochar application to the environment and economy. *Annals of Agricultural Sciences* **64**(2), 222–236.
28. Yaashikaa, P. R., Kumar, P. S., Varjani, S. and Saravanan, A. (2020) A critical review on the biochar production techniques, characterization, stability and applications for circular bioeconomy. *Biotechnology Reports* 28, e00570.
29. Yuan, T., He, W., Yin, G. and Xu, S. (2020) Comparison of bio-chars formation derived from fast and slow pyrolysis of walnut shell. *Fuel* **261**, 116450.
30. You, S., Ok, Y. S., Tsang, D. C. W., Kwon, E. E. and Wang, C. H. (2018) Towards practical application of gasification: A critical review from syngas and biochar perspectives. *Critical Reviews in Environmental Science and Technology* **48**(**22–24**), 1165–1213.
31. Zhang, Y., Minaret, J., Yuan, Z., Dutta, A. and Xu, C. (2018) Mild hydrothermal liquefaction of high water content agricultural residue for bio-crude oil production: A parametric study. *Energies (Basel)* **11**(11), 3129.
32. Lee, J., Lee, K., Sohn, D., Kim, Y. M. and Park, K. Y. (2018) Hydrothermal carbonization of lipid extracted algae for hydrochar production and feasibility of using hydrochar as a solid fuel. *Energy* **153**, 913–920.
33. Chen, D., Li, J., Zhang, T., Li, S., Wang, J., Niu, W., Liu, Y., Zheng, A. and Zhao, Z. (2022) Advancing biomass pyrolysis by torrefaction pretreatment: Linking the productions of bio-oil and oxygenated chemicals to torrefaction severity. *Fuel* **330**, 125514.
34. Zheng, A., Zhao, Z., Chang, S., Huang, Z., Wang, X., He, F. and Li, H. (2013) Effect of torrefaction on structure and fast pyrolysis behavior of corncobs. *Bioresource Technology* **128**, 370–377.
35. Chew, J. J. and Doshi, V. (2011) Recent advances in biomass pretreatment: Torrefaction fundamentals and technology. *Renewable and Sustainable Energy Reviews* **15**(8), 4212–4222.
36. Chen, D., Li, J., Zhang, T., Li, S., Wang, J., Niu, W., Liu, Y., Zheng, A. and Zhao, Z. (2022) Advancing biomass pyrolysis by torrefaction pretreatment: Linking the productions of bio-oil and oxygenated chemicals to torrefaction severity. *Fuel* 330, 125514.

37. Zhu, X., Luo, Z., Diao, R. and Zhu, X. (2019) Combining torrefaction pretreatment and co-pyrolysis to upgrade biochar derived from bio-oil distillation residue and walnut shell. *Energy Conversion and Management* **199**, 111970.
38. Duku, M. H., Gu, S. and Hagan, E. B. (2011) Biochar production potential in Ghana: A review. *Renewable and Sustainable Energy Reviews* **15**(8), 3539–3551.
39. Bridgwater, A. V. (2012) Review of fast pyrolysis of biomass and product upgrading. *Biomass and Bioenergy* **38**, 68–94.
40. Mondal, M., Goswami, S., Ghosh, A., Oinam, G., Tiwari, O. N., Das, P., Gayen, K., Mandal, M. K. and Halder, G. N. (2017) Production of biodiesel from microalgae through biological carbon capture: a review. *3 Biotech* **7**(2), 99.
41. Lee, J., Sarmah, A. K. and Kwon, E. E. (2019) Production and formation of biochar. In Ok, Y. S. et al. (eds.), *Biochar from Biomass and Waste* (pp. 3–18). Amsterdam: Elsevier.
42. Brown, A. E., Adams, J. M. M., Grasham, O. R., Camargo-Valero, M. A. and Ross, A. B. (2020) An assessment of different integration strategies of hydrothermal carbonisation and anaerobic digestion of water hyacinth. *Energies (Basel)* **13**(22), 5983.
43. Hagemann, N., Spokas, K., Schmidt, H.-P., Kägi, R., Böhler, M. and Bucheli, T. (2018) Activated carbon, biochar and charcoal: Linkages and synergies across pyrogenic carbon's ABCs. *Water (Basel)* **10**(2), 182.
44. Chatterjee, R., Sajjadi, B., Chen, W.-Y., Mattern, D. L., Hammer, N., Raman, V. and Dorris, A. (2020) Effect of pyrolysis temperature on physicochemical properties and acoustic-based amination of biochar for efficient $CO_2$ adsorption. *Frontiers in Energy Research* 8, 85.
45. Cantrell, K. B., Hunt, P. G., Uchimiya, M., Novak, J. M. and Ro, K. S. (2012) Impact of pyrolysis temperature and manure source on physicochemical characteristics of biochar. *Bioresource Technology* **107**, 419–428.
46. Zhang, M., Shu, L., Shen, X., Guo, X., Tao, S., Xing, B. and Wang, X. (2014) Characterization of nitrogen-rich biomaterial-derived biochars and their sorption for aromatic compounds. *Environmental Pollution* **195**, 84–90.
47. Peñafiel, M. E., Matesanz, J. M., Vanegas, E., Bermejo, D., Mosteo, R. and Ormad, M. P. (2021) Comparative adsorption of ciprofloxacin on sugarcane bagasse from Ecuador and on commercial powdered activated carbon. *Science of the Total Environment* **750**, 141498.
48. Zazycki, M. A., Godinho, M., Perondi, D., Foletto, E. L., Collazzo, G. C. and Dotto, G. L. (2018) New biochar from pecan nutshells as an alternative adsorbent for removing reactive red 141 from aqueous solutions. *Journal of Cleaner Production* **171**, 57–65.
49. Murtaza, G., Ahmed, Z., Dai, D.-Q., Iqbal, R., Bawazeer, S., Usman, M., Rizwan, M., Iqbal, J., Akram, M. I., Althubiani, A. S., Tariq, A. and Ali, I. (2022) A review of mechanism and adsorption capacities of biochar-based engineered composites for removing aquatic pollutants from contaminated water. *Frontiers in Environmental Science* **10**, 1035865.
50. Ambaye, T. G., Vaccari, M., van Hullebusch, E. D., Amrane, A. and Rtimi, S. (2021) Mechanisms and adsorption capacities of biochar for the removal of organic and inorganic pollutants from industrial wastewater. *International Journal of Environmental Science and Technology* **18**(10), 3273–3294.

51. Amen, R., Yaseen, M., Mukhtar, A., Klemeš, J. J., Saqib, S., Ullah, S., Al-Sehemi, A. G., Rafiq, S., Babar, M., Fatt, C. L., Ibrahim, M., Asif, S., Qureshi, K. S., Akbar, M. M. and Bokhari, A. (2020) Lead and cadmium removal from wastewater using eco-friendly biochar adsorbent derived from rice husk, wheat straw, and corncob. *Cleaner Engineering and Technology* **1**, 100006.
52. Gong, H., Chi, J., Ding, Z., Zhang, F. and Huang, J. (2020) Removal of lead from two polluted soils by magnetic wheat straw biochars. *Ecotoxicology and Environmental Safety* **205**, 111132.
53. Kokab, T., Ashraf, H. S., Shakoor, M. B., Jilani, A., Ahmad, S. R., Majid, M., Ali, S., Farid, N., Alghamdi, R. A., Al-Quwaie, D. A. H. and Hakeem, K. R. (2021) Effective removal of Cr(VI) from wastewater using biochar derived from Walnut Shell. *International Journal of Environmental Research and Public Health* **18**(18), 9670.
54. Ambaye, T. G., Vaccari, M., van Hullebusch, E. D., Amrane, A. and Rtimi, S. (2021) Mechanisms and adsorption capacities of biochar for the removal of organic and inorganic pollutants from industrial wastewater. *International Journal of Environmental Science and Technology* **18**(10), 3273–3294.
55. Ajmal, Z., Muhmood, A., Dong, R. and Wu, S. (2020) Probing the efficiency of magnetically modified biomass-derived biochar for effective phosphate removal. *Journal of Environmental Management* **253**, 109730.
56. Fan, R., Chen, C., Lin, J., Tzeng, J., Huang, C., Dong, C. and Huang, C. P. (2019) Adsorption characteristics of ammonium ion onto hydrous biochars in dilute aqueous solutions. *Bioresource Technology* **272**, 465–472.
57. Peñafiel, M. E., Matesanz, J. M., Vanegas, E., Bermejo, D. and Ormad, M. P. (2020) Corncobs as a potentially low-cost biosorbent for sulfamethoxazole removal from aqueous solution. *Separation Science and Technology* **55**(17), 3060–3071.
58. Nartey, O. D. and Zhao, B. (2014) Biochar preparation, characterization, and adsorptive capacity and its effect on bioavailability of contaminants: An overview. *Advances in Materials Science and Engineering* **2014**, 1–12.
59. Lee, S. S., Lim, J. E., El-Azeem, S. A. M. A., Choi, B., Oh, S.-E., Moon, D. H. and Ok, Y. S. (2013) Heavy metal immobilization in soil near abandoned mines using eggshell waste and rapeseed residue. *Environmental Science and Pollution Research* **20**(3), 1719–1726.
60. Ahmad, M., Ok, Y. S., Kim, B.-Y., Ahn, J.-H., Lee, Y. H., Zhang, M., Moon, D. H., Al-Wabel, M. I. and Lee, S. S. (2016) Impact of soybean stover- and pine needle-derived biochars on Pb and As mobility, microbial community, and carbon stability in a contaminated agricultural soil. *Journal of Environmental Management* **166**, 131–139.
61. Abbas, T., Rizwan, M., Ali, S., Zia-ur-Rehman, M., Farooq Qayyum, M., Abbas, F., Hannan, F., Rinklebe, J. and Sik Ok, Y. (2017) Effect of biochar on cadmium bioavailability and uptake in wheat (*Triticum aestivum L.*) grown in a soil with aged contamination. *Ecotoxicology and Environmental Safety* **140**, 37–47.
62. Ghosh, D., Masto, R. E. and Maiti, S. K. (2020) Ameliorative effect of Lantana camara biochar on coal mine spoil and growth of maize (Zea mays). *Soil Use and Management* **36**(4), 726–739.

63. Saxena, J., Rawat, J. and Kumar, R. (2017) Conversion of biomass waste into biochar and the effect on mung bean crop production. *Clean (Weinh)* **45**(7), 1501020.
64. Shakoor, A., Arif, M. S., Shahzad, S. M., Farooq, T. H., Ashraf, F., Altaf, M. M., Ahmed, W., Tufail, M. A. and Ashraf, M. (2021) Does biochar accelerate the mitigation of greenhouse gaseous emissions from agricultural soil? A global meta-analysis. *Environmental Research* **202**, 111789.
65. Ghassemi-Golezani, K. and Farhangi-Abriz, S. (2022) Improving plant available water holding capacity of soil by solid and chemically modified biochars. *Rhizosphere* **21**, 100469.
66. Jafri, N., Wong, W. Y., Doshi, V., Yoon, L. W. and Cheah, K. H. (2018) A review on production and characterization of biochars for application in direct carbon fuel cells. *Process Safety and Environmental Protection* **118**, 152–166.
67. Yu, J., Zhao, Y. and Li, Y. (2014) Utilization of corn cob biochar in a direct carbon fuel cell. *Journal of Power Sources* **270**, 312–317.
68. Suman, S., & Gautam, S. (2017) A comparative study between time, temperature, and fixed carbon using different biochar reductants as an alternate source of energy. *Energy Sources, Part A: Recovery, Utilization, and Environmental Effects*, **39**(10), 1029–1035. https://doi.org/10.1080/15567036.2017.1284959
69. Seo, J. Y., Tokmurzin, D., Lee, D., Lee, S. H., Seo, M. W. and Park, Y. K. (2022) Production of biochar from crop residues and its application for biofuel production processes: An overview. *Bioresource Technology* **361**, 127740.
70. Sun, X., Atiyeh, H. K., Li, M. and Chen, Y. (2020) Biochar facilitated bioprocessing and biorefinery for productions of biofuel and chemicals: A review. *Bioresource Technology* **295**, 122252.
71. Gupta, R. K., Hussain, A., Yadvinder-Singh, Sooch, S. S., Kang, J. S., Sharma, S. and Dheri, G. S. (2020) Rice straw biochar improves soil fertility, growth, and yield of rice–wheat system on a sandy loam soil. *Experimental Agriculture* **56**(1), 118–131.
72. Pandey, V., Patel, A. and Patra, D. D. (2016) Biochar ameliorates crop productivity, soil fertility, essential oil yield and aroma profiling in basil (*Ocimum basilicum* L.). *Ecological Engineering* **90**, 361–366.
73. Oliveira, F. R., Patel, A. K., Jaisi, D. P., Adhikari, S., Lu, H. and Khanal, S. K. (2017) Environmental application of biochar: Current status and perspectives. *Bioresource Technology* **246**, 110–122.
74. Zhang, X., Zhang, S., Yang, H., Feng, Y., Chen, Y., Wang, X. and Chen, H. (2014) Nitrogen enriched biochar modified by high temperature $CO_2$–ammonia treatment: Characterization and adsorption of $CO_2$. *Chemical Engineering Journal* **257**, 20–27.
75. Khan, Z., Khan, M. N., Zhang, K., Luo, T., Zhu, K. and Hu, L. (2021) The application of biochar alleviated the adverse effects of drought on the growth, physiology, yield and quality of rapeseed through regulation of soil status and nutrients availability. *Industrial Crops and Products* 171, 113878.
76. He, Y., Yao, Y., Ji, Y., Deng, J., Zhou, G., Liu, R., Shao, J., Zhou, L., Li, N., Zhou, X. and Bai, S. H. (2020) Biochar amendment boosts photosynthesis and biomass in C 3 but not C 4 plants: A global synthesis GCB. *Bioenergy* **12**(8), 605–617.

Chapter 19

# Building resilience to airborne pathogens

## An indoor air quality model

*Ashwani Kumar*

### 19.1 INTRODUCTION

The SARS-COV-2 variant that caused the COVID-19 pandemic led to a high fatality rate of approximately 2%–3% worldwide [1,2]. Coronavirus disease (COVID-19, SARS-CoV-2, Severe Acute Respiratory Syndrome Coronavirus 2) originated in December 2019 in Wuhan, China. As of January 2022, the World Health Organization (WHO) documented 360.57 million (36.05 crore) confirmed cases of COVID-19, including 5,620,865 deaths reported worldwide as of 27 January 2022. For vaccination record as of 25 January 2022, a total of 9.679,721,754 vaccine doses have been administered. More than 210 countries have been affected by SARS-CoV-2, and they have confirmed COVID-19 cases. Given the severity of the situation, the WHO declared COVID-19 a global health emergency [3–6]. In India, the first COVID-19 case was reported on 30 January 2020 in Kerala. The number of COVID-19 cases increased exponentially thereafter. To slow the spread, India implemented various measures, including a one-day campaign called "Janta Curfew" on 22 March 2020, followed by a 21-day nationwide lockdown starting on 25 March 2020, which was further extended to 17 May 2020 [7–10].

In India, out of 28 states, 26 states were affected by COVID-19, along with six union territories. The Health Ministry (Government of India) classified districts in red, orange, and green zones based on the severity. A total of 130 districts were identified as red zones, 284 districts as orange zones, and 319 districts as green zones. The red zones had the highest number of SARS-CoV-2 infections, orange zones has fewer SARS-CoV-2 infections, and green zones had no SARS-CoV-2 infections [11–14]. In Uttar Pradesh, 19 districts were designated as hot spots. Complete lockdowns were enforced in red zones to prevent community transmission Face masks were made mandatory across all zones to prevent nose and mouth exposure. Additionally, more than 300 government laboratories and over 100 private laboratories were involved in COVID-19 testing [15–18].

COVID-19 originates from the family of viruses called "coronaviridae." By June 2021, there were 36.05 crore COVID-19 cases worldwide, including


DOI: 10.1201/9781003521341-19

39.2 lakh deaths reported to the WHO. India, with a population of 1.34 billion, has been severely impacted by the COVID-19 pandemic. As of 27 January 2022, India reported 40,371,500 confirmed cases of COVID-19, with 491,700 deaths. By 17 January 2022, a total of 1,581,796,355 vaccine doses had been administered in the country. Maharashtra is the most affected state in India, with 60.2 lakh cases and 1.2 lakh deaths as of June 2021 [19–23]. In contrast, Mizoram is the least affected state of India, with 19,091 cases and 89 deaths reported. Despite multiple strategies to prevent COVID-19, the number of cases increased significantly from March 2020 and June 2021, and again from December 2021 to January 2022. The central and state governments implemented a 110-day lockdown as part of their efforts to curb the spread. This chapter highlights the main features of the coronavirus, its impact, prevention methods, and a proposed smart room management model to prevent indoor COVID-19 airborne transmission in India. Overall, India has been severely affected by COVID-19, and it has a long economic impact. The most effective strategy to stop the transmission in hospitals, shops, public places, schools, offices, conferences, elevators, and public transport includes following social distancing, hand sanitization, home isolation, and quarantine measures [24].

Air quality in a particular region is assessed by monitoring stations that collect large amounts of data on pollutant concentration levels. However, these bulky data are not easily readable by the general public and officials. This problem led to the development of the Air Quality Index (AQI), which compares the composition of local air with a standard composition defined by the government [25]. A typical AQI is an interpretive technique that converts complex air pollutant data into a single, easy-to-understand number. Globally, the AQI is referred to by different terminologies, such as the Pollutant Standard Index (PSI) [26], Air Pollution Index [27–30], and Oak Ridge AQI [31]. Several methodologies for calculating AQI have been proposed, including Aggregate AQI [32], Common AQI (CAQI) [33], New AQI, National AQI (NAQI) [34], Pollution index, PI [35], Aggregate Risk Index, and ARI [36]. Recent developments in AQI forecasting include fuzzy aggregation [37], coupled principal component analysis, and artificial neural network (ANN) [38], and combined IoT and cloud computing technology [39]. The AQI effectively highlights alarming pollutants concentration levels. Statistical analysis [40] concluded that AQI a major impact on respiratory diseases, with higher AQI levels indicating an increased risk of respiratory diseases in older people.

To deal with air quality, various types of air purifiers are available on the market, with efficiency rates up to 99%. These air purifiers remove pollutants from the air, thus improving indoor air quality. Among them, HEPA filters are the most common,. removing 99.97% of $PM_{2.5}$ pollutants. However, HEPA filters are expensive. Electrostatic precipitators can be used in the industrial sector to remove particulates of any size just by providing suitable voltage and current supply, but they are not feasible for home applications due to large setups [41–45].

The COVID-19 pandemic has significantly impacted the physical, mental, and economic well-being of people worldwide. The aim of this study is to highlight the importance of indoor air quality and its effect on human health. This chapter reviews various air quality enhancement techniques to improve the health of patients in COVID-19 hospitals [46–51]. Figure 19.6 consolidates the findings and presents a universal model for preventing airborne COVID-19 transmission in hospitals.

The chapter highlights the risk factors for adverse physical health during the COVID-19 pandemic across different age groups, as shown in Figure 19.1. Frontline workers, including medical staff, doctors, and sanitization department workers, are more prone to COVID-19 transmission. Additionally, the

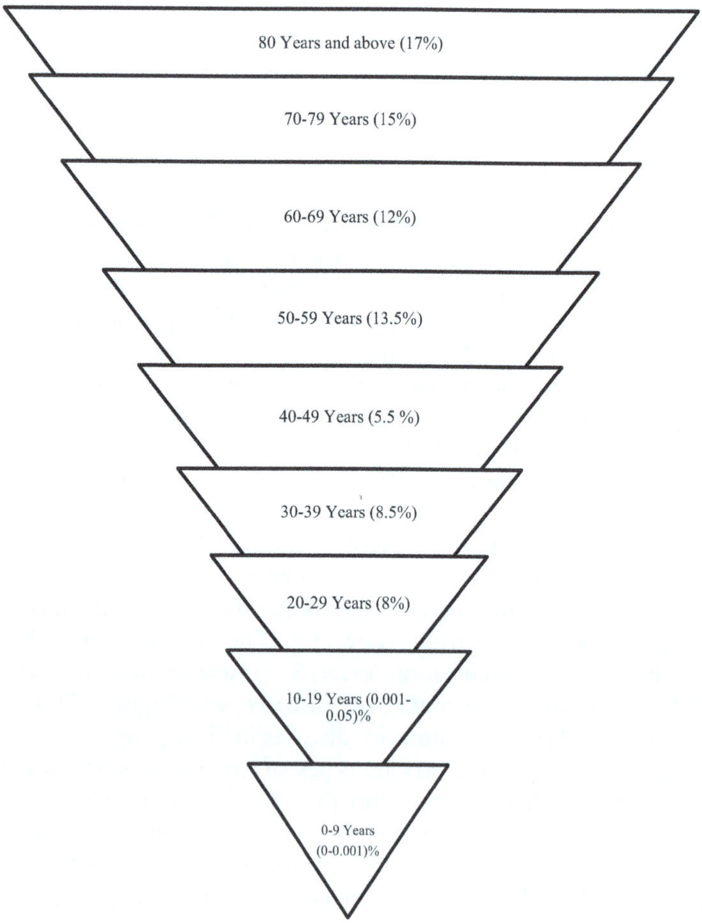


*Figure 19.1* Pyramidal approach for impact analysis of COVID-19 on different age groups (2020).

chapter addresses concerns and precautions related to COVID-19, identifying both protective and risk factors associated with physical health outcomes.

## 19.2 METHODOLOGY ADOPTED

The literature review started by defining the scope of the review and formulating a problem statement. After analyzing the literature, gaps and inconsistencies were figured out. Then, key advances in the field were highlighted, with a focus on reviewing recent papers from 2019 to 2021 that could have a significant impact. Finally, potential research areas for future exploration were highlighted. The conclusions drawn from the present study are strongly highlighted in the analysis of existing work.

## 19.3 ASSESSMENT OF AIR QUALITY

Table 19.1 shows the different names of AQI. AQI is an information tool that helps assess the quality of air in our surroundings [52]. The Environmental Protection Agency (EPA) developed the PSI by considering pollutants such as $PM_{10}$, $NO_2$, $O_3$, CO, and $SO_2$. The PSI is divided into five equal parts on a scale of 0–500. Although PM5 pollutants were not included in the PSI, they have a severe impact on human health [53]. To address this, the USEPA added these pollutants and renamed the PSI as AQI. The following equation was used for AQI formulation:

$$I_P = \frac{I_{\mathrm{Hi}} - I_{\mathrm{Lo}}}{\mathrm{BP}_{\mathrm{Hi}} - \mathrm{BP}_{\mathrm{Lo}}}\left(C_P - \mathrm{BP}_{\mathrm{Lo}}\right) + I_{\mathrm{Lo}}$$

$$\mathrm{AQI} = \mathrm{Max}\left(I_p\right)$$

*Table 19.1* Air quality measurement

| *Author/organization* | *Year* | *Name of index* | *Parameters* |
|---|---|---|---|
| Chelani et al. | 2002 | AQI | RSPM, $SO_2$, and $NO_2$ |
| CPCB, India | 2014 | National AQI | $PM_{10}$, $NO_2$, $PM_{2.5}$, $O_3$, CO, and $SO_2$ |
| Bishoi et al. | 2009 | New AQI | $PM_{10}$, $NO_2$, $PM_{2.5}$, $O_3$, CO, and $SO_2$ |
| Vann den Elshout et al. | 2008 | Common AQI | $PM_{10}$, $O_3$, $NO_2$ (main) and CO, $SO_2$ (auxillary) |
| US Environmental Protection Agency | 1976 | Pollution Standard Index | $PM_{10}$, $NO_2$, $O_3$, CO, and $SO_2$ |

As per AQI values, six different segments were created for health concerns. AQI values are mapped as follows:

$$0-50 \rightarrow \text{good};\ \ 51-100 \rightarrow \text{moderate};$$

$$101-150 \rightarrow \text{unhealthy for sensitive groups};$$

$$151-200 \rightarrow \text{unhealthy};\ \ 201-300 \rightarrow \text{very unhealthy};$$

$$301-500 \rightarrow \text{hazardous}$$

Currently, hybrid statistical models are used for high-precision measurement of air quality. Some of these statistical models are multiple linear models, ARIMA model, Support Vector Regressions (SVRs), ANNs, and hybrid models. Kumar and Goyal [54] proposed modifications to these models in their study on assessing AQI for Delhi. Zhu et al. [55] used hybrid models EMD-SVR-Hybrid and EMD-IMFs-Hybrid, integrating EMD and S-ARIMA for their study.

Due to the tropical nature and climatic conditions of India, the pollution problems encountered differ from those in other nations. Increased anthropogenic activities have led to high pollution levels, thereby degrading air quality and impacting human health. To assess air quality and implement mitigation and control measures for air pollution, the Central Pollution Control Board (CPCB) proposed the National Air Quality Monitoring Programme, a nationwide air quality monitoring system. In 2014, the CPCB came up with the NAQI [56].

## 19.4 AIR QUALITY AND COVID-19

Clean air is essential for human health as it impacts the metabolism of the human body. As discussed earlier in the Introduction, air pollution caused by the release of harmful gases has exacerbated the spread of COVID-19. Exposure to fine particles, particulate matter (PM), and nitrogen compounds significantly increases the risk of lung cancer and other pulmonary pathologies [57].

Household/indoor air pollution has increased health risks. The WHO Guidelines for Indoor Air Quality [58] indicate that poor air quality inside offices, houses, private buildings, and public buildings affects human health and can cause health issues. Indoor air pollutants can cause drowsiness, sore throat, or acute respiratory problems. Research studies report that particles $<300$ nm in diameter can penetrate the alveolar region of the lungs [59], while those $<200$ nm in diameter can reach the brain [60]. Additionally, particles $<240$ nm can cross the placental barrier, directly impacting the fetus [61]. Particles around 200 nm in size can cause airborne for a longer period of time due to their low depositional velocities [62]. Particles $<400$ nm in contribute significantly to the total particle mass observed in different residential areas in Hong Kong [63–68].

## 19.5 COVID-19 SOCIOECONOMIC IMPACT IN INDIA

Both SARS and COVID-19 are caused by coronaviruses. The virus affects the human respiratory tract and causes significant bodily damage. SARS-CoV is responsible for SARS, while SARS-CoV-2 causes COVID-19. Medical reports and research suggest that the human body does not have immunity to SARS-CoV-2, making it highly transmissible through surface or physical contact with COVID-19-positive patients. In terms of transmission, SARS-CoV-2 is more dangerous than SARS-CoV. The medical infrastructure of India was inadequate to contain COVID-19 spread and provide treatment for COVID-19. In 2020, there were only 49,000 ventilators in India for a population of 1.34 billion, which is a bare minimal amount. Therefore, it was required to manage COVID-19 with multiple strategies, including computational modeling and quantitative analysis for developing drugs and vaccines.

India's neighboring countries, including like Nepal, Pakistan, China, and Bangladesh, along with major countries in Asia, were severely affected by SARS-CoV-2. To prevent the spread of COVID-19, the Indian government implemented measures such as health checkups at airports, railway stations, and bus stations [10]. Compared with its neighboring countries, India's capacity to counter COVID-19 situation was lower and needed improvement. It was also observed that COVID-19-positive individuals who had not traveled across borders tested positive due to community transmission or physical contact with any positive person. The mobilization of Labor following the nation-wide lockdown was a major factor contributing to the transmission and spread of COVID-19 in India.

The spread of the COVID-19 virus has significantly impacted the social and economic conditions in India. Reduced purchasing power has adversely affected various factors, including manufacturing, tourism, hospitality, electronics, transportation, logistics, civil construction, and pharmaceuticals. With a shift from spending to saving, the economy experienced a negative growth rate of –23% GDP. This decline reflects the interdependent world economy. On the international stage, the reduced value of currencies will likely have a negative impact, influencing oil prices [11,12].

India's production industry has been directly affected by high labor mobility across the country and reduced global demand due to decreased consumption/purchase power, travel restrictions, and job losses [13]. It is anticipated that employment opportunities will decline in India, and the whole socioeconomic impact may take 2–3 years to recover.

## 19.6 COVID-19 IMPACT ON DIFFERENT AGE GROUPS AND SYMPTOMS IDENTIFICATION

Figure 19.1 shows age group variations among COVID-19-positive patients. The least affected group was children aged 0–9 years, while the most affected age group was 40–90 years. Individuals aged 0–30 years were less affected due to higher immune capacity.

As age increases, the likelihood of contracting COVID-19 increases significantly. However, during the second wave of COVID-19 in India, the above pyramidal graph changed, with the most affected age group being 30 years and above. In the third wave of COVID-19, the symptoms remained the same, but their impact was milder. Only patients with severe or critical conditions such as high BP, diabetes, or low immunity were affected by COVID-19.

Figure 19.2 shows a pyramidal model of COVID-19 symptoms in India. At the base of the pyramid is a group of persons who are not affected by the SARS-CoV-2 virus due to their strong immunity. Above this group is a group of persons diagnosed with COVID-19 but no symptoms. In India, a significant number of patients are asymptomatic. Among symptomatic cases, common symptoms include mild fever, cough, tiredness, headache, sore throat, and loss of taste (Figure 19.3). Serious symptoms of COVID-19 include shortness of breath, chest pain, and loss of speech and movement. Figure 19.3 shows symptoms progression at different levels in COVID-19 patients.

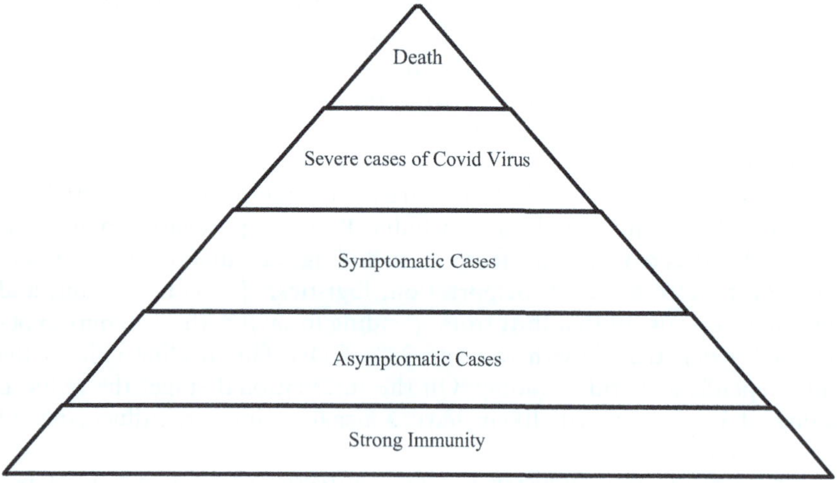


*Figure 19.2* Pyramidal model of COVID-19 symptoms in India.

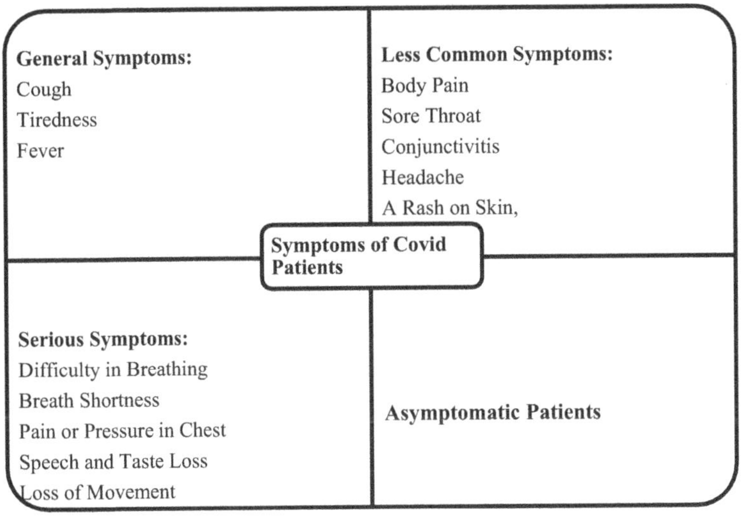


*Figure 19.3* Symptoms generation in patients.

## 19.7 PREVENTIVE MEASURES FOR COVID-19 IN INDIA

The simplest way of SARS-CoV-2 infection prevention is to use a face shield or face mask in public places and avoid surface contact/physical contact with infected individuals. As part of preventive measures, people should avoid visiting places with a high likelihood of public gatherings, such as public buses, schools, malls, food courts, jogging parks, cinema halls, and cross-border travel. It is also important to follow proper sanitization practices and other general preventive measures, as outlined in Figure 19.5. The Ministry of Health and Family Welfare in India has made significant efforts to raise public awareness by providing slogans and establishing a 24×7 helpline (+91-11-23978046 and 1800-180-1104). The Prime Minister of India introduced the famous slogan "Jab tak dawai nahi, tab tak dhilai nahi."

Figure 19.4 shows the U.S. government's traditional infection control pyramid, adopted by the Centers for Diseases Control and Prevention, highlighting the role of engineering and administrative control in managing infections. In India, local administration (police and medical teams) has done tremendous work for public. The Union Health Ministry of India launched "Aarogya Setu" app to establish an awareness network and build a database. This app is available on both iOS and Android platforms [14].

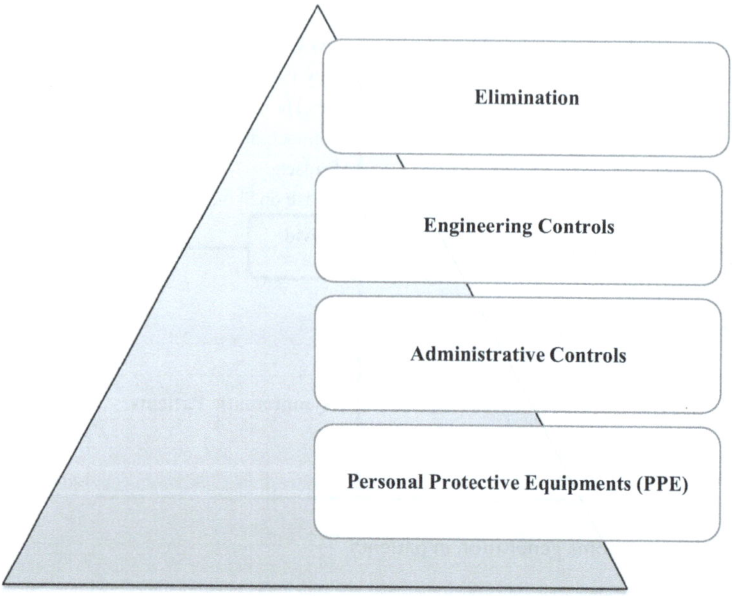


*Figure 19.4* Infection prevention/control model, adopted by the Centers for Diseases Control and Prevention, USA.

## 19.8 SMART ENGINEERING CONTROL MODEL ROOM FOR COVID-19 PATIENTS

This study presents a model for an engineering control room designed to prevent airborne transmission of COVID-19. Figure 19.6 shows the engineering control model room for controlling COVID-19. As the COVID-19 virus typically takes 5–14 days to show symptoms, maximizing patient protection against the SARS-CoV-2 virus in indoor environments is crucial. This model is particularly applicable to apartments, hospitals, and houses, where healthy indoor air is most important for the indoor environment. It can be obtained by using two windows and avoiding the recirculation of air in the room.

The room should have a direct sunlight facility. Germicidal ultraviolet (UV) and portable air filters are required for patients with mild symptoms of COVID-19. In cities with pollution levels, portable air filters are important for purifying the air. For critical patients, the room should be equipped with a ventilator and an all-axis moving trolley for life-saving interventions and medicine administration. As recirculation of air can transport airborne contaminants, it must be avoided.

## 19.9 CONCLUSIONS/ANALYSIS OF EXISTING WORK AND FUTURE POSSIBILITIES

From an Indian perspective, it can be summarized that until vaccines or medicine for COVID-19 are widely available, we should follow the prevention model (Figure 19.5) and ventilation (Figure 19.6) model. The key recommendations are as follows:

- Engineering-controlled rooms should be used as they have the potential to reduce SARS-CoV-2 airborne infection.
- Recirculation of air should be avoided, ensuring patients have access to fresh outdoor air.
- Using general PPE kits with hand sanitizers for COVID-19 prevention.
- Using portable air filters and germicidal ultraviolet (UV).
- Avoiding over-crowding in public transport, cinemas, hospitals, schools, offices, etc. as it reduces the chances of airborne transmission.
- Improving indoor air quality by using air-purifying filters is beneficial for reducing airborne COVID-19. UV germicidal filters, ionization filters, PCO, or newly emerging solid oxygen purifying can be effective in inactivating the virus in residential and public buildings These measures are incorporated into the smart room management model (Figure 19.6).

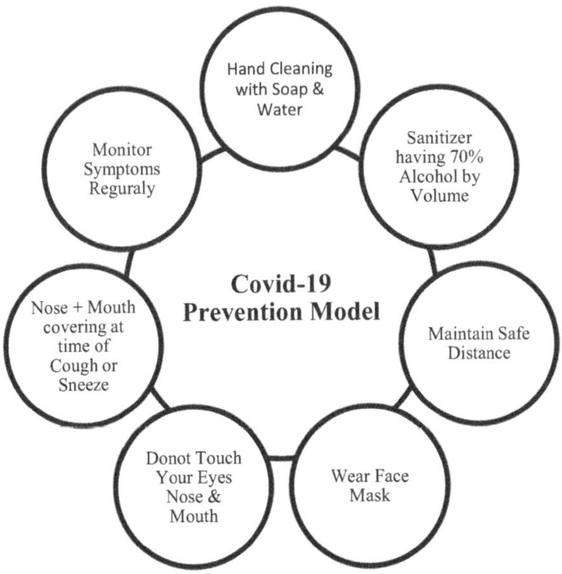


*Figure 19.5* COVID-19 prevention models in the context of India.

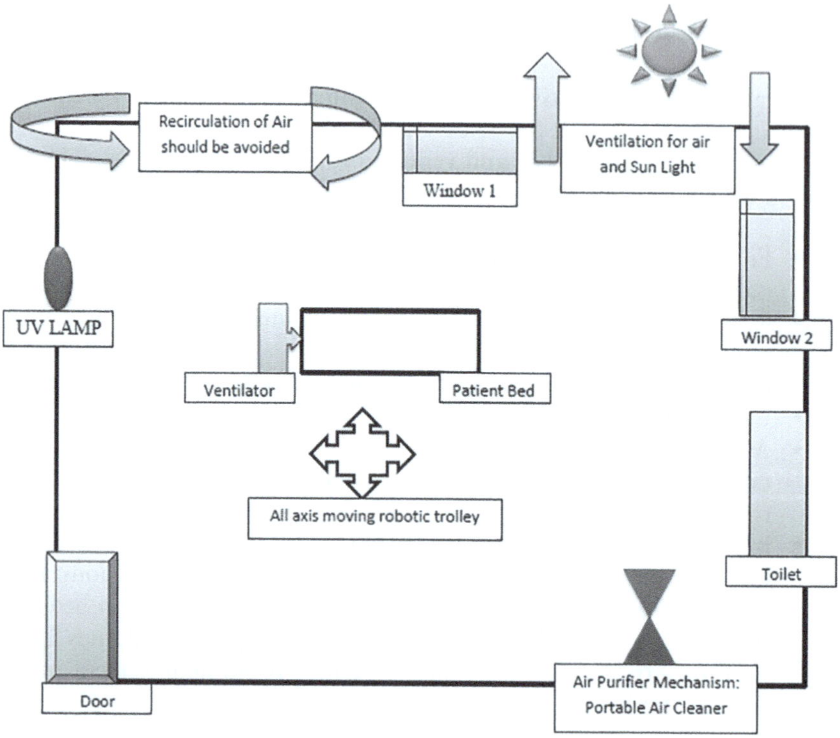


*Figure 19.6* Engineering control model room design for controlling airborne transmission of COVID-19.

- Pollutant concentration measured by AQI plays a crucial role in raising awareness among the public. The accuracy of AQI forecasting has been improved by using statistical methods. Using AQI can reduce pollutant concentration levels, thereby decreasing infection cases and mortality rates. The smart room management model (Figure 19.6) incorporates all these suggestions for optimal prevention of airborne transmission.

If these suggestions are implemented correctly, they have the potential to reduce airborne transmission of SARS-CoV-2. Additionally, adhering to general guidelines (Figure 19.5) is needed. These engineering controls will also protect healthcare warriors. It is suggested that using all the possible preventive measures to combat the COVID-19 is the most effective solution. Air pollution has been identified as a contributing factor to the spread of SARS-CoV-2. While a fully effective vaccine or medication remains the most promising long-term solution, mitigating air pollution can significantly help curb the virus's transmission in the interim.

## REFERENCES


1. Rodriguez-Morales, A.J.; Bonilla-Aldana, D.K.; Tiwari, R.; Sah, R.; Rabaan, A.A; Dhama, K., COVID-19, an emerging coronavirus infection: current scenario and recent developments: An overview. *Journal of Pure and Applied Microbiology*, 2020, 14, 5–12.
2 Sonawane, A.A.; Shastri, J.; Bavdekar, S.B., Respiratory pathogens in infants diagnosed with acute lower respiratory tract infection in a Tertiary Care Hospital of Western India using multiplex real time PCR. *Indian Journal of Pediatrics*, 2019, 86, 433–438. DOI: 10.1007/s12098-018-2840-8
3. Vaidyanathan, G., People power: How India is attempting to slow the coronavirus. *Nature*, 2020, 580, 442. DOI: 10.1038/d41586-020-01058-5.
4. Teixeira da Silva, J.A., Convalescent plasma: A possible treatment of COVID-19 in India. *Medical Journal Armed Forces India*, 2020. DOI: 10.1016/j.mjafi.2020.04.006.
5. Elfiky, A.A., Anti-HCV, nucleotide inhibitors, repurposing against COVID-19. *Life Sciences*, 2020, 248, 117477. DOI: 10.1016/j.lfs.2020.117477.
6. Wee, S.L. Jr; McNeil, D.G. Jr; Hernández, J.C., W.H.O. declares global emergency as Wuhan Coronavirus spreads. *The New York Times*, 2020.
7. Rathi, S.; Ish, P.; Kalantri, A.; Kalantri, S., Hydroxychloroquine prophylaxis for COVID-19 contacts in India. *The Lancet Infectious Diseases*, 2020. DOI: 10.1016/S1473-3099(20)30313-3.
8. Bureau, O., PM Modi calls for 'Janata curfew' on March 22 from 7 AM-9 PM. @businessline, 2020.
9. Novel Coronavirus -MOHFW, Home | Ministry of Health and Family Welfare | GOI, 2020.
10. Bhatnagar, T.; Murhekar, M.V.; Soneja, M.; Gupta, N.; Giri, S.; Wig, N., et al., Lopinavir/ritonavir combination therapy amongst symptomatic coronavirus disease 2019 patients in India: Protocol for restricted public health emergency use. *Indian Journal of Medical Research*, 2020, 151, 184–189. DOI: 10.4103/ijmr.IJMR_502_20.
11. Agarwal, N.; Meena, C.S.; Raj, B.P.; Saini, L.; Kumar, A.; Gopalakrishnan, N.; Kumar, A.; Balam, N.B.; Alam, T.; Kapoor, N.R.; Aggarwal, V., Indoor air quality improvement in COVID-19 pandemic: Review. *Sustainable Cities and Society*, 2021, 70, 102942. https://doi.org/10.1016/j.scs.2021.102942.
12. Sudhir, K.; Priester, J.; Shum, M.; Atkin, D.; Foster, A.; Iyer, G., et al., Research opportunities in emerging markets: An in-ter-disciplinary perspective from marketing, economics, and psychology. *Customer Needs and Solutions*, 2015, 2, 264–276. DOI: 10.1007/s40547-015-0044-1.
13. Albulescu, C., Coronavirus and oil price crash. *SSRN Journal*, 2020, 1–13. DOI: 10.2139/ssrn.3553452.
14. OECD Economic Outlook. Coronavirus: The world economy at risk, 2020. Available online at: https://www.oecd.org/economic-outlook/.
15. MyGov.in., Aarogya Setu Mobile App., 2020. Available online at: https://mygov.in/aarogya-setu-app/.
16. Jain, J.; Kumar, A.; Narayanan, V.; Ramaswamy, R.S.; Sathiyarajeswaran, P.; Shree Devi, M.S.; et al., Antiviral activity of ethanolic extract of Nilavembu Kudineer against dengue and chikungunya virus through in vitro evaluation. *Journal of Ayurveda and Integrative Medicine*, 2019. DOI: 10.1016/j.jaim.2018.05.006.

17. Mackay, I.M.; Arden, K.E., MERS coronavirus: Diagnostics, epidemiology and transmission. *Journal of Virology*, 2015, 12, 222. DOI: 10.1186/s12985-015-0439-5.
18. Peeri, N.C.; Shrestha, N.; Rahman, M.S.; Zaki, R.; Tan, Z.; Bibi, S., et al., The SARS, MERS and novel coronavirus (COVID-19) epidemics, the newest and biggest global health threats: What lessons have we learned? *International Journal of Epidemiology*, 2020. DOI: 10.1093/ije/dyaa033.
19. Tai, W.; He, L.; Zhang, X.; Pu, J.; Voronin, D.; Jiang, S.; et al., Characterization of the receptor-binding domain (RBD) of 2019 novel coronavirus: Implication for development of RBD protein as a viral attachment inhibitor and vaccine. *Cellular & Molecular Immunology*, 2020. DOI: 10.1038/s41423-020-0400-4.
20. Smith, R.D., Responding to global infectious disease outbreaks: Lessons from SARS on the role of risk perception, communication and management. *Social Science & Medicine*, 2006, 63, 3113–3123. DOI: 10.1016/j.socscimed.2006.08.004.
21. Broor, S.; Dawood F.S., Pandey B.G., Saha, S.; Gupta, V.; Krishnan, A.; et al., Rates of respiratory virus-associated hospitalization in children aged <5 years in rural northern India. *Journal of Infection*, 2014, 68, 281–289. DOI: 10.1016/j.jinf.2013.11.005.
22. Zou, L.; Ruan, F.; Huang, M.; Liang, L.; Huang, H.; Hong, Z.; et al., SARS-CoV-2 viral load in upper respiratory specimens of infected patients. *The New England Journal of Medicine*, 2020, 382, 1177–1179. DOI: 10.1056/NEJMc2001737.
23. McCloskey, B.; Heymann, D.L., SARS to novel coronavirus–old lessons and new lessons. *Epidemiology and Infection*, 2020, 148, e22. DOI: 10.1017/S0950268820000254.
24. Qiu, W.; Chu, C.; Mao, A.; Wu, J., The impacts on health, society, and economy of SARS and H7N9 outbreaks in China: A case comparison study. *International Journal of Environmental Research and Public Health*, 2018, 2018, e2710185. DOI: 10.1155/2018/2710185.
25. Beig, G.; Ghude, S.D.; Deshpande, A., Scientific evaluation of air quality standards and defining air quality index for India, Indian Institute of Tropical Meteorology, India, 2010.
26. Ott, W.R.; Hunt Jr, W.F., A quantitative evaluation of the pollutant standards index. *Journal of the Air Pollution Control Association*, 1976, 26(11), 1050–1054.
27. Shenfeld, L., Ontario's air pollution index and alert system. *Journal of the Air Pollution Control Association*, 1970, 20(9), 612–612.
28. Thom, G.C., Ott, W., *Air Pollution Indices: A Compendium and Assessment of Indices Used in the United States and Canada*. The Council, 1975.
29. Ott, W.R.; Thorn, G.C., Air pollution index systems in the United States and Canada. *Journal of the Air Pollution Control Association*, 1976, 26(5), 460–470.
30. Murena, F., Measuring air quality over large urban areas: Development and application of an air pollution index at the urban area of Naples. *Atmospheric Environment*, 2004, 38(36), 6195–6202.
31. Babcock Jr, L.R.; Nagda, N.L., Indices of air quality. In Thomas, W.A. (ed.), *Indicators of Environmental Quality*, pp. 183–197. Springer US: New York, 1972.

32. Kyrkilis, G.; Chaloulakou, A.; Kassomenos, P.A., Development of an aggregate air quality index for an urban medi-terranean agglomeration: Relation to potential health effects. *Environment International*, 2007, 33(5), 670–676.
33. Van den Elshout, S.; Léger, K.; Nussio, F., Comparing urban air quality in Europe in real time: A review of existing air quality indices and the proposal of a common alternative. *Environment International*, 2008, 34(5), 720–726.
34. Bishoi, B.; Prakash, A.; Jain, V.K., A comparative study of air quality index based on factor analysis and US-EPA methods for an urban environment. *Aerosol and Air Quality Research*, 2009, 9(1), 1–17.
35. Cannistraro, G.; Ponterio, L., Analysis of air quality in the outdoor environment of the city of Messina by an application of the pollution index method. *International Journal of Civil & Environmental Engineering*, 2009, 1(4), 214.
36. Sicard, P.; Lesne, O.; Alexandre, N.; Mangin, A.; Collomp, R., Air quality trends and potential health effects–development of an aggregate risk index. *Atmospheric Environment*, 2011, 45(5), 1145–1153.
37. Mandal, T.; Gorai, A.K.; Pathak, G., Development of fuzzy air quality index using soft computing ap-proach. *Environmental Monitoring and Assessment*, 2012, 184(10), 6187–6196.
38. Kumar, A.; Goyal, P., Forecasting of air quality index in Delhi using neural network based on principal component analysis. *Pure and Applied Geophysics*, 2013, 170(4), 711–722.
39. Jo, J.H.; Jo, B.; Kim, J.H.; Choi, I., Implementation of IoT-based air quality monitoring system for investigating particulate matter (PM10) in subway tunnels. *International Journal of Environmental Research and Public Health*, 2020, 17(15), 5429.
40. Ikram, M.; Yan, Z.J., Statistical analysis of the impact of AQI on respiratory disease in Beijing: Application case 2009. *Energy Procedia*, 2017, 107, 340–344.
41. Aditya, R., A review of general and modern methods of air purification. *Journal of Thermal Engineering*, 2018, 5(2), 22–28.
42. Heyder, J., Deposition of inhaled particles in the human respiratory tract and consequences for regional targeting in respiratory drug delivery. *Proceedings of the American Thoracic Society*, 2004, 1(4), 315–320.
43. Maher, B.A.; Ahmed, I.A.; Karloukovski, V.; MacLaren, D.A.; Foulds, P.G.; Allsop, D.; Mann, D.M.; Torres-Jardón, R.; Calderon-Garciduenas, L., Magnetite pollution nanoparticles in the human brain. *Proceedings of the National Academy of Sciences*, 2016, 113(39), 10797–10801.
44. Wick, P.; Malek, A.; Manser, P.; Meili, D.; Maeder-Althaus, X.; Diener, L.; Diener, P.A.; Zisch, A.; Krug, H.F.; von Mandach, U., Barrier capacity of human placenta for nanosized materials. *Environmental Health Perspectives*, 2010, 118(3), 432–436.
45. Lai, A.C.K., Particle deposition indoors: A review. *Indoor Air*, 2002, 12(4), 211–214.
46. Chao, C.Y.; Wong, K.K.; Cheng, E.C., Size distribution of indoor particulate matter in 60 homes in Hong Kong. *Indoor and Built Environment*, 2002, 11(1), 18–26.

47. Kumar, A. et al., *Advanced Computational Methods in Mechanical and Materials Engineering*. CRC Press: Boca Raton, FL, 2021. https://doi.org/10.1201/9781003202233.
48. Prasad, A. et al., *Biodegradable Composites for Packaging Application*. CRC Press: Boca Raton, FL, 2022. https://doi.org/10.1201/9781003227908.
49. Gajrani, K.K. et al., *Advances in Sustainable Machining and Manufacturing Processes*. CRC Press: Boca Raton, FL, 2022. https://doi.org/10.1201/9781003284574.
50. Singh, S. et al., *Nanomanufacturing and Nanomaterials Design: Principles and Applications*. Publisher Taylor & Francis (CRC Press): Boca Raton, FL, 2022. https://doi.org/10.1201/9781003220602.
51. Sharma, V.K. et al., *Additive Manufacturing in Industry 4.0: Methods, Techniques, Modeling, and Nano Aspects*. CRC Press: Boca Raton, FL, 2022. https://doi.org/10.1201/9781003360001.
52. Kurt, A.; Oktay, A.B., Forecasting air pollutant indicator levels with geographic models 3 days in advance using neural networks. *Expert Systems with Applications*, 2010, 37, 7986–7992.
53. Guo, Z.D.; Wang, Z.Y.; Zhang, S.F.; Li, X.; Li, L.; Li, C.; Cui, Y.; Fu, R.B.; Dong, Y.Z.; Chi, X.Y.; Zhang, M.Y., Aerosol and surface distribution of severe acute respiratory syndrome coronavirus 2 in hospital wards, Wuhan, China, 2020. *Emerging Infectious Diseases*, 2020, 26(7), 10–3201.
54. Kumar, A.; Goyal, P., Forecasting of daily air quality index in Delhi. *Science of the Total Environment*, 2011, 409(24), 5517–5523.
55. Zhu, S.; Lian, X.; Liu, H.; Hu, J.; Wang, Y.; Che, J., Daily air quality index forecasting with hybrid models: A case in China. *Environmental Pollution*, 2017, 231, 1232–1244.
56. Beig, G.; Ghude, S.D.; Deshpande, A., Scientific evaluation of air quality standards and defining air quality index for India, Indian Institute of Tropical Meteorology, India, 2010.
57. Heal, M.R.; Kumar, P.; Harrison, R.M., Particles, air quality, policy and health. *Chemical Society Reviews*, 2012, 41(19), 6606–6630.
58. Vijayan, V.K.; Paramesh, H.; Salvi, S.S.; Dalal, A.A.K., Enhancing indoor air quality: The air filter advantage. *Lung India: Official Organ of Indian Chest Society*, 2015, 32(5), 473.
59. Kumar, A. et al., *Thermal Energy Systems: Design, Computational Techniques and Applications*. CRC Press: Boca Raton, FL, 2023. https://doi.org/10.1201/9781003395768.
60. Kumar, A. et al., *Laser Based Technologies for Sustainable Manufacturing*. CRC Press: Boca Raton, FL, 2023. https://doi.org/10.1201/9781003402398.
61. Kumar, R. et al., *Sustainable Smart Manufacturing Processes in Industry 4.0*. Taylor & Francis (CRC Press): Boca Raton, FL, 2023. https://doi.org/10.1201/9781003436072.
62. Awasthi, M.K. et al., *Computational Fluid Flow and Heat Transfer: Advances, Design, Control & Applications*. CRC Press: Boca Raton, FL, 2024. https://doi.org/10.1201/9781003465171.
63. Singh, V.P. et al., *Energy Efficient Vehicles: Technologies and Challenges*. CRC Press: Boca Raton, FL, 2024. https://doi.org/10.1201/9781003464556.

64. Shah, S. et al., *Sustainability in Smart Manufacturing: Trends, Scope, and Challenges*. CRC Press: Boca Raton, FL, 2024. https://doi.org/10.1201/9781003467496.
65. Verma, V. et al., *Highly Efficient Thermal Renewable Energy Systems: Design, Optimization and Applications*. CRC Press: Boca Raton, FL, 2024. https://doi.org/10.1201/9781003472629.
66. Kumar, A. et al., *Advanced Materials for Biomedical Applications*. Taylor & Francis (CRC Press): Boca Raton, FL, 2022. https://doi.org/10.1201/9781003344810.
67. Kumar, A. et al., *Applications of Biotribology in Biomedical Systems*. Springer Nature, 2024. https://link.springer.com/book/9783031583261.
68. Kumar, A. et al., *Advanced Material for Biomechanical Applications*. Taylor & Francis (CRC Press): Boca Raton, FL, 2022. https://doi.org/10.1201/9781003286806.

Chapter 20

# Urban air quality

## A comparative study of pollution levels in Dewas versus Bhopal and a case study in Sainath Nagar near Kolar Road construction site, Bhopal

*Sachin Kumar Banait, Sakshi Sarathe, Sujeet Kesharvani, Gaurav Dwivedi, Prashant Baredar, and Puneet Verma*

### 20.1 INTRODUCTION

In contemporary urban environments, air quality has emerged as a pivotal concern owing to its profound implications for public health, environmental sustainability, and overall quality of life. The rapidly expanding urbanization, industrialization, and vehicular emissions continue to shape the atmospheric composition of urban areas, making it imperative to investigate deeper into understanding and comparing their air quality profiles of urban areas. In this context, comparing air quality between two cities provides valuable insights into the spatial and temporal variations of air pollution, as well as factors influencing these differences. The two cities represent contrasting urban landscapes, characterized by diverse demographic, geographical, and industrial characteristics.

According to Dutta et al. (2021), the study investigates rising air pollution in major Indian cities due to urbanization, population growth, vehicle proliferation, and industrial expansion, Focusing on Delhi, Kolkata, and Chennai, it analyzes pollution trends with meteorological parameters and compares pollution levels to national standards, with Delhi and Kolkata emerging as most affected, particularly during winter.

As per Agarwal et al. (2020), air quality in six Indian megacities and six cities in Hubai, China, during strict lockdowns. Real-time $PM_{2.5}$ and $NO_2$ concentrations were monitored over 3 months, with data converted to AQI based on U.S. Environmental Protection Agency standards. Both countries saw immediate drops in AQI for $NO_2$ after one week of lockdown, with China experiencing a gradual decline in AQI for $PM_{2.5}$. Overall, China achieved a 42.5% reduction in AQI for $PM_{2.5}$ and a 37.42% drop in AQI for $NO_2$, while India recorded a 64.65% decrease in AQI for $PM_{2.5}$ and 65.80% decrease in AQI for $NO_2$.

 DOI: 10.1201/9781003521341-20

Chiritescu et al. (2024) study urban air pollution levels in 33 cities across Romania during 2019 and 2020, with a particular focus on understanding the impact of the COVID-19 pandemic on pollutant concentrations. Analyzing time-series data of major air pollutants such as $PM_{10}$, $PM_{2.5}$, $NO_2$, CO, $SO_2$, and $O_3$, the research aimed to assess variations in pollutant levels across different seasons and under varying levels of anthropogenic activity. The results indicate significant reductions in pollutant concentrations during the spring of 2020, coinciding with the implementation of strict lockdown measures. Spatial analysis reveals regional variations in pollutant concentrations, highlighting specific hot spots and cold spots across Romania.

Zhou et al. (2024) examine the impact of environmental complaint reporting on atmospheric environmental quality using panel regression and simultaneous equation models. Analyzing data from 295 Chinese cities from 2018 to 2020, it finds that such reporting significantly improves air quality, particularly in cities with higher pollution and economic development. The study highlights the effectiveness of complaint reporting in industrial areas and emphasizes its role in reducing air pollution from major emission sources.

Liu et al. (2024) examine the influence of urban forests and parks on air quality and microclimate in Jinan, China. The results show that urban forests significantly reduce $PM_{2.5}$ and ozone concentrations during warm seasons, with levels up to 6.5 and 23 $\mu g/m^3$ lower than in downtown areas. Conversely, urban parks reduce $PM_{2.5}$ but have minimal impact on gaseous pollutants. Both environments lower temperatures by 1.36°C–6.8°C and increase relative humidity, enhancing comfort for residents. This underscores the vital role of urban vegetation in enhancing air quality and creating a favorable living environment.

Zheng et al. (2024) investigate the potential of using plants as biomonitors to assess air pollution levels. Utilizing an air pollution tolerance index derived from various physiological and biochemical parameters, the study reveals *Cedrus deodara* and *Sabina chinensis* have the highest air pollution tolerance, with *Cedrus deodara* showing superior performance. Conversely, *Ligustrum lucidum* and *Euonymus japonicas* were found to be the most sensitive species. Overall, *Ligustrum lucidum* and *Euonymus japonicas* could serve as effective bioindicators, while *Cedrus deodara* is recommended for green belt development in Xi'an, China.

Verma et al. (2023) examine air pollution in Brisbane City, Queensland, Australia, and its impacts on respiratory health life expectancy. While governments have implemented strategies to improve air quality, natural events also affect pollutant levels. The study investigates changes in criteria pollutant concentrations during the 2020 COVID-19 lockdown compared to pre-lockdown and previous years. It also considers post-lockdown scenarios to assess the impact of behavioral changes. The results indicate COVID-19 restrictions influenced pollutants levels, but the locations of monitoring stations also played significant role.

Singh et al. (2024) investigate the impact of unplanned urbanization on key environmental factors such as land surface temperature (LST), vegetation cover, and $PM_{2.5}$ in Bhopal, Madhya Pradesh, India. It examines their spatial–temporal changes and assesses their interrelationships, culminating in an Environmental Quality Index spanning 2000–2020. Trend analyses reveal a notable increase in LST and $PM_{2.5}$ levels across different city zones, with rises of 1°C–2°C and 23–25 μg/m$^3$, respectively, underscoring challenges in environmental quality management.

Chaurasia Sadhana et al. (2013) evaluate ambient air quality across industrial, commercial, and residential areas of Bhopal city using the New National Ambient Air Quality Standards. Conducted in February and March 2012, the study assessed pollutants including $SO_2$, $NO_x$, $PM_{10}$, and $PM_{2.5}$, with PM levels consistently exceeding permissible limits at all sites. The Air Quality Index (AQI) indicated severe pollution levels, with significant positive correlations observed for all parameters during the study period.

Jha Pragya et al. (2023) examine air pollution variations ($PM_{10}$, $PM_{2.5}$, $SO_2$, $NO_2$, $NH_3$, and ozone) during distinct lockdown phases in Dewas, M.P., India: pre-lockdown (January 1 to March 24), lockdown (March 25 to May 31), and post-lockdown (June 1 to December 31). The study analyzes data spanning 2018–2022, revealing significant reductions in particulate matter (PM) and gaseous pollutant concentration during the lockdown, except for ozone.

Agrawal et al. (2024) compare air quality dynamics in Bhopal, India, and Adelaide, Australia, amid COVID-19 restrictions, analyzing $PM_{2.5}$, $PM_{10}$, $NO_2$, and $O_3$ concentrations across pre-lockdown, lockdown, and post-lockdown phases from 2019 to 2022. While significant reductions in PM levels were observed during lockdowns in Indian and Chinese cities due to various factors, Bhopal and Adelaide exhibited nuanced trends influenced by local conditions, including bushfires in Australia. Bhopal displayed decreased $PM_{10}$ and $PM_{2.5}$ levels but varied patterns in $NO_2$ and CO, while Adelaide experienced minor changes. The study undergoes the necessity for tailored, sustainable air quality management strategies considering regional specifics and emphasizes the role of international collaboration in addressing broader atmospheric challenges.

Kori Reeta et al. (2019) assess the ambient AQI in Bhopal, Madhya Pradesh, India, from 2017 to 2018 across eighteen locations. It study focuses on seven pollutants, namely, $PM_{10}$, $PM_{2.5}$, $NO_2$, $SO_2$, $O_3$, $NH_3$, and Pb. The results indicated that gaseous pollutants were within standard limits, while $PM_{10}$ and $PM_{2.5}$ exceeded National Ambient Air Quality Standards at all locations. The AQI ranged from poor to satisfactory, with an overall moderate rating for Bhopal city during the study period.

Rahul & Charan (2022) investigate air pollution dynamics in Mandideep, India, amid rapid urbanization. Data collected from 23 monitoring stations between August and October 2021 reveal exceedances of $PM_{2.5}$ and $PM_{10}$

concentrations over Indian and WHO guidelines. Trends showed decreasing $PM_{10}$, $PM_{2.5}$, $SO_2$, and CO levels from 2014 to 2016, with stable $NO_2$ and increasing $O_3$ concentrations, notably affected by industrial and vehicular emissions. Seasonal variations indicated highest pollutant levels in winter, with haze formation linked to static conditions, while ozone pollution correlated with higher temperatures.

Road construction is a vital component of infrastructure development, facilitating economic growth, connectivity, and urban expansion. However, alongside its benefits, road construction activities also pose significant challenges to environment quality, particularly air quality. The process involved in road construction, such as excavation, paving, and traffic diversion, can generate a variety of pollutants that are released into the atmosphere, contributing to air pollution these pollutants include PM, nitrogen oxides ($NO_x$), volatile organic compounds (VOCs), carbon monoxide (CO), and others, which have known detrimental effects on human health and environment.

The impact of road construction on air quality extends beyond the immediate vicinity of construction sites, affecting surrounding areas and communities. Construction-related emissions can lead to increased levels of air pollutants, exacerbating existing air quality issues in urban and peri-urban regions. Additionally, the duration of road construction projects, often spanning several months or years, prolongs the exposure of nearby populations to harmful pollutants, potentially resulting in adverse health outcomes.

Understanding the effects of road construction on air quality is essential for effective environmental management and public health protection. By identifying the sources and mechanisms of air pollution associated with road construction, policymakers, urban planners, and construction stakeholders can implement mitigation measures and regulatory strategies to minimize environmental impacts and safeguard air quality.

The high need to monitor and control the air pollution during road construction is highlighted by the literature survey:

Font et al. (2019) compare air quality trends and policy initiatives in Paris and London, focusing on $NO_2$ and $PM_{10}$ and $PM_{2.5}$ concentrations from 2005 to 2016. Despite efforts like Euro emission standards and low emission zones, both cities experienced an initial increase in roadside $NO_2$, followed by a gradual decrease starting from 2010, associated with the introduction of Euro V heavy vehicles. However, achieving European limit values for $NO_2$ remains a distant goal, with estimated timelines of 20 years for Paris and 193 years for London. Additionally, while Euro V light diesel vehicles correlated with decreased roadside $PM_{10}$ levels, an increase in motorcycles in London hindered significant reductions in $PM_{2.5}$ concentrations from 2010 to 2016.

Giunta (2020) examines air quality impacts associated with road construction and operation, focusing on a motorway project. The study estimates emissions of CO, $NO_x$, and $PM_{10}$ during both phases, highlighting comparable releases of CO and $NO_x$ but more significant $PM_{10}$ emissions during construction. Findings reveal that storage, transit on unpaved roads, and crushing contribute to 85% of $PM_{10}$ emissions during construction, while tunnel portals exhibit higher pollutant concentrations during operation. Additionally, the dispersion model suggests that topography significantly influences Co concentrations.

Faber et al. (2015) investigate the impact of aerosol emissions from construction activities on local air quality, focusing on earthworks and road works. Using advanced aerosol and trace gas instruments, fuel-based emission factors were determined, with earthwork activities and compacting showing highest median emission factors for $PM_{10}$, Construction activities were found to contribute significantly to total $PM_{10}$ and traffic-related NOx emissions in Germany, with earthworks alone contributing about 44% of total $PM_{10}$ emissions from construction activities in the country.

Font et al. (2014) examine the impact of road widening projects in urban areas on air quality, focusing on $PM_{10}$ and $PM_{2.5}$, $NO_x$, and $NO_{2.}$ During construction, $PM_{10}$ concentrations rose by up to 15 μg/m$^3$, with post-completion increases observed during rush hours: 2–4 μg/m$^3$ for $PM_{10}$, 1 μg/m$^3$ for $PM_{2.5}$, and 40 and 8 μg/m$^3$ for $NO_x$ and $NO_2$, respectively. Breaches of the $NO_2$EU limit value post-development indicated a significant decline in residential air quality. Moreover, an increase in glutathione-dependent oxidative potential of $PM_{10}$ suggested a rise in pro-oxidant components linked to vehicle abrasion. These findings underscore the exacerbation of air pollution due to increased traffic following road expansion.

Park & Lee (2024) investigate the impact of highway capping as a sustainable urban development strategy, focusing on Bundang–Suseo Highway in South Korea. Employing diverse methods including on-road monitoring, in situ measurements, and Unmanned Aerial Vehicle (UAV) assessments. The study reveals a shift toward more stable pollutant levels post-cap installation. In situ monitoring demonstrates significant reductions in noise and pollutants like ultra-fine particles and black carbon (BC) near the cap, while UAV monitoring captures changes in pollutant levels at various altitudes. Despite ground-level increases in $PM_{2.5}$, $PM_{10}$, and $NO_2$, the cap leads to decreased levels of pollutants, highlighting crucial role of intentional cap design in enhancing urban air quality.

Guo et al. (2024) present a novel methodology for creating city-level vehicle emission inventories, demonstrated in Changzhou in 2022. The analysis reveals that light-duty passenger vehicles contribute significantly to carbon monoxide and volatile organic compounds emissions, while heavy-duty trucks are prominent in nitrogen oxides and fine PMs emissions. Gasoline vehicles dominate carbon monoxide and volatile organic compounds emissions, while diesel vehicles are key sources of nitrogen oxides and the fine PMs emissions. Notably, China IV vehicles exhibit the highest emission contribution rates. Emissions peak in winter and are lowest in April, with higher intensity observed in northeast of Changzhou.

In the transportation sector, achieving sustainability is imperative due to significant adverse environmental and health effects of road construction and operation (Giunta, 2023). The construction phase, including earthmovings, materials production, and machinery operation, contributes substantially to global greenhouse gas (GHG) emissions, with particular attention to air quality concerns like $PM_{10}$, CO, and $NO_x$ emissions (Giunta, 2023). This study focuses on analyzing emissions from the construction of an Italian road, employing a meticulous process involving data collection, emission estimation, identification of critical construction periods, and investigation of high-impact sites and activities (Giunta, 2023). The spatial and temporal analysis of $PM_{10}$, CO, and $NO_x$ emissions highlights the need for comprehensive understanding and monitoring during the construction phase (Giunta, 2023).

Alshetty & Shiva Nagendra (2022b) investigate the influence of urban factors such as traffic patterns, land use, and meteorological conditions on road dust resuspension and ambient PM concentrations in urban areas, focusing on 18 locations in Chennai and 10 in Delhi. The results reveal PM concentration peaked at construction sites due to fugitive emissions and heavy-duty vehicle movement ($PM_{10}$=183.3±25 μg/m$^3$, $PM_{2.5}$=80.7±6.7 μg/m$^3$), especially at dumpsites with a higher proportion of heavy commercial vehicles. Size distribution analysis indicates that 13%–18% of dust is <10 μm, 6%–9% is <2.5 μm, and 4%–6% is <1 μm, highlighting the impact of traffic, land use, and seasonal variations on road dust dynamics.

Angelevska et al. (2021) address pressing issue of air pollution in urban areas of North Macedonia, particularly attributed to road transport development. It highlights continuous exceedance of PM and nitrogen dioxide limits, emphasizing the lack of comprehensive reduction measures in national plans. By categorizing short- to long-term mitigation strategies into five subcategories, the study offers a structured framework for local authorities to tackle air pollution effectively. This framework aids in selecting and evaluating suitable measures to manage road transport pollution, facilitating improved urban air quality planning and policymaking. The study serves as a crucial step toward implementing air pollution management strategies in urban areas, aiming to mitigate adverse effects of road transport on air quality.

Xu et al. (2023) delve into the relationship between urban congestion, air pollution, and road area ratio, a crucial metric in road design and city planning. Utilizing panel data from Chinese cities spanning 2006 to 2018, the research examines how increasing the relative capacity of urban roads impacts air pollution levels. Additionally, study explores underlying mechanisms of pollution from congestion by decomposing the road area ratio. The results indicate that a higher urban road area ratio correlates with reduced air pollution, irrespective of influence of surrounding cities.

Abhijith & Kumar (2019) investigate the impact of roadside green infrastructure (GI) on pollutant concentrations, including $PM_{10}$, $PM_{2.5}$, and $PM_1$, BC, and particle number concentration, within three GI configurations: hedges only, trees only, and a mix of trees and hedges/shrubs, both close (<1 m) and away (>2 m) from roads. Measurements were conducted over 30 days for 3 months, capturing morning and evening traffic peaks. The results show reductions in pollutant concentrations behind GI in away-road conditions, with the 'hedges only' configuration exhibiting maximum reductions of up to 63% for BC. In close-road conditions, reductions varied, with the 'trees only' configuration reporting increases in most pollutants, while combination of trees and hedges led to reduced pollutant concentrations. Wind direction significantly influenced reductions, with along-road winds allowing assessment of their influence, resulting in reductions of up to 52% for BC. SEM-EDS analysis revealed a dominance of natural particles (50%) and a reduction in vehicle-related particles behind GI. These findings provide insights into air quality modifications under different GI configurations, supporting the formulation of guidelines for GI design to mitigate air pollution exposure near busy roads.

Robinah N. et al. (2022) investigate the impact of road infrastructure equipment on ambient air pollution, noise, and vibration in Mganda. Six equipment types (excavator, roller, grader, concrete mixer, tamper, and wheel loader) from a Kampala case study project were analyzed. GHG emissions, including carbon dioxide ($CO_2$), carbon monoxide (CO), nitrogen dioxide ($NO_2$), hydrocarbon (HC), and PM, were assessed. $CO_2$ was predominant emission (84.1% of total), with grader emitting highest $CO_2$ at 1531.5 g/hour (37.1%), Nitrogen dioxide was the lowest at 1.43 g/hour for the concrete mixer (1.4%). Overall, GHGs accounted for 88.8% of emissions, while air criteria pollutants constituted 11.2%. The grader emitted the highest PM (100.5 g/hour). Most emissions met Environmental Protection Agency standards, except for PM (100.5 g/hour). Carbon monoxide and nitrogen dioxide concentrations exceeded safe ambient air quality limits. Noise levels were above recommended 70.00 decibels, except for the wheel loader. The excavator exceeded permissible vibration limits by 4%.

Alshetty & Shiva Nagendra (2022a) examine ambient PM, resuspendable road dust, and PM mass deposition in the human respiratory tract during preconstruction and construction phases. PM emission rates were estimated and compared between construction roads (CR) and non-construction roads for both phases. Maximum PM concentrations ($PM_{10}$=270.1 µg/m$^3$, $PM_{2.5}$=71.8 µg/m$^3$, $PM_1$=56.3 µg/m$^3$) occurred at night on CRs due to heavy-duty vehicles moving excavated earth. Within 0–100 m from construction sites, average $PM_{10}$ concentrations exceeded 250 µg/m$^3$. Between 100–200 m, 200–400 m, and 400–500 m, $PM_{10}$ values ranged from 200 to 250 µg/m$^3$, 150 to 200 µg/m$^3$, and 100 to 150 µg/m$^3$, respectively. Road dust resuspension, particularly from heavy-duty trucks, significantly influenced PM concentrations. The study reveals $PM_{10}$ deposition fractions in construction workers airways during construction (74%–78%), followed by $PM_{2.5}$ (23%–54%) and $PM_1$ (20%–25%). Implementing sustainable practices, pollution control technologies, and local-scale policies is crucial for mitigating construction-related pollution.

Thammasaroj & Jinsart (2019) examine roadside fine PM levels, specifically $PM_{2.5}$ and $PM_{10}$, in construction and traffic-congested areas of Bankok. Two eastern routes, Pattanakan and Srinagarindra, were selected for study. $PM_{2.5}$ and $PM_{10}$ were sampled using mini-volume and cyclone air samplers, respectively, over 5 days in November to December 2018. Dust samples were analyzed for five heavy metals (Fe, Cu, Zn, Cd, and Pb) using graphite furnace atomic absorption spectroscopy. Traffic conditions were monitored via a video-camera, showing higher congestion in construction areas ($PM_{2.5}$: 39.38 μg/m$^3$, $PM_{10}$: 55.08 μg/m$^3$) exceeded those in control areas ($PM_{2.5}$: 36.15 μg/m$^3$, $PM_{10}$: 48.87 μg/m$^3$), with lowest levels observed at background sites ($PM_{2.5}$: 17.28 μg/m$^3$, $PM_{10}$: 27.32 μg/m$^3$). The study highlights the need for long-term monitoring to assess pollution trends and health impacts post-construction.

Kim et al. (2012) investigates GHG emission from onsite construction equipment, offering a comparative analysis across various equipment types used in different construction activities. Twenty-four cases from a road construction project in Korea were examined, estimating GHG emissions based on final design documents. The study provides emission ranges according to equipment productivity and site conditions, with major activities generating emissions ranging from 256.52 to 376.70 t $CO_2$e, with 282.17 for fair site conditions. A focus group interview yielded guidelines for improving equipment productivity to mitigate GHG emissions, emphasizing proper equipment selection based on specific site conditions. These findings offer valuable insights for reducing GHG emissions from onsite equipment usage.

PM particles, primarily generated through a mechanical process such as the break-up of larger solid particles and wind-blown dust, such as from road traffic emission and industrial activities, residential and commercial combustion, fossil fuel burning, agricultural activities, and various non-combustion process. By analyzing key air quality indicators such as $PM_{2.5}$ and $PM_{10}$, this study seeks to identify patterns of pollution concentration within study. This study is conducted on two cities Dewas and Bhopal, showing comparison between pollution concentration levels to show that now in terms of better quality of life, we need to focus on monitoring and mitigation of air pollution also in smaller cities. This chapter also contains some of the mitigation techniques to minimize air pollution. In this study, we are also focusing on air quality in nearby locality of Kolar Road construction site, Mahabalipuram, Bhopal, we have installed a device for 4 days from 19 January 2024 to 22 January 2024. Here, we have recorded readings for $PM_{2.5}$ concentration levels, after analyzing data, we have drawn some plots and we have also given some techniques for reducing PM concentrations at construction sites.

## 20.2 STUDY AREA

For urban air quality comparison, the study area includes two industrial cities of Madhya Pradesh, India. Their names are Bhopal and Dewas. Dewas – a bustling urban center in Malwa region of Madhya Pradesh, India – is recognized for its industrialization and is home to a government-operated banknote printing facility. Dewas is located at coordinates 22.96°N latitude and 76.06°E longitude, spanning an area of ~50 $km^2$. Bhopal, capital city of Madhya Pradesh, recognized as a pivotal financial and economic hub in state. Bhopal's economy is predominantly industrial, with key sectors including pharmaceuticals, automobiles, textiles, and electrical goods production. Notable industrial areas within and around Bhopal include the expansive Govindpura Industrial Area, industrial suburb of Mandideep, and emerging zones like Bagroda AKVN, Tamot Plastic Park, Acharpura Industrial Area, Badiyakhedi Industrial Area, and Jamuniya Khejda. Bhopal spans an area of approximately 648.24 $km^2$. Bhopal city is positioned at coordinates 23°15′N latitude and 77°24′E longitude (Figure 20.1).

For the road construction site, Sainath Nagar is the locality nearby the Kolar Road, Bhopal. We chose this site because it is near the Kolar Road construction site. It will be feasible to record the air pollution levels in the nearby locality of road construction site. Its coordinate is 23.18N, 77.41E. The population in areas around the Kolar Road is rapidly growing. Many commercial markets, hospital, residential colonies, and schools are situated here.

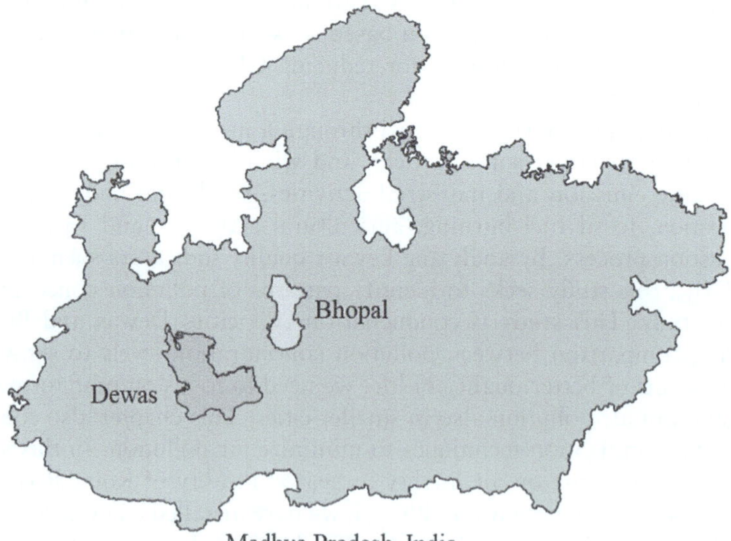


*Figure 20.1* Study area map of Bhopal and Dewas for comparison of particulate matter in the air.

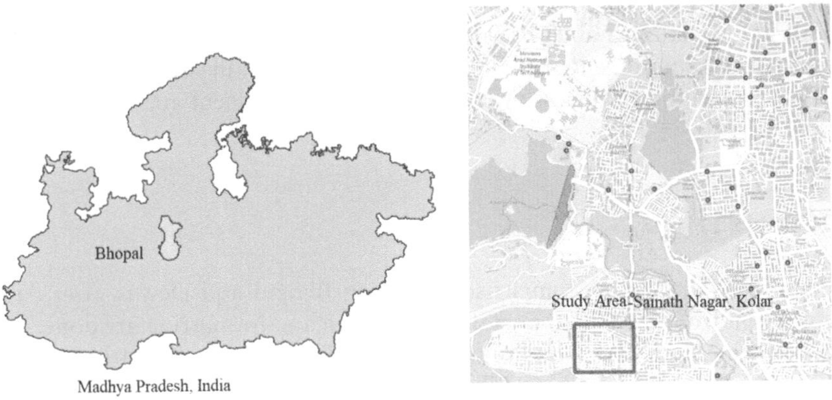


*Figure 20.2* Study area location near the Kolar Road construction site.

## 20.3 DATA COLLECTION

Data are collected from the CPCB site for TT Nagar station in Bhopal and Bhopal Chauraha Dewas. Data is collected for year 2019 to 2023 We would like to extend our sincere gratitude to the Central Pollution Control Board (CPCB) for providing access to their valuable data through their portal. Information sourced from the Central Pollution Control Board's website (https://airquality.cpcb.gov.in/ccr/#/caaqm-dashboard-all/caaqm-landing/data) has been instrumental in enriching our research and enhancing our understanding of air quality dynamics. We acknowledge efforts of the Central Pollution Control Board in collecting, collating, and disseminating crucial air quality data, which serves as a cornerstone for various environmental studies and policy formulations (Figure 20.2).

## 20.4 DEVICE USED

A DustTrak™II Aerosol monitor, model 8530, is a versatile instrument designed to provide real-time aerosol mass readings while also allowing for gravimetric sampling. It is suitable for a wide range of applications including occupational hygiene, indoor air quality assessment, outdoor environmental monitoring, emissions tracking, remote monitoring, and process monitoring. Equipped with a desktop design and battery operation, this monitor offers convenience and portability. Its single-channel, light-scattering laser photometer delivers accurate and reliable aerosol mass readings in real time. The instrument utilizes a sheath air system to isolate aerosols in the optics chamber, ensuring optimal performance and minimal maintenance requirements. Key features include the ability to measure aerosol concentrations corresponding to different particle size fractions, including $PM_1$, $PM_{2.5}$ respirable particles, $PM_{10}$, or custom size fractions. This functionality enables users to tailor their monitoring efforts to specific regulatory requirements or research objectives, making the DustTrak™II Aerosol monitor a valuable tool for air quality assessment and management (TSI, 2024).

## 20.5 METHOD

For urban air quality comparison between Bhopal and Dewas after data collection from the CPCB portal, data sorting and formatting are done. We have taking data of 15-minute frequency and converted it to 24-hour average frequency. By using *r* and Python programming, we filled missing data with the forward fill method to analyze data and plotting graphs.

For road construction site data, we have downloaded data from the DustTrak™II which is at the frequency of 15 minutes. By using r and Python programming, we have analyzed data and plot graphs.

## 20.6 MITIGATION TECHNIQUES FOR AIR POLLUTION

- Mitigating air pollution globally requires a multifaceted approach involving various strategies at local, national, and international levels. Here are some key mitigation techniques:
  - **Transition to Clean Energy Sources**: Shifting from fossil fuels to renewable energy sources like solar, wind, and hydroelectric power can significantly reduce air pollution. This involves promoting policies that incentivize clean energy production and discourage the use of coal and other high-emission fuels (Edenhofer et al., 2011; Peng et al., 2020; Yuan et al., 2022).
  - **Improving Industrial Process**: Implementing cleaner production technologies and enforcing stricter emissions standards for industries can help reduce air pollution. This includes adopting technologies such as scrubbers, catalytic converters, and particulate filters to minimize the emission of pollutants like sulfur dioxide ($SO_2$), nitrogen oxides ($NO_x$), and PM (Castells-Quintana et al., 2021; Grigonyte-lopez Rodriguez, 2020).
  - **Promoting Sustainable Transportation**: Encouraging the use of public transportation, walking, and cycling along with adoption of electric vehicles (EVs) can help reduce emissions from the transportation sector. Developing infrastructure for EV charging stations and implementing policies such as fuel efficiency standards and vehicle emissions testing are crucial (Dioha et al., 2022; Liang et al., 2023; Skipper et al., 2023).

  - **Urban Planning and Design:** Designing cities with an emphasis on compact, mixed-use developments, greenspaces, and pedestrian-friendly infrastructure can reduce the need for long commutes and improve air quality. Implementing zoning regulations that limit industrial activities near residential areas can also mitigate exposure to pollutants (Fan et al., 2022; Jung & Yoon, 2022; Santos et al., 2022).
  - **Afforestation and Green Space:** Planting trees and creating green spaces not only sequester carbon dioxides ($CO_2$) but also help filter air pollutants and mitigate the urban heat island effect. Urban forestry programs and green roof initiatives can contribute to improving air quality in urban areas (Hewitt et al., 2020; Jayasooriya et al., 2017; Liu et al., 2021).
  - **International Cooperation and Policy Frameworks:** Collaborative efforts among nations through international agreements such as Paris Agreement can set targets for reducing GHG emissions, which contribute to air pollution and climate change, implementing and strengthening regulations under frameworks like Kyoto Protocol can also drive global action on air quality (Bauer & Menrad, 2019; Leahy et al., 2020; Lima et al., 2020).
- Here are some of the techniques to control air pollution at construction site of road that should be practiced strictly:
  - Employing covering mechanisms for dump trucks that transport excavated material (Giunta, 2023).
  - Implementing speed control measures for driving during loading operations (Giunta, 2023).
  - Utilizing appropriate methods for material stacking and stockpiling, including regulating heights of stacks (Giunta, 2023).
  - Encouraging behavioral changes among workers to enhance safety and efficiency (Giunta, 2023).
  - Adopting modern construction methods such as pre-fabrication and modular construction (Giunta, 2023).
  - Hardening ground surfaces to minimize dust and soil erosion (Giunta, 2023).
  - Installing sealed roads within construction sites to improve accessibility and minimize environmental impact (Giunta, 2023).

## 20.7 RESULTS

In the Results, we have shown plots for $PM_{2.5}$ and $PM_{10}$, based on the reading of the CPCB Station of Bhopal Chauraha, Dewas, TT Nagar, Bhopal, and also for the Sainath Nagar, Kolar Road construction site, Bhopal.

## 20.7.1 Comparison for $PM_{2.5}$ levels between Bhopal and Dewas

Figure 20.3a and b shows that $PM_{2.5}$ levels in both the cities exceed their prescribed safe limit of 60 μg/m$^3$ for $PM_{2.5}$ (Das et al., 2020). Data provided by the CPCB show the highest levels of $PM_{2.5}$ as 354.48 and 139.56 μg/m$^3$ for Bhopal and Dewas, respectively, which is way above the limits.

Figure 20.4a and b shows that $PM_{10}$ levels in both cities are higher than the prescribed limit of 100 μg/m$^3$ (Das et al., 2020). Data collected from the CPCB station show the highest levels of $PM_{10}$ as 436.64 and 412.18 μg/m$^3$ for Bhopal and Dewas, respectively. Data show that small cities are also getting higher pollution levels.

After analyzing data from the road construction site, we have plotted several graphs, which help us draw some valuable insights. Let us examine the following plot:

Figure 20.3 shows that readings for $PM_{2.5}$ exceeded the 60μg/m$^3$ prescribed safe limit by the Central Pollution Control Board (Das et al., 2020). The maximum level of $PM_{2.5}$ reaches up to 1,040 μg/m$^3$. It can create serious problems like cardiopulmonary problems, asthma, bronchitis, and pneumonia in older people, workers working in such environments, and

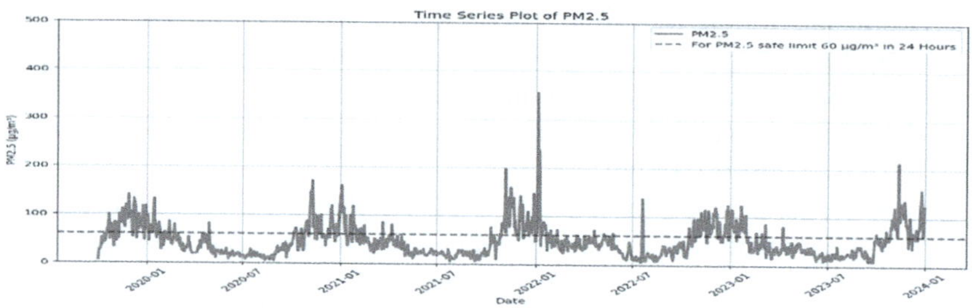


*Figure 20.3a* Time-series plot of $PM_{2.5}$ for Bhopal.

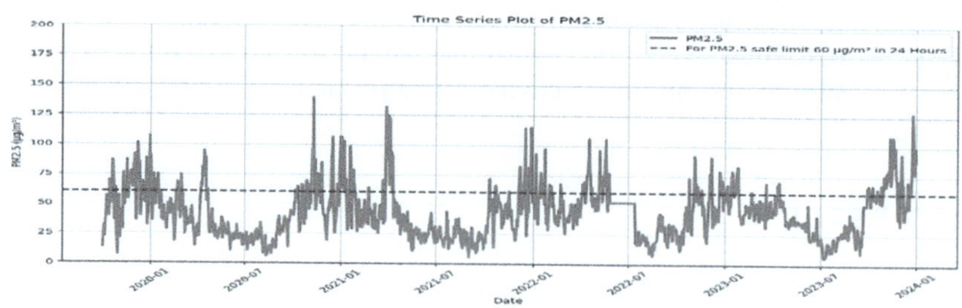


*Figure 20.3b* Time series plot of $PM_{2.5}$ for *Dewas.*

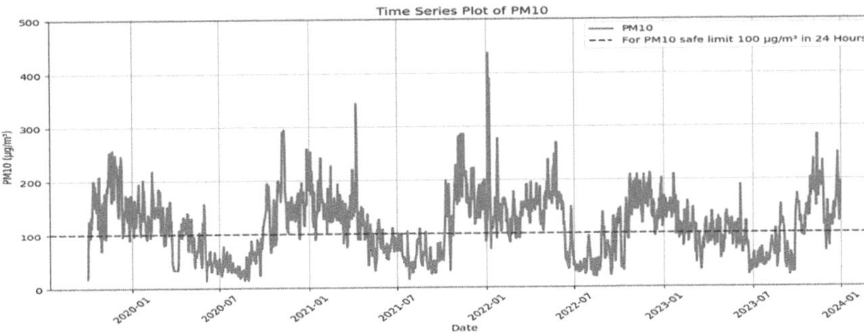


*Figure 20.4a* Time series plot of $PM_{10}$ for Bhopal.

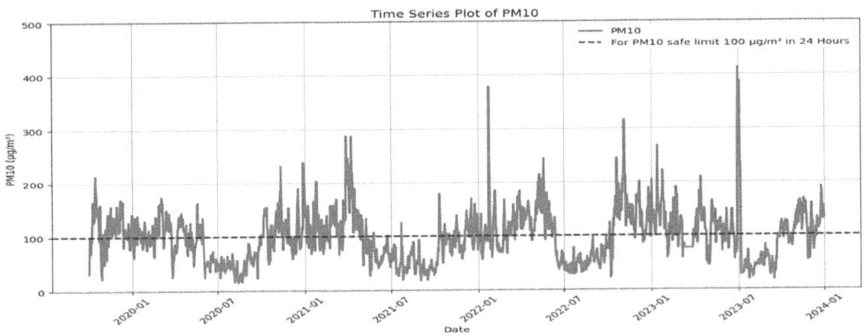


*Figure 20.4b* Time series plot of $PM_{10}$ for Dewas.

people living in nearby areas. It also reduces visibility, which can cause serious accidents.

Figure 20.4 shows that $PM_{2.5}$ levels typically increase after 6 PM to 12 PM. Higher levels in the evening may be due to traffic, higher levels at night could result from both construction and traffic. In the morning, it is due to construction activity and traffic associated with school, college, and office (Figures 20.5 and 20.6).

## 20.8 CONCLUSIONS AND FUTURE SCOPE

As shown in the above plots, PM levels are significantly beyond the prescribed limits of the CPCB. To ensure a clean and safe environment, we should focus on the monitoring and mitigation of pollution, even in smaller

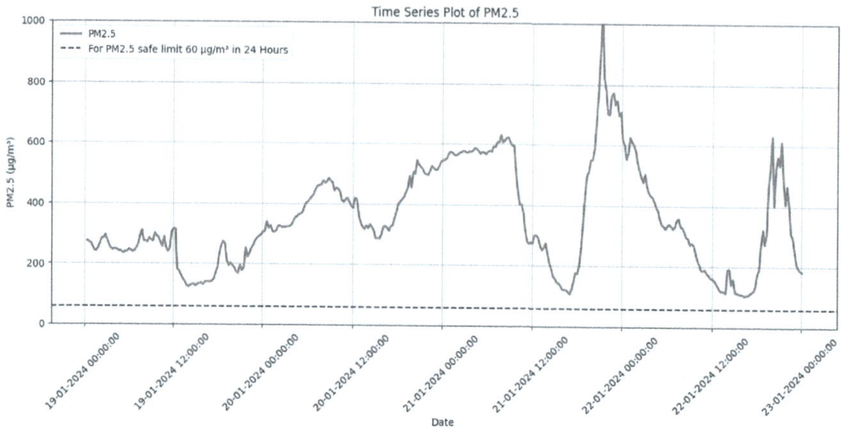


*Figure 20.5* Time-series plot for $PM_{2.5}$ concentration at Sainath Nagar.

cities. This study, conducted between Dewas and Bhopal, aims to highlight the need for monitoring and mitigation of air pollution. Proper steps must be taken to address this issue.

- Rising levels of PM are a serious concern, leading to various health issues and reduced visibility.
- To reduce the high concentration of PM, we should focus on the techniques suggested in this chapter.
- We should improve discipline in transportation and industrial sectors by ensuring construction material and waste transportation are covered during transport.
- Using scrubbers, catalytic converters, electrostatic precipitators, etc. can help reduce direct air pollution from transportation and industrial activities.
- Creating green spaces withing and around cities can serve as a natural sink for air pollution.
- To reduce traffic at construction sites, we should promote the use of public transport and consider options like carpooling for commuting to offices or any other working places.

This study focused solely on PM and has not considered any other pollution and meteorological data. Hence, future studies could include other pollutants and meteorological data in the analysis.

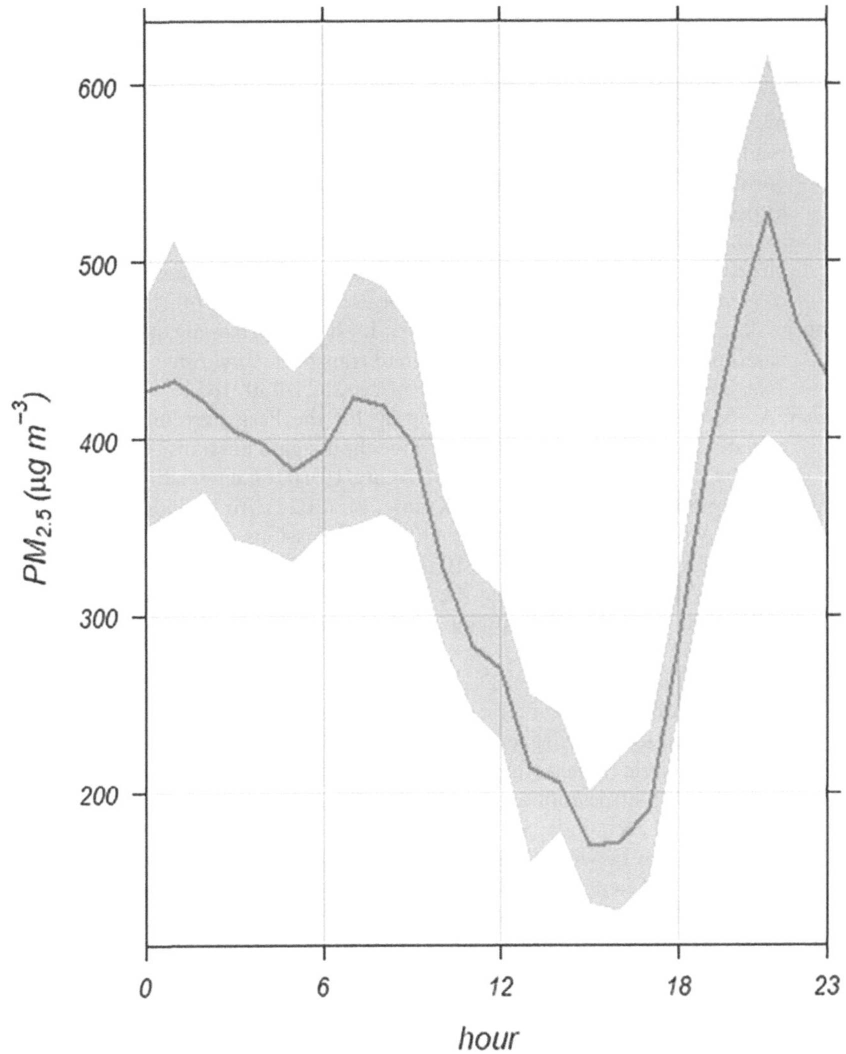


*Figure 20.6* $PM_{2.5}$ concentration pattern for 24 hours.

## REFERENCES

Abhijith, K. V., & Kumar, P. (2019). Field investigations for evaluating green infrastructure effects on air quality in open-road conditions. *Atmospheric Environment*, *201*, 132–147. https://doi.org/10.1016/j.atmosenv.2018.12.036

Agarwal, A., Kaushik, A., Kumar, S., & Mishra, R. K. (2020). Comparative study on air quality status in Indian and Chinese cities before and during the COVID-19 lockdown period. *Air Quality, Atmosphere & Health*, *13*(10), 1167–1178. https://doi.org/10.1007/s11869-020-00881-z

Agrawal, A., Kesharvani, S., Dwivedi, G., Choudhary, T., Verma, R., & Verma, P. (2024). Quantifying the impact of lockdown measures on air pollution levels: A comparative study of Bhopal and Adelaide. *Science of the Total Environment*, *909*, 168595. https://doi.org/10.1016/j.scitotenv.2023.168595

Alshetty, D., & Shiva Nagendra, S. M. (2022a). Impact of vehicular movement on road dust resuspension and spatiotemporal distribution of particulate matter during construction activities. *Atmospheric Pollution Research*, *13*(1), 101256. https://doi.org/10.1016/j.apr.2021.101256

Alshetty, D., & Shiva Nagendra, S. M. (2022b). Urban characteristics and its influence on resuspension of road dust, air quality and exposure. *Air Quality, Atmosphere & Health*, *15*(2), 273–287. https://doi.org/10.1007/s11869-021-01102-x

Angelevska, B., Atanasova, V., & Andreevski, I. (2021). Urban air quality guidance based on measures categorization in road transport. *Civil Engineering Journal*, 7(2), 253–267. https://doi.org/10.28991/cej-2021-03091651

Bauer, A., & Menrad, K. (2019). Standing up for the Paris agreement: Do global climate targets influence individuals' greenhouse gas emissions? *Environmental Science & Policy*, *99*, 72–79. https://doi.org/10.1016/j.envsci.2019.05.015

Castells-Quintana, D., Dienesch, E., & Krause, M. (2021). Air pollution in an urban world: A global view on density, cities and emissions. *Ecological Economics*, *189*, 107153. https://doi.org/10.1016/j.ecolecon.2021.107153

Chiritescu, R. V, Luca, E., & Iorga, G. (2024). Observational study of major air pollutants over urban Romania in 2020 in comparison with 2019. *Romanian Reports in Physics*, *76*, 702.

Das, M. S., Shukla, V. K., Tarun, D., Sharma, S. K., Fasiur, R., & Prashant, G. D. (2020). National ambient air quality standards status and trends 2019.

Dioha, M. O., Lukuyu, J., Virgüez, E., & Caldeira, K. (2022). Guiding the deployment of electric vehicles in the developing world. *Environmental Research Letters*, 17(7). https://doi.org/10.1088/1748-9326/ac765b

Dutta, S., Ghosh, S., & Dinda, S. (2021). Urban air-quality assessment and inferring the association between different factors: A comparative study among Delhi, Kolkata and Chennai Megacity of India. *Aerosol Science and Engineering*, *5*(1), 93–111. https://doi.org/10.1007/s41810-020-00087-x

Edenhofer, O., Pichs-Madruga, R., Sokona, Y., Seyboth, K., Matschoss, P., Kadner, S., Zwickel, T., Eickemeier, P., Hansen, G., Schlömer, S., & von Stechow, C. (eds.) (2011). *IPCC*, 2011: *Summary for policymakers*. In: *Renewable Energy Sources and Climate Change Mitigation*. IPCC Special Report. Cambridge University Press: Cambridge, pp. 3–160.

Faber, P., Drewnick, F., & Borrmann, S. (2015). Aerosol particle and trace gas emissions from earthworks, road construction, and asphalt paving in Germany: Emission factors and influence on local air quality. *Atmospheric Environment*, *122*, 662–671. https://doi.org/10.1016/j.atmosenv.2015.10.036

Fan, P. Y., Chun, K. P., Mijic, A., Tan, M. L., Liu, M. S., & Yetemen, O. (2022). A framework to evaluate the accessibility, visibility, and intelligibility of green-blue spaces (GBSs) related to pedestrian movement. *Urban Forestry & Urban Greening*, *69*, 127494. https://doi.org/10.1016/j.ufug.2022.127494

Font, A., Baker, T., Mudway, I. S., Purdie, E., Dunster, C., & Fuller, G. W. (2014). Degradation in urban air quality from construction activity and increased traffic arising from a road widening scheme. *Science of the Total Environment*, *497–498*, 123–132. https://doi.org/10.1016/j.scitotenv.2014.07.060

Font, A., Guiseppin, L., Blangiardo, M., Ghersi, V., & Fuller, G. W. (2019). A tale of two cities: is air pollution improving in Paris and London? *Environmental Pollution*, *249*, 1–12. https://doi.org/10.1016/j.envpol.2019.01.040

Giunta, M. (2020). Assessment of the impact of CO, $NO_x$ and $PM_{10}$ on air quality during road construction and operation phases. *Sustainability*, *12*(24), 10549. https://doi.org/10.3390/su122410549

Giunta, M. (2023). Sustainable practices in road constructions: Estimation and mitigation of impact on air quality. *Transportation Research Procedia*, *69*, 139–146. https://doi.org/10.1016/j.trpro.2023.02.155

Grigonyte-lopez Rodriguez, J. (2020). Reduction of fine particle emissions from small-scale wood combustion using a novel heat exchanger system. Report series in aerosol science no. 231. http://www.atm.helsinki.fi/FAAR

Guo, M., Ning, M., Sun, S., Xu, C., Zhang, G., Zhang, L., Zhang, R., Zheng, J., Chen, C., Jia, Z., Liu, Y., & Bo, Y. (2024). Estimation and analysis of air pollutant emissions from on-road vehicles in Changzhou, China. *Atmosphere*, 15(2), 192. https://doi.org/10.3390/atmos15020192

Hewitt, C. N., Ashworth, K., & MacKenzie, A. R. (2020). Using green infrastructure to improve urban air quality (GI4AQ). *Ambio*, *49*(1), 62–73. https://doi.org/10.1007/s13280-019-01164-3

Jayasooriya, V. M., Ng, A. W. M., Muthukumaran, S., & Perera, B. J. C. (2017). Green infrastructure practices for improvement of urban air quality. *Urban Forestry & Urban Greening*, *21*, 34–47. https://doi.org/10.1016/j.ufug.2016.11.007

Jung, S. J., & Yoon, S. (2022). Effects of creating street greenery in Urban Pedestrian roads on microclimates and particulate matter concentrations. *Sustainability*, *14*(13), 7887. https://doi.org/10.3390/su14137887

Kim, B., Lee, H., Park, H., & Kim, H. (2012). Greenhouse gas emissions from onsite equipment usage in road construction. *Journal of Construction Engineering and Management*, *138*(8), 982–990. https://doi.org/10.1061/(ASCE)CO.1943-7862.0000515

Leahy, S., Clark, H., & Reisinger, A. (2020). Challenges and prospects for agricultural greenhouse gas mitigation pathways consistent with the Paris agreement. *Frontiers in Sustainable Food Systems*, 4. https://doi.org/10.3389/fsufs.2020.00069

Liang, Y., Kleijn, R., & van der Voet, E. (2023). Increase in demand for critical materials under IEA Net-Zero emission by 2050 scenario. *Applied Energy*, *346*, 121400. https://doi.org/10.1016/j.apenergy.2023.121400

Lima, M. A., Mendes, L. F. R., Mothé, G. A., Linhares, F. G., de Castro, M. P. P., da Silva, M. G., & Sthel, M. S. (2020). Renewable energy in reducing greenhouse gas emissions: Reaching the goals of the Paris agreement in Brazil. *Environmental Development*, *33*, 100504. https://doi.org/10.1016/j.envdev.2020.100504

Liu, H., Kong, F., Yin, H., Middel, A., Zheng, X., Huang, J., Xu, H., Wang, D., & Wen, Z. (2021). Impacts of green roofs on water, temperature, and air quality: A bibliometric review. *Building and Environment*, *196*, 107794. https://doi.org/10.1016/j.buildenv.2021.107794

Liu, K., Li, J., Sun, L., Yang, X., Xu, C., & Yan, G. (2024). Impact of urban forest and park on air quality and the microclimate in Jinan, Northern China. *Atmosphere*, 15(4), 426. https://doi.org/10.3390/atmos15040426

Park, K., & Lee, J. (2024). Mitigating air and noise pollution through highway capping: The Bundang-Suseo Highway Cap Project case study. *Environmental Pollution*, *346*, 123620. https://doi.org/10.1016/J.ENVPOL.2024.123620

Peng, W., Dai, H., Guo, H., Purohit, P., Urpelainen, J., Wagner, F., Wu, Y., & Zhang, H. (2020). The critical role of policy enforcement in achieving health, air quality, and climate benefits from India's clean electricity transition. *Environmental Science & Technology*, *54*(19), 11720–11731. https://doi.org/10.1021/acs.est.0c01622

Pragya, J., Singh, R. J., & Kumar, A. A. (2023). Assessment of air status pre-lockdown, lockdown and post-lockdown in Dewas, MP (India). *International Journal of Innovative Science and Research Technology*, *8*(12), 1634–1640.

Rahul, M. K., & Charan, T. S. (2022). Analysis of pollution of mandideep region in Bhopal. *International Journal of Scientific Research in Civil Engineering*, 6(2), 34–46.

Reeta, K., Alok, S., Nisha, U., Shrivastava, K., Ankita, K., Roy, A., Saket, M., & Poonam, D. (2019). Recent status of ambient air quality of Bhopal City, Madhya Pradesh, India. *International Journal of Research & Review*, 6(11), 365–371.

Robinah, N., Safiki, A., Thomas, O, & Annette, B. (2022). Impact of road infrastructure equipment on the environment and surroundings. *Global Journal of Environmental Science and Management*, *8*(2), 251–264.

Sadhana, C., Pragya, D., Ravindra, S., & Dev, G. A. (2013). Assessment of ambient air quality status and air quality index of Bhopal city (Madhya Pradesh), India. *International Journal of Current Science*, *9*, 96–101.

Santos, T., Ramalhete, F., Julião, R. P., & Soares, N. P. (2022). Sustainable living neighbourhoods: Measuring public space quality and walking environment in Lisbon. *Geography and Sustainability*, *3*(4), 289–298. https://doi.org/10.1016/j.geosus.2022.09.002

Singh, S., Shukla, A., & Jain, K. (2024). Assessing the urbanization-induced impact on environmental parameters of a city from a remote-sensing perspective. *Remote Sensing Applications: Society and Environment*, *34*, 101169. https://doi.org/10.1016/J.RSASE.2024.101169

Skipper, T. N., Lawal, A. S., Hu, Y., & Russell, A. G. (2023). Air quality impacts of electric vehicle adoption in California. *Atmospheric Environment*, *294*, 119492. https://doi.org/10.1016/j.atmosenv.2022.119492

Thammasaroj, P., & Jinsart, W. (2019). Effects of overcrowded traffic and road construction activities in bangkok on $PM_{2.5}$, $PM_{10}$ and heavy metal composition. *EnvironmentAsia, 12*(special issue), 28–35. https://doi.org/10.14456/ea.2019.60

TSI. (2024). DustTrak™ II Aerosol Monitor 8530. https://tsi.com/products/aerosol-and-dust-monitors/aerosol-and-dust-monitors/dusttrak%e2%84%a2-ii-aerosol-monitor-8530/ Date accessed 14/04/2024.

Verma, P., Sisodiya, S., Banait, S. K., Chowdhury, S., Dwivedi, G., & Zare, A. (2023). The impact of coronavirus disease of 2019 (COVID-19) lockdown restrictions on the criteria pollutants. *Processes, 11*(1). https://doi.org/10.3390/pr11010296

Xu, S., Sun, C., Wei, H., & Hou, X. (2023). Road construction and air pollution: Analysis of road area ratio in China. *Applied Energy*, *351*, 121794. https://doi.org/10.1016/j.apenergy.2023.121794

Yuan, R., Ma, Q., Zhang, Q., Yuan, X., Wang, Q., & Luo, C. (2022). Coordinated effects of energy transition on air pollution mitigation and $CO_2$ emission control in China. *Science of the Total Environment*, *841*, 156482. https://doi.org/10.1016/j.scitotenv.2022.156482

Zheng, J., Qu, X., Hou, R., Tang, X., Xu, Z., Huang, Z., Wang, Z., Zhang, W., Yang, C., & Li, T. (2024). A comparative study of air pollution tolerance capabilities of four tree species in Xi'an city, China. *International Journal of Environmental Science and Technology*, *21*(1), 665–674. https://doi.org/10.1007/s13762-023-04970-1

Zhou, X., Cao, G., Peng, B., Xu, X., Yu, F., Xu, Z., Yan, Y., & Du, H. (2024). Citizen environmental complaint reporting and air quality improvement: A panel regression analysis in China. *Journal of Cleaner Production*, *434*, 140319. https://doi.org/https://doi.org/10.1016/j.jclepro.2023.140319

# Index

For Product Safety Concerns and Information please contact our
EU representative GPSR@taylorandfrancis.com Taylor & Francis
Verlag GmbH, Kaufingerstraße 24, 80331 München, Germany